Microplastics Pollution and Worldwide Policies on Plastic Use

Microplastics Pollution and Worldwide Policies on Plastic Use discusses microplastic pollution and global policies developed to tackle the problem. It details the mechanisms of microplastics occurrence, sources, and impacts. It then offers a comprehensive overview of the various policies created by specific countries in Asia, Europe, the Americas, and Africa to address plastics use and minimize its effects.

- Describes microplastics pollution found worldwide in drinking water and food chains.
- Addresses policies implemented in Asia, Europe, the Americas, and Africa and details local policies for various countries within each region, including requirements and penalties for non-compliance.
- Explains the mission and vision of global organization such as the United Nations, G7, World Economic Forum, World Bank, and Lisbon Treaty.

This book is aimed at academics, industrial professionals, policy makers, and general readers interested in the mitigation of microplastic pollution.

Lee Tin Sin is a researcher, professional engineer, and associate professor. He graduated with a Bachelor of Engineering (Chemical-Polymer), First Class Honours, and a PhD in Polymer Engineering from Universiti Teknologi Malaysia. Dr Lee has been involved in rubber processing, biopolymer, nanocomposite, and polymer synthesis with more than 100 publications of journal papers, book chapters, and conferences. He was the recipient of the Society of Chemical Engineers Japan Award for Outstanding Asian Researcher and Engineer, 2018, for his contribution to polymer research. He was also conferred the Meritorious Service Medal by the Sultan of Selangor, Malaysia in 2019.

Bee Soo Tueen is a researcher, professional engineer, and associate professor. She graduated with a Bachelor of Engineering (Chemical-Polymer), a Master of Engineering (Polymer Engineering), and a PhD in Polymer Engineering from Universiti Teknologi Malaysia. Dr Bee has published more than 70 journal papers on nanocomposites, flame retardants, and biopolymers.

Green Chemistry and Chemical Engineering
Series Editor
Dominic C.Y. Foo
University of Nottingham, Malaysia

Managing Biogas Plants
A Practical Guide
Mario Alejandro Rosato

The Water-Food-Energy Nexus
Processes, Technologies, and Challenges
I. M. Mujtaba, R. Srinivasan, and N. O. Elbashir

Hemicelluloses and Lignin in Biorefineries
Jean-Luc Wertz, Magali Deleu, Séverine Coppée, and Aurore Richel

Materials in Biology and Medicine
Sunggyu Lee and David Henthorn

Resource Recovery to Approach Zero Municipal Waste
Mohammad J. Taherzadeh and Tobias Richards

Hydrogen Safety
Fotis Rigas and Paul Amyotte

Nuclear Hydrogen Production Handbook
Xing L. Yan and Ryutaro Hino

Water Management
Social and Technological Perspectives
Iqbal Mohammed Mujtaba, Thokozani Majozi and Mutiu Kolade Amosa

Efficiency and Sustainability in the Energy and Chemical Industries
Scientific Principles and Case Studies, Third Edition
Krishnan Sankaranarayanan

Microplastics Pollution and Worldwide Policies on Plastic Use
Lee Tin Sin and Bee Soo Tueen

For more information about this series, please visit: www.routledge.com/Green-Chemistry-and-Chemical-Engineering/book-series/CRCGRECHECHE

Microplastics Pollution and Worldwide Policies on Plastic Use

Lee Tin Sin and Bee Soo Tueen

CRC Press
Taylor & Francis Group
Boca Raton London New York

CRC Press is an imprint of the
Taylor & Francis Group, an **informa** business

Designed cover image: Shutterstock

First edition published 2024
by CRC Press
2385 NW Executive Center Drive, Suite 320, Boca Raton FL 33431

and by CRC Press
4 Park Square, Milton Park, Abingdon, Oxon, OX14 4RN

CRC Press is an imprint of Taylor & Francis Group, LLC

ISBN: 978-1-032-48202-6 (hbk)
ISBN: 978-1-032-48203-3 (pbk)
ISBN: 978-1-003-38786-2 (ebk)

DOI: 10.1201/9781003387862

Typeset in Times New Roman
by Apex CoVantage, LLC

Dedication

For our dearest children

Chen Chen and Yuan Yuan, with love forever

Contents

Preface

Plastics are widely used in every moment of our lives. However, substantial uses of plastics and improper disposal have caused serious environmental problems. Many countries experience negative impacts brought about by plastic pollution for decades. As a result, many countries have been developing governance policies to control the consumption of plastics, such as plastic tax, extended producer responsibilities (EPR), penalties, etc. This book covers the discussions on the policies and regulations of the Asian, European, African, and American continents to address the plastic issues in over 100 countries. In the Chapter 1, the impacts of microplastics are discussed; in Chapter 2 through Chapter 5, the policies and regulations about plastics use in Asian, Europe, America, and Africa, respectively, are discussed. Finally, Chapter 6 is about the policies of plastic use implemented by global economic organizations. Although a variety of approaches are being engaged to tackle the plastic pollution problems, education retains its central importance in driving the successful implementation of all the policies and regulations. In fact, many of the policies and regulations are made up based on "reduce, reuse, and recycle" (the 3Rs) of plastics. Thus, the authors are optimistic this book can provide comprehensive information to the policy makers, researchers, educators, and environmentalists on the foundational knowledge about plastic use policies. In the opinion of the authors, all the policies, regulations, and commitments need sustained efforts. In remembrance of the words of Winston Churchill: "Now is not the end. It is not even the beginning of the end. But it is, perhaps, the end of the beginning." Combatting plastic pollution calls for prolonged and diligent work before achieving success in the near future.

Lee Tin Sin and Bee Soo Tueen

Authors

Lee Tin Sin is a researcher, professional engineer, and associate professor. He graduated with a Bachelor of Engineering (Chemical-Polymer), First Class Honours, and PhD in Polymer Engineering from Universiti Teknologi Malaysia. Dr Lee has been involved in rubber processing, biopolymer, nanocomposite, and polymer synthesis, with more than 120 publications of journal papers, book chapters, and conferences. He was the recipient of the Society of Chemical Engineers Japan Award for Outstanding Asian Researcher and Engineer, 2018, for his contribution to polymer research. He was also conferred the Meritorious Service Medal by the Sultan of Selangor, Malaysia in 2019. He co-authored "Polylactic Acid" (Elsevier), the first and second editions which were published in 2012 and 2019, respectively. In 2023, he co-authored "Plastics and Sustainability," published by Elsevier. He is dedicated to making meaningful contributions to both the plastics industry and the formulation of policies aimed at improving the quality of human life.

Bee Soo Tueen is a researcher, professional engineer, and associate professor. She graduated with a Bachelor of Engineering (Chemical-Polymer), a Master of Engineering (Polymer Engineering), and a PhD in Polymer Engineering from Universiti Teknologi Malaysia. Dr Bee has published more than 80 journal papers on nanocomposites, flame retardant, and biopolymer works. She co-authored "Polylactic Acid" (Elsevier), the second edition of which was published in 2019. In the year 2023, she collaborated as a co-author on "Plastics and Sustainability," a publication released by Elsevier. Her unwavering commitment lies in making substantial contributions to both the plastic industry and the development of policies geared towards enhancing the overall quality of human life.

1 Microplastics Pollution: Serious Issues

1.1 INTRODUCTION

The current serious environmental pollution problem has gained huge attention from various parties. Without a doubt, plastic pollution is considered as one of the serious environmental pollution problems nowadays. The huge generation of plastics waste has severely contributed to serious plastic pollution, and these issues have attracted much attention and awareness from various countries in controlling the usage and production of goods made of plastics. Improper handling of plastics waste can become more serious when the disposed plastics wastes flow into natural water sources such as river, lakes, and oceans. These plastic wastes are mainly non-biodegradable polymers, and they are unable to be fully decomposed into harmless, basic components under natural environmental conditions. However, these non-biodegradable plastics wastes partially degrade into smaller fragments, known as microplastics. According to a study reported by Briggs (2021), scientists had found that more than approximately 12 million pieces of the litter found in natural aquatic environments are plastics waste. The huge amounts of plastics wastes are mainly contributed by the large production of plastics products. Plastic is durable, light in weight, low in cost, and has excellent gas and liquid permeability resistance, good mechanical properties, and physical appearance. These excellent characteristics have caused favourability towards the usage of plastics being produced commercially in various industries and applications worldwide, especially in food packaging and biomedical applications. By referring to the report investigated by Plastic Europe (2020), the global production of plastics had reached 368 million tonnes in 2019. Besides, they also mentioned that the generation of plastics wastes in the year of 2018 was a massive, huge amount due to massive production of plastics products. According to Hopewell et al. (2009), the generation of large amount of plastics waste is mainly attributed to the disposal plastics items and goods which are one-time usage products such as food packaging products, biomedical devices, etc. These plastics items will be disposed after one use and contribute to the severe increment of plastics waste globally. Currently, plastic wastes have been recovered as energy fuel, and some plastic wastes were further recycled into other usages due to increasing of public awareness of the importance of green environmental issues. However, a lot of plastic wastes were disposed in landfills and caused the severe utilization of a lot of lands just for plastic waste disposal. Furthermore, the disposal of plastic wastes in landfill could have some detrimental effects on our environment. The disposal of plastic waste in landfills was found to cause serious contamination of the groundwater by leaching through the soil. Furthermore, the leaching of plastic wastes through soil is also found to contaminate the other water sources when the partially degraded plastic wastes are leached

DOI: 10.1201/9781003387862-1

easily into the surface water of other sources. One of the severe problems caused by the improper of plastic wastes disposal is marine plastic pollution. Marine plastic pollution is a severe environmental problem which has threatened the survival of aquatic animals, and thus, further influences the food safety and quality for humans. Furthermore, marine plastic pollution could affect coastal tourism and as well as contributing to and accelerating climate change. In a study conducted by Jambeck et al. (2015), they estimated that an approximately 4.80 to 12.70 million tonnes of plastic waste were disposed into the ocean per year based on the data before 2014. According to International Union for Conservation of Nature (2021), they reported at least 14 million tonnes of plastic wastes were discharged into the ocean per annum in the year 2021. This observation indicates that the disposal of plastic wastes into the ocean had severely increased since 2014. This indicates that the generation of plastic wastes gradually increased from year to year. These plastic wastes were not handled properly and caused these inappropriately handled wastes to leach directly or indirectly into the marine environment through either the drainage system or the wastewater treatment plant's (WWTP) effluent. The leaching of plastics wastes could cause the plastics to be transported for a long way, from residential areas to the aquatic environment, and then accumulate in the ocean and on land surfaces (Reisser et al., 2013; Ryan et al., 2009). That plastic wastes can be transported easily for a long way from the disposed areas to the ocean is mainly attributed to the low-density behaviour of plastic materials compared to other materials such as glass and metal; this behaviour can cause the plastic wastes to flow with the water into the aquatic environment (Ryan et al., 2009).

1.2 ORIGIN OF MICROPLASTICS: PRIMARY AND SECONDARY MICROPLASTICS

The environmental pollution caused by plastic wastes becomes worse when these plastic wastes are degraded into smaller fragments during the transportation process into aquatic environments. Plastics can be classified in accordance with their different particle sizes, which are macroplastics, microplastics, and nanoplastics (see Table 1.1). Currently, the pollution of microplastics have become a main concern to environments; especially, microplastics have caused severe environmental issues the marine ecosystem. In general, microplastics are categorized into two main groups, which are primary microplastics and secondary microplastics.

The plastics with particle size approximately less than 5 mm in length are known as microplastics (Neetha et al., 2020). In general, microplastics are divided into

TABLE 1.1
Types of Plastics in Terms of Different Particle Sizes

Types of plastics	Particle size	References
Macroplastics	> 25 mm	Romeo et al., 2015
Microplastics	1 mm < sizes < 5 mm	Neetha et al., 2020
Nanoplastics	1 nm < sizes < 1000 nm	Gigault et al., 2018

two major groups, which are primary microplastics and secondary microplastics. Primary microplastics are the plastics that are intentionally designed small in size for commercial usages such as cosmetics products, personal care products, microfibres from the textiles industries, etc. (Ryan et al., 2009). The primary microplastics are produced to be primarily used in the commercial products and industrial raw materials. There are various types of primary microplastics in industries such as (Boucher and Friot, 2017; Issac and Kandasubranmanian, 2021):

a. The appearance of microbeads in cosmetics products and personal care products.
b. The presence of microplastics in abrasive blasting agents and paints.
c. The production of polymer resins for manufacturing purposes in the form of powder or pellets.
d. The paint coating of ships or other marine transportation and building construction.

Primary microplastics can enter the environment directly, through various water channel systems such as household wastewater treatment systems and industrial wastewater treatment systems. The microplastics used in the personal care products such as facial scrub wash, body scrub shower gel, etc., and cosmetics products are typically known as microbeads (Boucher and Friot, 2017; Sundt et al., 2014). According to Rochman et al. (2015), microbeads are commonly produced from different types of plastics such as polypropylene (PP) and polyethylene (PE), and the particles' sizes range of 1 mm to 5 mm. The application of microplastics in personal care products, especially facial scrub wash, gained much attention in the personal care products industries after the patent of using the microbeads made of plastics in such applications was published (Sundt et al., 2014). The favourability of using microplastics as microbeads in personal care products is mainly attributed to the fact that the plastic-made microbeads could be precisely manufactured in microscopic sizes (Cole et al., 2014; European Chemicals Agency, 2019). This caused the natural, exfoliated materials that were used in facial scrubs, which were higher cost in production, to be gradually replaced by the usage of plastic-made microbeads (Fendall and Sewell, 2009; Issac and Kandasubramanian, 2021; Napper et al., 2015). The primary microplastics used as microbeads in cosmetics and personal care products are being washed with domestic water into household wastewater treatment systems and then, further, flushed into the ocean or other aquatic environments (Issac and Kandasubramanian, 2021; Kalcikova et al., 2017). Table 1.2 summarizes the types of polymers which are commonly used as microbeads, the average size and shape of the microbeads used inside the personal care products. The spillage of polymer resins during the manufacturing process or transportation also could lead to the leaching of polymer resins, either in pellet form or powder form, into the aquatic environment as well. The wastewater containing the primary microplastics will be treated in the wastewater treatment system, and then, the treated wastewater will eventually discharge into the ocean. The treatment process in the wastewater treatment system is unable to filter the microplastics inside the wastewater due to their tiny particle size. All these microplastics would accumulate in the ocean and persist for many years without being chemically degraded into residues with simple

TABLE 1.2
Microplastics Information About Microbeads in Personal Care Products (Sundt et al., 2014)

Microplastics information	Personal Care Products
Types of polymers used	Polypropylene (PP), Polyethylene (PE); PTFE, PET, and nylon
Particles size	0.001 mm to 0.8 mm
Particles shape	Irregular, spherical shape

chemical structures. On the other hand, primary microplastics have also been used in the coating paints for the bodies of marine transportation vehicles such as ships. The application of the microplastics in coating paints for ships can further contribute to the release of primary microplastics directly into the aquatic environment, when the paint coatings on the ships are subjected to the abrasive blasting of sea water. Besides, the degradation of the paint coating also could cause the paint to be removed from the ships' bodies and split into the ocean. On the other hand, primary microplastics were found to be released to the ocean during maintenance, when the coating paints on the ships are being removed by using abrasive blasting techniques (Sundt et al., 2014).

Secondary microplastics are the form of microplastics that are generated from the breakdown of plastics products with a larger size into microplastics with a smaller particle size of less than 5 mm in length (Boucher and Friot, 2017; Ryan et al., 2009). The formation of secondary microplastics is generally attributed to the exposure of larger plastic goods to weathering effects such as sunlight, ultraviolet radiation, mechanical abrasion effect from wind, wave action, etc., and further degraded into smaller plastics fragments (Ryan et al., 2009; Shahnawaz et al., 2019). Besides, the plastic microfibres that are released from synthetic textiles such as clothing, fishing nets, etc., during the laundry process and washing process are also considered as secondary microplastics. This is due to the presence of an abrasion effect: during the process of washing and manufacturing in the laundering of synthetic textiles, microfibres from clothing can be released, and these microfibres, made of plastics, are being washed into the industrial or household wastewater treatment systems (Boucher and Friot, 2017; De Falco et al., 2019; OECD, 2020). These secondary microplastics, produced in landfill areas due to fragmentation by weathering and the presence of secondary microplastics in the treated wastewater system, would eventually leach into the ground water and flow into the oceans. The secondary microplastics generated from the degradation of larger plastic parts would float on the surface water of ocean and eventually lead to the marine plastic pollution, which is harmful for the aquatic environment (Ramkissoon, 2020). Similarly, microfibres (secondary microplastics) will be detached from the synthesis textiles during the washing and manufacturing process. This had led to the disposal of the secondary microplastics wastewater, discharged from the synthetic textiles industries into the wastewater treatment systems and, further, into the ocean marine environment (Sundt et al., 2014). Besides, the secondary microplastics not only can cause marine plastic pollution; it also could lead to air pollution. Secondary microplastics, lightweight and with low density, would flow together with air and cause secondary microplastics to be found in dust and airborne, fibrous particles. The health effects of the presence of microplastics in air dust and airborne, fibrous particles on human body through inhalation remains unknown.

1.2.1 Sources of Microplastics in Drinking Water

As mentioned in previously, most microplastics are discharged into the ocean or other water sources due to lack of proper planning and management in handling plastic wastes. According to Boucher and Friot (2017), the total amounts of primary microplastics that are released into the ocean are estimated to be 1.50 million tonnes per year globally. This also indicates that marine microplastics is one of the main pollutants in the marine environment. Various researchers have observed that most of the microplastics discovered in the marine environment mainly come from micro-plastics wastes that were originally used by consuming products and formed from the degradation of plastics goods (Boucher and Friot, 2017; Browne et al., 2011; Qiu et al., 2020). The microplastics that are commonly found in various water sources are microbeads in personal care products such as scrubbing agents in facial scrub foam, etc., cosmetics products, synthetic clothing, residues of rubber used to make tyres, and other types of polymer wastes that are dumped in various wastewater systems and around lakes, rivers, and beaches (Laskar and Kumar, 2019). Currently, there are a few sources of microplastics that have been identified to severely cause marine plastic pollution. Microplastics could contaminate the drinking water sources through atmospheric fallout, surface runoff, the wastewater treatment plant's effluent and sludge, and the leaching of microplastics from soil and landfilling.

1.2.1.1 Atmospheric Fallout

Air is one of the most important substances that is needed by humans and all living animals to survive. However, air nowadays has been severely polluted, and this could further harm the survival of humans and living animals. These air pollution issues have currently gained huge attention globally due to their threat to humans and all living animals. Airborne microplastics have been newly identified as one of the harmful air pollutants among all air pollutants in the atmosphere. Airborne microplastics have also been identified as one of microplastic pollution sources in marine plastics pollution. According to studies conducted by Boucher and Friot (2017) and Prata (2017), the presence of airborne microplastics is estimated to be approximately 7% of total marine pollution. The occurrence of microplastics in the atmospheric environment is mainly considered as a resultant of anthropogenic activities due to the fact that airborne microplastics can be easily transported from one area to another area by wind and atmospheric deposition processes due to their light weight and low-density behaviour (Qiu et al., 2020). Atmospheric deposition is defined as the process where microplastic particles, aerosols, gases, etc., deposit themselves on the surface of earth from the atmosphere. The deposition of micro-plastics in various natural water sources could pollute the drinking water source, and the effect of microplastics on the human body when ingested remains unknown. According to Klein and Fischer (2019), there are two main types of atmospheric deposition mechanisms, which are dry deposition and wet deposition. Dry deposition is also known as the gravitational sedimentation process, where the particles that float in the atmosphere settle on any surfaces due to gravitational forces. On the other hand, wet deposition is referred to as the process of precipitating all atmospheric particles with atmospheric hydrometeors such as rainout, snow, etc. (Klein and Fischer, 2019; Schlesinger and Bernhardt, 2013). Microplastics that are found in various marine environments might be contributed by the atmospheric fallout (Cai et al., 2017). According to Cai et al. (2017), the degradation mechanism of the microplastics obtained through the atmospheric fallout process is observed to pose similar degradation actions with microplastics collected in the surface of water. This observation indicates that some of the microplastics may be possibly settled on the terrestrial surface with dry deposition, and some may be transported into the aquatic environments through wet deposition (Cai et al., 2017; Chen et al., 2019; Prata, 2017). The presence of microplastics due to atmospheric fallout is found to be in the dynamic cycle of the atmosphere, the earth, and the marine environment. According to a study conducted by Dris et al. (2016), they found that atmospheric fallout is one of the microplastic pollutant sources that contributed to the presence of microplastics in various water systems. The concentration and deposition of airborne microplastics is mainly affected by various factors such as meteorological conditions, the growth of the human population, the activities of humans, and local environments (Chen et al., 2019). According to two research studies conducted by Dris et al. (2015) and Dris et al. (2016), they observed that the number of microplastics collected during a rainy day was found to be several times higher than the amount of microplastics collected under standard atmospheric conditions. According to Table 1.3, the particles in sediment collected during wet weather was found to be much higher than the particles in the sediment collected during dry weather. This indicates that wet deposition is the main contributor to atmospheric fallout due to the low density of microplastics, which requires rain precipitation be deposited into the marine environment. Besides, the number of microplastics is also found to be affected by wind direction. A study conducted by Browne et al. (2010) found that the number of microplastics was high at downwind sites, which are eight fragments per 50 mL of sediment, while the number of microplastics found in the other sites was only three fragments per 50 mL of sediment. This observation also indicates that wind movement could help the microplastics with low-density to travel and reach the marine environment easily. This is because the low density of microplastic makes it able to suspend easily in the atmosphere and deposit in the marine environment (Chen et al., 2019; Klein and Fischer, 2019).

TABLE 1.3
The Numbers of Microplastics Collected during Dry Weather and Wet Weather

Wet weather/after raining	Dry weather	References:
280.00 particles/m^2/day	29.00 particles/m^2/day	Dris et al. (2015)
11 to 355 particles/m^2/day	2 to 34 particles/m^2/day	Dris et al. (2016)

TABLE 1.4
The Amount of Suspended, Airborne Microplastics Collected in Shanghai and Paris

Population in city	The collected suspension airborne microplastics	References
Shanghai	1.42 microplastics per m^3 of air	Chen et al., 2019; Liu et al., 2019c
Paris (half population of Shanghai)	0.93 microplastics per m^3 of air	Liu et al. (2019a)

Besides, the population of humans in the area is also found to relate and affect the amount of microplastics in that area. In a study reported by Chen et al. (2019) and Liu et al. (2019a), they compared the collection of airborne microplastics in Shanghai and Paris, respectively, to perform the comparison on the microplastics collected in relating the population numbers. By referring to Table 1.4, the amount of suspension airborne microplastics that collected in Shanghai was determined with 1.4 microplastics per m^3 of air, while the amount of suspended microplastics collected in Paris is approximately 0.93 microplastics per m^3 of air. The microplastics collected in Shanghai was observed to be significantly higher than the airborne microplastics collected in Paris. This is mainly attributed to the population of Shanghai being two times the population of Paris. This also indicates that the size of a city population could influence the generation of airborne microplastics in a particular area (Liu et al., 2019a). This is because anthropogenic activities are influenced by population densities; higher population density could lead to a severe increment in generating plastics wastes, and thus, induce the release of the airborne microplastics (Chen et al., 2019). For example, the drying of clothes and blankets made of synthetic textiles under sunlight and airflow could contribute to increasing the discharge of airborne microplastics in Shanghai. The exposure of clothes and blankets consisting of synthetic textiles to sunlight and ultraviolet radiation is more likely to degrade into microfibres (Chen et al., 2019; Liu et al., 2019a). According to Qiu et al. (2020), the activities of humans on land could also contribute to the release of higher concentrations of airborne microplastics through wear and tear on the tyres of all automobiles and the grinding and cutting of synthetic plastics parts in industries. Besides, the transferring mechanisms of microplastics in the atmosphere is one of the important concerns that caused the presence of microplastics in the aquatic environment. This is because the transferring mechanisms of microplastics by the moving air flow (or wind) could release microplastics to the soil and travel into the surface water (Rezaei et al., 2019). In a study conducted by Rezaei et al. (2019), they investigated the effect of wind in transferring microplastics into the soil by collecting sediments involved in the wind-erosion activities at 11 different location sites. The original soil samples for these 11 locations sites were initially collected. The original soil was tested in a wind tunnel at a velocity of 12 m/s to prepare the wind-eroded soil samples. Rezaei et al. (2019) concluded that the wind erosion played an important role in transporting the microplastics in the atmosphere. Wind erosion action is a medium that led to microplastics being deposited in the surface waters.

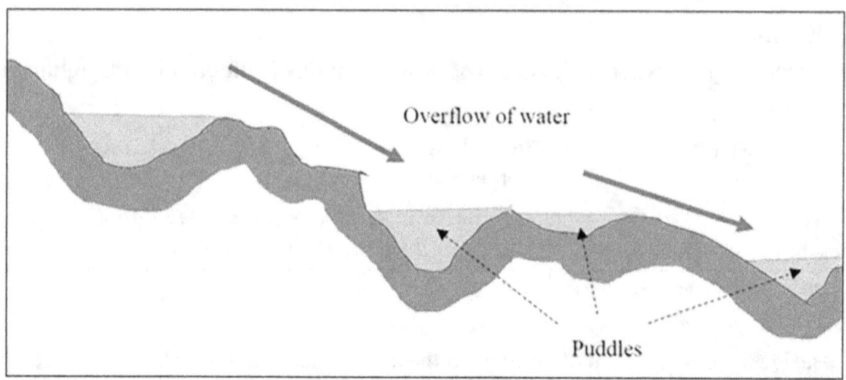

FIGURE 1.1 The formation of puddles when the excessive rainwater begins to overflow during the surface runoff phenomenon.

1.2.1.2 Surface Runoff

One of the main sources that contributes to the presence of microplastics in drinking water sources is surface runoff. Surface runoff is basically referred to the phenomenon of open channel-flow over the ground surface. The occurrence of surface runoff is due to the overflow or retention of rainwater, storm water, or other rainfall water sources which cause the infiltration capacity to no longer be able to withstand the ground surface. In other words, the happening of surface runoff is attributed to the rainfall water's exceeding the maximum capacity limit of the rainwater that can be absorbed by the soil (Horton, 1993; Pilgrim et al., 1978). This situation has further caused the water absorbability of soil to reach its saturation capacity by the excessive rainwater. Consequently, the excess of rainwater has caused the excess of rainwater to overflow, and this phenomenon has led to the formation of puddles, as shown in Figure 1.1. The formation of puddles is due to the collection of the overflow rainwater into a section of rough slope (Hillel, 2005). As shown in Figure 1.1, the open channel system is unable to further absorb the excess rainwater and causes this excess rainwater to overflow. The excessive rainwater flows down to the rough slope from a higher elevation to a lower elevation area and forms puddles.

Surface runoff has various types of runoff conditions; for example, rainfall runoff, stormwater runoff, etc. In general, surface runoff is categorized into two mechanisms, which are throughflow and saturated overland flow. Throughflow mechanisms refers to the infiltration mechanisms of rainwater into the soil through macropores; the excess rainwater would overflow and reach the lower stream channels such as ponds, creeks, etc. This phenomenon may lead to the leaching out of microplastics from the soil, and the leached out microplastics would be carried away by the rainwater to the groundwater (Kirkby and Chorley, 1967). Eventually, the microplastics will reach the stream channel during the percolation process (Horton, 1993; Pilgrim et al., 1978). On the other hand, the occurrence of saturated overland flow is mainly due to the saturated soil horizon being unable to withstand the intensity of rainfall. Then, the saturated overland would transport the microplastics in the soil into the surface water (Horton, 1993).

According to Liu et al. (2019a), a huge number of land-based microplastics is flowing into the marine environment through urban surface runoff and highway stormwater runoff. Urban areas are usually found to experience the occurrence of urban surface runoff and highway stormwater runoff. In urban areas, most plastic wastes were disposed of in landfills, and these plastics waste slowly degrade into smaller fragments to form microplastics. Land-based microplastics mainly originate from microplastics that leach out from poorly managed landfills and the disposal of plastic debris from industries. The leaching of microplastics from landfills and disposed microplastics debris could be transferred to the marine environment through the method of surface runoff (Horton and Dixon, 2017; Lutz et al., 2021). Besides, the accumulated microplastics on the road surface could be flushed and washed by the rainwater running through urban roads. Then, the flushed microplastics travelled into the marine environment through various surface runoff methods and pathways such as natural reservoirs and artificial open channel systems (Lutz et al., 2021). On the other hand, one of the most pervasive microplastics in urban areas is microplastic residues torn from the tyres of various vehicles on roads (Qiu et al., 2020). In urban areas, the large of amounts of vehicles on the road daily have caused the formation of microplastics due to the frictional effect between the tyres of the vehicles and the surface of road. The frictional effect has caused the microplastics to be torn from tyres and deposited on the road surface during the driving process. Some of the deposited microplastics from vehicles' tyres are trapped in the rough surface of the road, while some of the microplastics that stain the road surface are washed away by rainfall and flow into various freshwater channel environments through surface runoff. Eventually, these microplastics flow directly into freshwater channel reservoirs in residential areas, which is the freshwater source for drinking water (Kole et al., 2017).

During heavy rainfall, the occurrence of stormwater runoff would carry and wash away microplastics into the aquatic environment. This is mainly attributed to the fact that rainwater that carried microplastics will flow directly into surface water or other sewage collection channel systems in accordance with the local situation. Sewage collection channel systems are mainly divided into two different types of systems, which are combined systems and separate systems (Kole et al., 2017). Combined systems collect all the water inflow from the natural water channel into wastewater treatment plants (Kole et al., 2017). On the other hand, the separated system is referred to as the system that only collects the wastewater discharged from households and industrial areas into wastewater treatment plants. The separated systems are widely applied due to the expected increment in the rainfall intensity which is contributed by climate change. Thus, the usage of separated sewerage systems minimized the duty load of the rainwater treatment process by reducing the rainwater amount that was being transported into wastewater treatment plants. The microplastics discharged from the urban areas such as microplastics from landfills, micro debris from industrial areas, microplastics torn from the vehicles' tyres, etc., are carried by stormwater runoff from the land. This has further caused microplastics to be unable to be effectively collected because the microplastics in stormwater would directly flow into the surface water with stormwater. This basically caused 12% of 1,043 tonnes of microplastics that were torn from the tyres of vehicles were finally discharged and ended up in surface water (Kole et al., 2017).

The stormwater management process, such as stormwater retention ponds, is a cost-effective method that is used to perform stormwater runoff treatment. Stormwater retention ponds collect the stormwater runoff flow from various water sources. The collected stormwater in the stormwater retention ponds will be further stored in these ponds for several days or weeks. The stormwater runoff carries pollutants such as microplastics when draining into the stormwater retention ponds. The pollutants in the stormwater runoff will be further removed with the sedimentation process under gravitational effects. The gravitational sedimentation process can effectively remove most of the microplastics by forming a thin biofilm layer on the surface of microplastics. The formation of biofilm could promote the deposition of microplastics through the sedimentation of microplastics with low-density behaviour (Harrison et al., 2018; Liu et al., 2019b). As a results, the deposition of microplastics on biofilm can be accumulated in the bottom of stormwater retention ponds for several years, until the sediments of microplastics are discharged from the ponds. The stored stormwater will be discharged from the stormwater retention ponds after the stormwater has been stored for a period of days or weeks (Liu et al., 2019b; Minelgaite et al., 2015; Vollertsen et al., 2009). That the microplastic sediments will be left in the bottom of stormwater retention ponds for several years is mainly attributed to the slow degradation of microplastics under the condition of lacking available exposure to sunlight. Therefore, the stormwater that was discharged from the retention pond was found to contain fewer microplastics than the inlet of the stormwater into the stormwater retention ponds. This is attributable to the fact that most of the microplastics have been removed by the sedimentation process via gravitational effects. However, the removal frequency of microplastic sediments depends on the numbers of microplastics in the stormwater inlet. However, the large amount of microplastics in the stormwater inlet will cause the microplastic sediments in the retention ponds to be removed from the stormwater retention ponds in accordance with the maintaining effect of removal efficiency (Liu et al., 2019b). The disposal of microplastic sediments without treatment in landfills could cause the possible leaching of microplastics into soil, and eventually, into the groundwater. Therefore, the existence of microplastics in stormwater retention ponds also indicates the carriage of microplastics on land by the stormwater runoff into the marine environment through the wastewater treatment plant or the stormwater retention ponds.

According to the studies conducted by Liu et al. (2019b) and Liu et al. (2019a), the presence of microplastics in stormwater retention ponds had implied that stormwater runoff could flow into the stormwater retention ponds by carrying the microplastics on land. In these studies, the sediments from several different stormwater retention ponds in Denmark had been collected and analysed. Basically, the stormwater that enter the stormwater retention ponds originated from residential areas, industrial areas, commercial areas, and discharge from highway (Liu et al., 2019a, 2019b). According to Liu et al. (2019b), they had extracted 2,232 microplastics from the sediments samples that were collected from three different stormwater retention

TABLE 1.5
The Amounts and Number Concentration of Microplastics in Accordance with Five Different Ranges of Particles Sizes

Particles size range, µm	Amount, item	Number concentration, item/kg
10–50	1,135	8,894
50–250	979	7,671
250–500	72	564
500–1,000	15	118
1,000–2,000	31	243

ponds. Among all the microplastics collected, 1,135 of the collected microplastics were found in the particle size-range between 10µm to 50µm, with a concentration of 8,894 per kg, as summarized in Table 1.5. Besides, the particles size in the range within 10µm to 50µm was observed to be the highest amount and number concentration, while the microplastics with a particles size between 500 µm to 1000 µm was found to be the lowest in terms of amount and number concentration, with values of 15 and 118 per kg, respectively (Liu et al., 2019b). From this observation, approximately 50.85% of microplastics found in the collected sediments are microplastics with particle sizes of 10 µm to 50 µm.

In another research study conducted by Liu et al. (2019a), they had collected sediments samples from seven different stormwater retention ponds and performed analysis on the presence of microplastics in every stormwater retention pond. According to Liu et al. (2019a), the total number of microplastics collected from all seven stormwater retention ponds was determined to be 3,436. They also observed that microplastics were found to be present in every stormwater retention pond. They also found and concluded that the sixth stormwater retention pond that received the stormwater runoff from the commercial area posed the highest microplastics concentration of 22,894 per m^3, as summarized in Table 1.6. Table 1.6 shows the particles' number per m^3 and mass of microplastic per m^3 studied by Liu et al. (2019a).

1.2.1.3 Wastewater Treatment Plant's Effluent and Sludge

Wastewater treatment plants are the primary source of microplastics in drinking water sources such as the oceans and rivers (Qiu et al., 2020). Primary and secondary microplastics mainly originate from cities and flow into the drainage system or sewage system by stormwater or surface fallout. A large number of microplastics are discharged into wastewater treatment plants daily with un-treated wastewater. One of the common examples is the discharge of microbeads from personal care products into wastewater treatment plants through the residential drainage system (Issac and Kandasubramanian, 2021; Kalcikova et al., 2017). Wastewater treatment plants also receive stormwater runoff that contains microplastics. Wastewater treatment plants achieve high efficiency in screening and trapping microplastics from flowing out through the wastewater treatment plants' effluent (Prata, 2018). However, there is a considerable amount of microplastic that is detected in wastewater treatment plants' effluent that was discharged from the wastewater treatment plant. This indicates

TABLE 1.6
The Particles' Number per Volume and Mass per Volume for the Microplastics Found in a Total of Seven Stormwater Retention Ponds

Stormwater retention pond	Particles number per volume, particles/m³	Mass per volume, µg/m³
1	796	186
2	490	128
3	1,409	231
4	5,249	521
5	11,348	664
6	22,894	1,143
7	494	85

Source: Data collected and compounded from a study conducted Liu et al. (2019a).

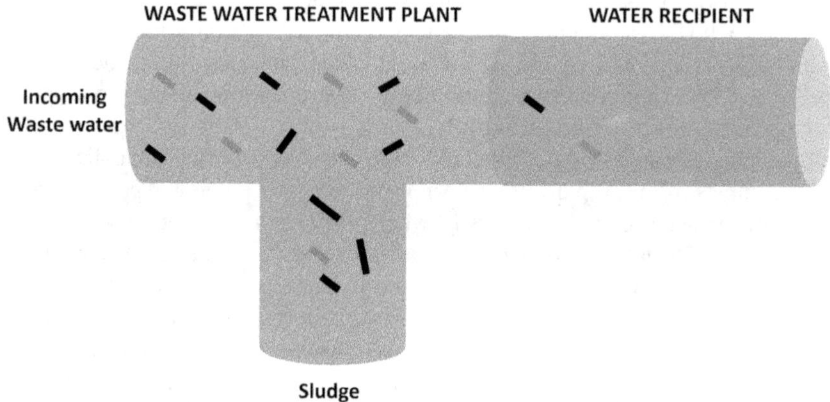

FIGURE 1.2 Schematic diagram of number of microplastics in different in WWTPs (Magnusson and Noren, 2014).

that the microplastics in stormwater runoff are unable to be fully removed, although the removal efficiency of the microplastics in wastewater treatment plants is high (Talvitie et al., 2017). The presence of microplastics in the discharge effluent from wastewater treatment plants could lead to the contamination of the marine environment (Murphy et al., 2016; Prata, 2018). Figure 1.2 illustrates the number of microplastics that present inside the raw wastewater and the treated effluent. According to Figure 1.2, a lower amount of microplastics can be observed to be present inside the wastewater treatment plant's effluent. Microplastics have been trapped in wastewater treatment plants, and thus, removed from the wastewater by undergoing a few treatment processes (Magnusson and Noren, 2014).

According to a few studies conducted by the researchers, approximately 78.34% to 95.00% of microplastics suspended in wastewater can be effectively removed during the preliminary and primary treatment of the wastewater. Most of the large microplastics are removed in the preliminary treatment, as these large microplastics are probably stuck on grit and thus removed (Freeman et al., 2020; Talvitie et al., 2016). Moreover, large microplastics are easily removed with the sedimentation process due to gravitational settlement and entrapment (Murphy et al., 2016; Prata, 2018). This observation was also proven by the studies conducted by Dris et al. (2015), Murphy et al. (2016) and Talvitie et al. (2016). Based on the results obtained by Murphy et al. (2016) and Talvitie et al. (2016), they found 78.34% and 98.40% of microplastics were removed from the wastewater in the preliminary and primary treatment, respectively. In the study conducted by Dris et al. (2015), the number of microplastics had been highly reduced from the range of 260×10^3 to 320×10^3 particles per unit m^3 in the untreated wastewater to the range of 50×10^3 to 120×10^3 particles per unit m^3 after the primary treatment process.

Dissolved air floatation (DAF) can also participate in primary treatment. It is mainly designed to remove low-density particles such as microplastics by creating bubbles to attach all suspended solids together. Then, the floc that containing the adhered microplastics are floated and removed by a skimming device (Bui et al., 2020; Sol et al., 2017). According to Talvitie et al. (2017), the overall percentage of microplastics removal that the dissolved air floatation can achieve was up to 95%. Moreover, approximately 7% to 20.10% of the microplastics were found to be further removed in the secondary treatment in wastewater treatment plants (Murphy et al., 2016; Prata, 2018; Talvitie et al., 2016). In the secondary treatment process, such as activated sludge treatment, microplastics are removed from wastewater by attaching the microplastics with the floc to form higher-density particles and then further settled down to the sludge at the bottom (Freeman et al., 2020; Talvitie et al., 2016). Table 1.7 summarizes the daily number of microplastics and the percentage of microplastics removal in four different locations of sample collected in wastewater treatment plants (Murphy et al., 2016).

Besides, some wastewater treatment plants have tertiary treatment sections. The main aim of the tertiary treatment system is to reduce the number of organic matters, pathogens, nutrients, and solids in which microplastics can be removed in this process (Freeman et al., 2020). In the tertiary treatment system, different technologies can be applied in tertiary treatment systems such as biologically active filter (BAF), disc filter (DF), membrane bioreactor (MBR), rapid sand filtration (RSF), and ozone

TABLE 1.7
The Number of Microplastics and the Removal Percentage of Microplastics in Four Sample Collection Locations at a Wastewater Treatment Plant

Location in wastewater treatment plant	Number of microplastics, 10^6 particles/day	Removal percentage, %
Influent	4,097 ± 1,365	0.00
Effluent of preliminary treatment	2,270 ± 406	44.59
Effluent of primary treatment	887 ± 74	78.34
Effluent of secondary treatment	65 ± 11	98.41

Source: Data collected from Murphy et al. (2016).

(Hidayaturrahman and Lee, 2019; Lares et al., 2018; Talvitie et al., 2016}. However, the application of a biologically active filter (BAF)-type tertiary treatment system was observed to provide an insignificant effect on reducing the concentration of microplastics (Talvitie et al., 2016). Talvitie et al. (2016) found that no decrement was observed in the concentration of microplastics in the effluent treated with a biologically active filter (BAF) when compared with the effluent before being treated with BAF. This observation might be due to the fact that the application of biologically active filter is mainly used to treat the effluent by reducing the nitrogen level in wastewater treatment plants (Talvitie et al., 2017; White et al., 2012). Contrastingly, the overall elimination percentages of microplastics after the tertiary treatment system, such as disc filter (DF), membrane bioreactor (MBR), rapid sand filter (RSF) and ozone treatment, were found to reach higher than 95 % (Hidayaturrahman and Lee, 2019; Lares et al., 2018; Talvitie et al., 2017).

A membrane bioreactor (MBR) is a treatment technology that combines membrane filtration and aeration tanks (Baresel et al., 2017; Bui et al., 2020; Liu et al., 2020b; Pico and Barcelo, 2019). The membrane filtration process and the biodegradation process in the aeration tank will effectively filter the large microplastics, but these processes are unable to effectively filter tiny microparticles and allow these tiny microplastics to pass through (Ngo et al., 2019). In the research conducted by Lares et al. (2018) and Talvitie et al. (2017), they found that the overall microplastics elimination percentages after the membrane bioreactor process had reached up to 99.90% and 99.40%, respectively. The high microplastics-removal percentage of the membrane bioreactor might be due to the fact that the pore size of the filter used is the smallest when compared to other technologies (Iyare et al., 2020; Ngo et al., 2019). In the membrane bioreactor technology, the range of pores sizes that are offered by these microfiltration and ultrafiltration membranes are 0.10 μm to 50 μm and 0.001 μm to 0.10 μ, respectively (Zhang et al., 2020a). The pores with smaller sizes could enable more tiny microplastics to be trapped by the membrane installed inside the membrane bioreactor (Ngo et al., 2019). However, the membrane fouling that may be occurring in membrane bioreactor after a long time of usage can significantly increase the maintenance cost (Bui et al., 2020).

On the other hand, a disc filter (DF) is a liquid filter device which is also used to filter microplastics. A disc filter contains an overlapping filter screen and flange rings by forming multiple layers of filter (Xu et al., 2021). According to Hidayaturrahman and Lee (2019), they found that disc filtration can remove approximately 79.40% of microplastics that are not settled down during the coagulation process. They also found this technology can achieve the overall microplastics elimination percentage up to 99.10%. Meanwhile, the overall microplastics elimination percentage by using the technology of a disc filter can achieve up to 98.5%, as obtained by Talvitie et al. (2017). In this technology, microplastics that have larger particle sizes than pore sizes of the disc filter are effectively retained (Hidayaturrahman and Lee, 2019). However, the efficiency of the disc filter was observed to be significantly reduced when the membranes are clogged by the microplastics. In this situation, the clogging effect of the microplastics further caused the occurrence of accelerated backwash frequency, and a certain amount of microplastics can permeate through the membrane. This had further caused the removal percentage of microplastics to be significantly reduced (Bui et al., 2020; Hidayaturrahman and Lee, 2019; Talvitie et al., 2017).

Rapid sand filtration is also one of the technologies used in wastewater treatment plants to remove microplastics. Anthracite and quartz sand are commonly utilized as the filter material in rapid sand filtration. Rapid sand filtration is mainly used to remove organic and inorganic substances (Bui et al., 2020; Xue et al., 2021). The sand grains in rapid sand filtration can be used to trap and adhere the microplastics effectively (Hidayaturrahman and Lee, 2019; Sembiring et al., 2021; Talvitie et al., 2017). In this technology, the removal of microplastics from wastewater relied on the hydrophilic interaction between sand particles inside the rapid sand filtration system and microplastics. The formation of aggregation between the microplastics and EPS secreted by microorganisms can be trapped by the rapid sand filtration easily (Xu et al., 2021).

By referring to the results obtained by Hidayaturrahman and Lee (2019), the application of ozone treatment can effectively remove 89.90% of microplastics, and the overall microplastics elimination of ozone treatment achieved up to 99.20%. In the ozone treatment process, microplastics are exposed to ozone treatment to cause the degradation of microplastics (Bui et al., 2020; Chen et al., 2018; Hidayaturrahman and Lee, 2019; Singh and Sharma, 2007; Sol et al., 2020). This is attributed to ozone molecules being a reactive species that can attack the structure of the microplastics and then initiate the degradation of the microplastics (Krystynik et al., 2021; Singh and Sharma, 2007). However, the efficiency of the elimination of microplastics would be lowered when the decomposition of macroplastics from large plastic particles into microplastics occurs due to the ozonation. Consequently, the ozonation process could highly increase the number of microplastics in wastewater treatment plants (Bui et al., 2020; Sol et al., 2020; Wang et al., 2020). If the ozonation treatment is incomplete, this could cause the effluent of the wastewater treatment plant to contain the intermediate product or by-product. Although a few studies conducted had proven that approximately 98% of microplastics can be removed from the wastewater treatment plant, a small amount of the microplastics still can escape from wastewater treatment plants and be discharged into the marine environment. The accumulation of the escaped microplastics over a period of time still can lead to significant contamination in the aquatic environment (Murphy et al., 2016; Prata, 2018; Talvitie et al., 2016).

1.2.1.4 Leaching from Soil and Landfilling

The microplastics that leach into soil are one of the main sources of microplastics contamination in drinking water. This is because the microplastics that leach in the soil can further leach into groundwater, which is one of the drinking water sources (Yu et al., 2019). The porous structure of the soil with a lot of micrometres size of pores could allow the leaching of microplastics into a deeper layer of soil (Zhou et al., 2020b). In general, the leachate from the landfill and the sewage sludge discharged from the wastewater treatment plant are the main contributing sources of microplastics that leach into the soil, and thus, further leach into the marine environment. Landfill leachate is one of the common sources of microplastics because some of the microplastics are disposed together with municipal solids-wastes in the landfill and the degradation of larger plastics parts into microplastics (Bhada-Tata and Hoornweg, 2012). The sewage sludge that functions to trap microplastics in wastewater treatment plants will be primarily used as fertilizer (Zhang and Liu, 2018).

According to Nizetto et al. (2016), Qiu et al. (2020) and Yu et al. (2019), approximately 21% to 42% of plastics waste is disposed of and stored in landfills globally. The plastic wastes that are buried in a landfill are subjected to an extreme environment, as the pH of the leachate ranges from 4.5 to 9.0, and thus, the salinity of the landfill is high (He et al., 2019; Qiu et al., 2020). The plastics wastes buried in landfills are subjected to microbial degradation by breaking down the large plastic parts into microplastics under extreme environmental conditions, although oxygen and sunlight are absent in landfills (He et al., 2019; Qiu et al., 2020; Silva et al., 2020). The leachate or wastewater that consists of the microplastics also can be formed from the physico-chemical and biological transformation of plastic wastes that are being landfilled (Silva et al., 2020). According to He et al. (2019), they found that microplastics were present in the leachate of landfills. He et al. (2019) conducted an investigation of six different municipal solid waste landfills from four different cities in China, which are Shanghai, Wuxi, Suzhou, and Changzhou. In this study, they collected the leachate from the collecting wells in 12 landfills, and these samples are considered as the landfill's leachate before the pre-treatment process. The microplastics in these collected leachates were extracted by filtering the sample with mesh sizes of 25, 45, 75, and 150 μm. After that, the collected particles were sorted after the cleaning process and then identified using Fourier Transform infrared Spectroscopy (FTIR) spectrometer.

In research conducted by He et al. (2019), they found the existence of microplastics in the leachates collected from 12 active and closed landfills (Table 1.8). They also found that the total concentration of microplastics in all 12 landfills had reached

TABLE 1.8
The Concentration of Microplastics Found in Leachate Collected from 12 Different Landfills

Landfill	Concentration of plastics (particles per m^3)
1	3,580
2	18,380
3	790
4	24,580
5	1,380
6	1,170
7	960
8	960
9	420
10	2,960
11	2,210

up to 24.58 particles per litre. In the collected leachate from landfills, different types of microplastics such as polyethylene (PE), polypropylene (PP), polyvinyl chloride (PVC), etc., were found in these collected leachates. The presence of microplastics that had been identified to occur in leachate samples was mainly in the form of lines, fragments, flakes, foams, and pellets (He et al., 2019). Lines are the elongated particles, while flakes are the particles in sheets. The fragments are irregular and thick pieces. Pellets are plastic particles with spherical shapes, while foams are spongy particles. According to He et al. (2019), they found that the microplastics in the form of lines, fragments, flakes, and foam occupied approximately 99.36% of the plastics found in the leachate landfills. However, the microplastics in the form of pellets was observed to occupy the lowest portion of microplastics that were in the leachate landfills. The high amount of microplastics found in leachate from landfills indicates the fragmentation process of microplastics is serious in the landfills (He et al., 2019).

On the other hand, microplastics are also found to be present in the sludge used in wastewater treatment plants. The microplastics removed from the treatment process in wastewater treatment plants are trapped in the sludge (Qie et al., 2020). According to Talvitie et al. (2016), microplastics removed from wastewater in the activated sludge process were recycled inside wastewater treatment plants, where the microplastics were not removed together with the effluent. There is an approximately 20% of the microplastics had been recycled in the treatment process. Approximately 80% of the microplastics remained in the dried sludge after dewatering process of the sludge (Talvitie et al., 2017). On the other hand, Mason et al. (2016) investigated and found the presence of 15,800 particles per kg of microplastics in the sludge. The presence of microplastics in the sludge from wastewater treatment plants is mainly due to the microplastics being unable to be degraded chemically, biologically, and thermally in wastewater treatment plants. During the treatment process of wastewater treatment plants, the microplastics that are removed from the wastewater are settled down at the bottom of equalization basin. The disposal of sludge from the wastewater treatment plant could cause the deposited microplastics on the sludge to travel to the marine environment (He et al., 2019). This is because the sludge from wastewater treatment plants will be treated and then the treated sludge can be further used as fertilizer in the agricultural field (Rolsky et al., 2019). Microplastics in the sludge that were used as fertilizers in agriculture leached into the soil and eventually leached into the groundwater. This caused the groundwater, one of the main sources of drinking water, to be contaminated by microplastics (Talvitie et al., 2017). Sludge is used as a soil amendment to increase soil fertility. This is because the favourable soil properties and nutrient recyclability are created due to the presence of several nutrients in sludge such as sodium and magnesium (Clarke and Smith, 2010; Rolsky et al., 2019).

Microplastics can enter groundwater bodies by the leaching process. The microplastics in landfills can leach into groundwater bodies because landfills in some developing countries do not have leachate collection systems and engineered liners (Kjeldsen et al., 2002; Parvin and Tareq, 2021). The absence of engineered liners in the landfills had further caused plastics waste to be unable to be appropriately stored in landfills, and eventually, this had led to the migration of microplastics into groundwater. On the other hand, the leachate from landfills can carry microplastics into surface water through hydraulic connections (Parvin and Tareq, 2021). The vertical, downward movement of microplastics is assisted with the occurrence of macropores that are present inside the soil. In contrast, the retention of microplastics is more favourable in soils that have a larger portion of micropores, instead of the leaching process. This is mainly attributed to the fact that the size of microplastics is smaller than the size of the pores in soil, and this causes the microplastics to be able to leach into the soil and enter the groundwater (Blasing and Amelung, 2017). Factors such as biota, features, aggregation, and cracking of soil can contribute to the vertical and horizontal distribution of microplastics in soil. The cracks found in soil are a pathway for microplastics to penetrate to a substantial depth and enter the deeper layers of soil. Ploughing is a method by which microplastics can enter deeper soil layers until reaching the depth of the plough. The depth of the soil that the microplastics can leach into mainly depends on conventional tillage practices. On the other hand, soil features such as the presence of macropores could further enhance the vertical movement of microplastics in soil. Macropores are pores with a length of approximately more than 0.08 mm and are generally made by earthworms and roots. Thus, microplastics can enter easily to deeper layers of soil through macropores (Rillig et al., 2017).

Moreover, soil biota is one of the factors that contributes to the leaching of microplastics into the soil. The small scale of transportation by soil biota can spread microplastics horizontally, thus further enhancing the entry of microplastics into the soil (Rillig et al., 2017). Rillig et al. (2017) investigated the migration of microplastics into the soil by using earthworms. In their research, they used four adult and healthy earthworms, and the soil used was initially sieved to remove stones and then steamed to remove other soil biotas that can affect the accuracy of the results. In this study, the microplastics used is polyethylene (PE), with the range of microplastics size within 710 μm to 2800 μm. Rillig et al. (2017) conducted their experiment in an air-conditioned greenhouse for 21 days. In this experiment, the bottoms of 40 pots were sealed with a permeable, black fleece to avoid the escape of earthworms and stagnant water. The same mass of microplastics was added to the surface of the soil for every pot. After 21 days, the results showed that the microplastics were moved by the earthworms into the soil. The smallest range of size of microplastics was observed to move the bottom layer of soil by earthworms, whereas other the size ranges of microplastics were moved to the middle soil layer. The polyethylene (PE) microplastics proved to be transported by earthworms down to a depth of 10 cm. Thus, the soil biota can move the microplastics into a deeper layer of soil, and eventually, leach into groundwater (Rillig et al., 2017).

1.2.1.5 Personal Care and Cosmetic Products

Personal care and cosmetic products are one of the main sources of microplastics that are generated by industrial activities. This is attributed to the fact that microplastics are added into personal care and cosmetic products during the manufacturing process. The majority of microplastics in personal care and cosmetic products are discharged into wastewater treatment plants through industrial wastewater and sewage wastewater systems during the washing process. Some of the personal care and cosmetic products will be disposed as rubbish when over the expiration date. Eventually, the disposed personal care and cosmetic products will end up at landfills. Microplastics are added as ingredients in personal care and cosmetic products such as toothpaste, facial cleansers, facial scrubs, shower gels, sunscreen, cosmetics, shampoo, etc. The microplastics that are added into personal care and cosmetic products during the manufacturing process are known as microbeads. These microbeads are manufactured, plastic particulates with an approximate particle size range of 1μm to 1000 μm, which is dependent on the performance of microplastics in the personal care and cosmetic products. Besides, the particle sizes of microbeads in personal care and cosmetic products also depend on the percentages of microbeads required in the products. For example, manufactured microbeads are usually applied in personal care and cosmetics products for exfoliating purposes. Most of the microbeads used in the production of personal care and cosmetic products are non-biodegradable polymers. Non-biodegradable polymers of microbeads can persist in the environment for a long period of time, which is attributed to the fact that plastic microbeads may need a very long time to complete degradation (Leslie, 2015). Microbeads made of non-degradable polymers that are used in personal care and cosmetic products can be divided into two sections, which are thermoplastics and thermoset. The thermoplastics used in personal care and cosmetic products include polypropylene (PP), polyethylene (PE), polytetrafluoroethylene, polystyrene (PS), etc., while thermosets mainly consist of polyethylene terephthalate, polyurethane, polymethyl methacrylate. The microbeads that are commonly used in personal care and cosmetic products are mostly made of polyethylene; up to 93% of the microbeads are made of polyethylene.

In personal care and cosmetic products, microbeads generally act as exfoliants, film formers, emulsifiers, viscosity controllers, and bulking agents. In fact, natural ingredients such as sea salts, oatmeal, and almond flour were added into personal care and cosmetics products during the production process instead of microbeads before the 1990s (Lassen et al., 2015). Since the 1990s, the application of microbeads in personal care and cosmetic products has sharply increased. This is attributed to the usage of microbeads having provided higher effectiveness in abrasive strength and qualities of the personal care and cosmetic products in comparison to natural materials. However, the discharge of microbeads used in personal care and cosmetic products has caused pollution in the marine environment. To obtain a suitable alternative to microbeads, various alternatives from natural plants such as crushed walnuts shell, oats, jojoba seeds, pumice, etc., with their micro abrasives effects and environment impacts have been assessed and discussed in terms of different types of materials (plastics, natural plants, and minerals) and the lifecycle of the personal care and cosmetic products with the use of these materials (Lassen et al., 2015). In research conducted by Hunt et al. (2003), they performed an investigation and assessment on the three different impact categories by using a lifecycle analysis method. In this study, the three different impact categories that were included in the assessment were the impacts on ecosystem damage or detriment impacts on environment, the resource scarcity or loss of non-renewable resources, and lastly, the impacts on human health. They had conducted a comparative lifecycle assessment to evaluate the impacts of several alternatives (such as microbeads from polymer materials, natural plants, minerals, etc.) for all the stages in their lifecycle. In this study, they were focused on the extraction process, processing, and transportation of every stage involved, and they also compared all the data to the microbeads.

According to Hunt et al. (2003), they found that almost all the plant's alternatives could contribute to the environmental problem due to land and water consumption. Among all the alternatives from plants, the usage of almond was found to cause severe, detrimental impacts on environments. This is mainly due to the fact that the plantation of almond required the heavy usage of water in water-scarce regions. Plant materials were evaluated to display relatively high environmental impacts due to their occupation of land usage during plantation. Besides, the application of mineral-type alternatives for personal care and cosmetics products was found to have bad impacts in the category of impact on the ecosystem, especially high global-warming impacts. Titanium oxide is the mineral-type alternative that was found to have significant detrimental impacts on the ecosystem. On the other hand, carboxymethyl cellulose and citric acid are mineral-type alternatives that can lead to high global-warming impacts, which is attributed to formation of fine particles and global-warming problems. This is because the manufacturing of carboxymethyl cellulose and citric acid as an alternative to minerals required higher levels of temperature during processing. However, minerals such as pumice and sand were found to have minor overall impacts on ecosystems. On the other hand, the impacts of plant-material alternatives such as apricot, bamboo, olive, orange, wheat, etc., were also found to have minor overall impacts on ecosystems. The impact of plastic microbeads on the ecosystem was observed to pose little demand for water usage and land occupation, in comparison to other alternatives for personal care and cosmetic products.

On the other hand, Hunt et al. (2003) also evaluated the effects of other alternatives in applying personal care and cosmetic products on impacts of the resource scarcity category. Impact in the category of resources scarcity is represented by the loss of non-renewable resources during the manufacturing and production process. They observed that the mineral-type alternative of vermiculite provided acceptable impact in the category of resources scarcity, in comparison to other mineral-type alternatives such as carboxymethylcellulose, citric acid, and titanium oxide. Carboxymethylcellulose, citric acid, and titanium oxide displayed high impacts in the categories of the resource's scarcity due to the high usage of fossil fuels. All the alternatives from plants were found to have a lower impact in the category of resource scarcity, except the alternative of almonds. Plastic microbeads were also found to provide significant impact on resource scarcity due to the usage of fossil fuels during the manufacturing and production process.

In the third category of human health impacts, the evaluation is conducted by measuring in the terms of disability-adjusted life years, which also represents the loss of the equivalent of one year of full health. By referring to the study conducted by Hunt et al. (2003), they observed that, in almost all the plant alternatives, they caused low impacts in the category of human health impacts, except for almonds. Almond was found to cause a high level of human health impacts, and this is mainly attributed to the non-carcinogenic toxicity of almonds. On the other hand, the minerals type of alternatives such as carboxymethylcellulose, citric acid, and titanium dioxide had displayed high impacts in terms of human health. The high impacts in the human health impacts category of carboxymethylcellulose, citric acid, and titanium dioxide are mainly attributed to the terrestrial acidification problem, the global warming issue, and the formation of fine particles, which are harmful to human health. Similarly, the application of microbeads made of plastics such as nylon and polymethyl methacrylate also displayed a high impact on human health. Most plastics microbeads can contribute to the global warming issue and the formation of fine particles. The high impact of the global warming issue of plastics in comparison to other materials is mainly due to high resource-scarcity. This is because most of the raw materials used to produce plastics are fossil fuels. Furthermore, the high impacts of plastic microbeads are also attributed to the formation of fine particles during manufacturing process, and it is harmful to human health. In terms of overall performance, alternatives of silica were observed to show lower environmental impacts than other alternatives of micro-abrasives. However, further investigation is needed on alternatives to micro-abrasives to replace the usage of microbeads in personal care and cosmetics products in terms of being safe to use by humans and having fewer environmental impacts.

1.2.1.6 Synthetic Textiles

Synthetic textile is another source contributing to the presence of microplastics in aquatic environments. In textile industries or the daily washing process in household areas, microplastic fibres were released and removed from synthetic textiles such as carpets, curtain, clothes, etc., due to abrasion in the washing machine during the clothes washing process (Lassen et al., 2015). In synthetic textile industries, the types of polymers commonly used in the manufacturing of synthetic textiles are acrylic, polystyrene, polyamide, and polyester (Sundt et al., 2014). The consumption of synthetic textiles is found to be approximately 69.7 million tonnes per annum globally, while the growth of the consumption of synthetic textiles fibres had increased by approximately 79.3 % in the period of 1992 to 2010. The increment in the consumption of microplastic fibres is mainly attributed to the increasing development of countries, social media, and technologies. Currently, clothes made with synthetic textiles can be easily purchased online via social media such as Facebook, Lazada, Instagram, etc., and can be bought from real shops. According to Boucher and Friot (2017), they found that the amount of microplastic fibres from synthetic textiles released into the ocean for India and South Asia is, significantly, the highest, with the releasing amount approximately 15.9%. After that, they found that this is followed by China, with a value of approximately 10.3%. In s study conducted by Lassen et al. (2015), they found the appearance of microplastic fibres in the water released after washing quilts, sweaters, and shirts, respectively, using a washing machine with a rotational speed of 600 rpm. In this study, the number of microplastic fibres released from the washing of quilts is determined to be approximately 900 microfibres, while the number of microplastics fibres released from the washing of sweaters and shirts are found to be approximately 1,900 microfibres and 1,160 microfibres, respectively. This observation also indicates that the releasing of microplastic fibres of sweaters was found to be significantly higher than quilts and shirts. In a study conducted in United Kingdom, they found that every litre of water discharged from laundry was detected with 20 mg of microplastic fibres or 200 microplastic fibres. In the worst condition, microplastic fibres can release up to approximately 280 mg or 1,900 microplastic fibres for every washing process of 0.2 kg synthetic textiles. In a study tested in the Netherlands (Lassen et al., 2015), the washing of 660g nylon-type synthetic textiles was found to release 260 mg of microplastic fibres. In a record of Sweden, the amount of microplastic fibres released from the washing process of synthetic textiles into the sewage system in one capita of a day is approximately 0.6 g or 4,000 microplastic fibres per day. In this study, the water discharged into the sewage system is mainly released from laundry. According to Sundt et al. (2014), the amount of microplastic fibres released per annum can be estimated and calculated by taking 280 kg of clothes are washed per capita per year, 50% of the clothes are made from synthetic textiles, and a total of 4,200 litres of water will be discharged from the washing machines, with a total of 5 million per capita. According to a study conducted by Sundt et al. (2014) as shown in Table 1.9, the total estimated volume of microplastic fibres that is released during the washing process of synthetic textiles in three different counties of the Netherlands, the United Kingdom, and Sweden are found to be in the range of 275 tonnes microplastic fibres per year to 1,095 tonnes of microplastic fibres per year with a capita of 5 million. The wastewater containing microplastic fibres was released from the washing machine and then discharged into the drainage system. Then, wastewater containing microplastic fibres was delivered into wastewater treatment plants. In wastewater treatment plants, the microplastic fibres that are collected in the sewage sludge are approximately 100 tonnes of microplastic fibres per year to 530 tonnes of microplastics fibres per year. By comparing the total microplastic fibres in water discharged from laundry and the total microplastic collected in the sewage sludge, a significant number of microplastic fibres are still unable to be removed effectively in wastewater treatment plants due to the smaller particle sizes of microplastics fibres, which are mainly in the range of 20 μm to 30 μm (Lassen et al., 2015). The un-captured microplastic fibres in wastewater treatment plants is estimated to be 6 tonnes of microplastic fibres per year to 60 tonnes of microplastic fibres per year, and these microplastic fibres would be released into the aquatic environment. However, most of the microplastic fibres are still able to be captured by the filtration system in wastewater treatment plants.

TABLE 1.9
The Total Amount of Microplastic Fibres Released after the Washing of Synthetic Textiles per Year

Country	Amount of microplastic fibres released annually	
Netherlands	$\frac{280\ kg}{year.capita} \times \frac{260\ kg}{0.66\ kg} \times 5\ million \times \frac{1\ tonne}{10^9\ mg} =$	275.76 tonnes/year
United Kingdom	$\frac{4,200\ L}{year.capita} \times \frac{30\ mg}{1\ L} \times 5 million \times \frac{1\ tonne}{10^9\ mg} =$	315 tonnes per year
Sweden	$\frac{0.6\ g}{day.capita} \times \frac{365\ day}{1\ year} \times 5\ million \times \frac{1\ tonne}{10^6\ mg} =$	1,095 tonnes per year

1.2.1.7 Plastic Mulching

Plastic mulching is one of the contributing sources to the occurrence of microplastics in agricultural lands. Plastic mulching is a technology that has been widely implemented in the global agricultural industries to ensure improved sustainability. In this technology, plastic mulches can help to improve the production yield of crops by reducing the growth of the weeds around the crops. Besides, plastic mulches also can reduce the evaporation of water to retain the moisture of soils by modifying and controlling the transpiration rate of crops, humidity, and the temperature of soils to create an optimum microclimate condition for the crops (Hayes, 2021; Qadeer et al., 2021). According to Ren et al. (2021), the global agricultural plastic mulches market in 2016 was estimated at about 4 million tonnes, with approximately $10.6 million. They also predicted the global market would achieve approximately 9.14 million tonnes by the year of 2030, with an average annual growth rate of 5.6% due to the wide application of plastic mulches. China is the largest plastic mulches user in the global market, accounting for approximately 60% to 80% of plastic mulches' global usage. Besides, approximately 19% of the agricultural lands in China are covered with plastic mulches. According to Meng et al. (2020), the application of plastic mulching in China has rapidly increased since the year of 1982, with 117,000 ha of agricultural lands had used a total of 6,000 tonnes of plastic film. In 2016, they found that approximately 18.4 million ha of agricultural lands had been covered with a total of 1.5 million tonnes of plastic film. Currently, most agriculture industries have used plastic mulching technology to ensure the amount of crop supply meets the global demand due to the increase of the worldwidc population.

Although plastic mulching technology has provided many benefits to our daily life, it is also found to contribute a detrimental effect on the environment. This is mainly attributed to the improper handling or management of the used plastic mulches, which could significantly contribute to the presence of microplastic on lands. According to Li et al. (2020), most of the globally conventional plastic mulches that are currently used in agricultural industries are made of low-density polyethylene (LDPE). Besides, the plastic polymers that are commonly used in traditional plastic mulching technology are polyethylene and polyvinyl chloride (PVC) (Ren et al., 2021). In plastic mulching technology, the usage of polyethylene has become the major choice of plastic polymer that used in the manufacturing of plastic film for agricultural purposes. This is because polyethylene has ease in the processing, good flexibility, high durability, high resistance to chemicals, and is non-toxic and odourless (Ren et al., 2021). However, the non-biodegradable behaviour of polyethylene has significantly contributed to environmental impacts due to its tendency to persist in the environment for a long period (Kasirajan and Ngouajio, 2012).

According to Hayes (2021), the common and popular method used to treat the plastic mulches in agricultural lands after the harvest process is to remove the plastic mulches from the soils and then follow up by covering the soils with the new plastic mulches for the following crops. Most agricultural land owners would conduct illegal, open burning of the plastic mulches, and some agricultural land owners would send the used plastic mulches to the landfilling centre after the harvest process. The mismanagement process of the plastic mulches in agricultural lands is mainly attributed to the high cost required to recycle the used plastic films with high contaminants (Qi et al., 2020). The high contaminants of used plastic mulches in agricultural lands with high contaminant weight up to approximately 40% to 50% is mainly due to the usage of fertilizers, pesticides, UV light, soil particles, and other contaminants. However, the plastic mulches that are accepted for recycling must containing less than 5% of the total contaminants' weight (Kasirajan and Ngouajio, 2012). Besides, cleaning or removing the plastic mulches from the soils is very challenging due to the fact that plastic film is very thin and hard to be fully remove from soil (Qi et al., 2020). If plastic mulches are not fully removed from the soil after the harvesting process, the leftover plastic mulches would further accumulate in the agricultural soils. The leftover plastic mulches that had accumulated in the soils would lead to fragmentation into smaller particles by the soils' loosening action during the preparation process in planting the new crops. The accumulated plastic mulches can also further fragment into microplastics or nanoplastics through the slow degradation process with the presence of UV light and temperature fluctuations (Li et al., 2020). As plastic polymers in the agricultural soils have the persistence to degrade, slow degradation would take place, and this would further cause the number of microplastics in the terrestrial environment to significantly increase year over year. This microplastic in the terrestrial environment can be further transferred to nearby areas or aquatic environments through runoff to streams and rivers. The microplastics originating from the degradation of plastic mulches that appear in both terrestrial areas and aquatic environments will harm the environment and threaten the health of consumers. The additives added in the plastic mulches will leach from the used plastic mulches during the fragmentation process. The additives may react with the existing agricultural chemicals in the soils and change the physical and chemical properties of the agricultural soils. Besides, the presence of microplastics in the agricultural soils would adsorb the heavy metals and hydrophobic, organic contaminants in the soils and become the carrier for harmful chemicals. Harmful chemicals inside microplastics will be carried along during the travelling of microplastics to the new environment and thus enter the body of the consumers. This would further cause the contaminants carried by the microplastics to be spread to the surroundings by entering the body of consumers and then bioaccumulate in the body of consumers and thus lead to health impacts (Qadeer et al., 2021).

Various experimental studies had been conducted to investigate the effect of plastic mulching activity on the occurrence of microplastics in the agricultural soils. According to Isari et al. (2021), they performed an investigation on the soil samples collected from agricultural lands to produce watermelon and tomato, which had an operating time of over 10 years in Elis, Western Greece. In this study, they selected watermelon and tomato fields that are far from industrial areas to prevent the contamination of microplastics from other sources that will affect the results of microplastics due to plastic mulching technology. Their results showed that the number of microplastics obtained in the overall five soil samples from the watermelon field was approximately 301 ± 140 particles/kg, while the overall five soil samples from the tomatoes field were approximately 69 ± 38 particles/kg. According to Isari et al. (2021), the number of microplastics obtained in the watermelon field was observed to be higher than the microplastics obtained in the tomato field. In another investigation conducted by Huang et al. (2020) on the soil samples collected from the

TABLE 1.10
The Number of Microplastics in Different Layers of Soils from 5 Years of Mulching a Cotton Field

Year of mulching	Depth (cm)	Numbers of microplastics, particles/kg	Average numbers of microplastics, particles/kg
5	0 to 5	62 ± 21	80.3 ± 49.3
	5 to 20	103 ± 69	
	20 to 40	68 ± 41	
15	0 to 5	200 ± 158	290 ± 49.3
	5 to 20	450 ± 163	
	20 to 40	220 ± 113	
24	0 to 5	1,055 ± 275	1,075.6 ± 346.8
	5 to 20	1,375 ± 269	
	20 to 40	825 ± 420	

cotton forms in Xinjing Uygur Autonomous Region, China to verify the relationship between the duration of application of plastic mulches in agricultural lands and the accumulation of microplastics in the soil, the plastic mulches were made from polyethylene with approximately 0.006 to 0.009 mm thickness, while the soil samples were collected from the cotton farms with different levels of agricultural soils. Table 1.10 summarizes the number of microplastics obtained in different layers of soils collected from samples and the total amount of microplastics in soil samples that were collected from cotton after 5, 15, 24 years for applying plastic mulching (Huang et al., 2020).

Several replacement materials and methods have been suggested to replace the use of plastic mulching in agricultural industries to avoid the accumulation of microplastics from the degradation of the polyethylene plastic mulches. The suggested replacement materials and methods such as the paper mulches, biodegradable plastic mulches, photodegradable plastic mulches, and recycling of plastic mulches have been investigated and compared with conventional plastics mulches. The paper-based film has been developed on agricultural lands to replace the usage of plastic mulching. However, the poor mechanical properties and water resistance effect has caused the paper-based film to be easily break when exposed to the environment, and the degradation duration is too short to do planting. According to Kasirajan and Ngouajio (2012), the addition of fibre mats and thicker paper are suggested to overcome the shortness of the paper-based film. On the other hand, photodegradable plastic mulches have been developed to replace the conventional plastic mulching. This is because the photodegradable plastics would break down into carbon dioxide and water when subjected to UV radiation. Although photodegradable plastic mulches are more practical than conventional polyethylene plastic mulching, the degradation process may be incomplete when the UV radiation is reduced or absent (Kasirajan and Ngouajio, 2021). Another method is suggested to reduce the usage of plastic mulches by applying biodegradable plastic mulches in agricultural industries. This is because biodegradable plastic mulches will completely breakdown into carbon dioxide, water, and microbial biomass through the aerobic bio-assimilated process with the presence of microorganisms and will not release any residue in the soil over time (Hayes, 2021). Although there is no residue in keeping the soil, which is relatively environmentally friendly, there are some important things that still need to be considered and further investigated such as the interaction between the biodegradable process, the factors that affect the bio-assimilated process, types of biodegradable polymer mulches to different types of crops, etc. (Kasirajan and Ngouajio, 2012; Zhang et al., 2022).

1.2.2 Mixed Source of Microplastics

Sludge from wastewater treatment which is commonly applied as fertilizer in agriculture is another source of microplastics that can consist of primary and secondary microplastics. The sewage wastewater from household and industrial areas contains several mixtures of different polymer types of microplastics such as microbeads in personal care cosmetic products, microfibres in textiles, fragments of plastic waste, etc., which will be transported to the wastewater treatment plants before being discharged into the rivers and oceans. The sewage wastewater will eventually undergo preliminary, primary, secondary, and tertiary treatment processes to remove microplastics. The treated wastewater will be released into the environment at the end of the treatment process. In the preliminary treatment, the operations involved are the screening and aerated grit chamber and are aimed to remove the large size of solid particles to prevent the coarse solid particles from destroying the operations after the preliminary treatment process. The removal efficiency of microplastics in preliminary treatment can reach up to the approximate range of 35% to 58%. After that, the water discharged from preliminary treatment would proceed to primary and secondary treatment systems to remove the settled solid particles, organic and inorganic compounds inside the wastewater. Besides, the primary and secondary treatment processes also function to remove the approximately 97.8% of the microplastics. The discharged wastewater from the secondary treatment process would enter the tertiary treatment process to remove the presence of heavy metals and other chemicals in the sewage. The overall efficiency of the wastewater treatment plant to remove the microplastics is approximately up to 99% (Golwala et al., 2021). As an example, a wastewater treatment plant in Italy has received approximately 1×10^9 microplastics per day from a population of 1.2 million capita. According to Golwala et al. (2021), the removal efficiency of the whole wastewater plant is approximately 84%, and about 8.4×10^8 particles per day and will be successfully captured in the wastewater treatment plant. For what remains, 1.6×10^8 microplastics will be discharged into the aquatic environment (Alvim et al., 2020).

Most of the microplastics' shapes that are obtained in the wastewater treatment plants are approximately 59% in fibre shape, followed by approximately 33% in fragment shape. Although most of the microplastics found in the wastewater treatment plant mainly originate from personal care and cosmetic products, these microplastics are mostly obtained in irregular shape. It is expected that approximately 1.09×10^9 of microplastics will enter terrestrial environments through the bio-solids pathway daily. According to Zhang et al. (2020a), approximately 30,000 tonnes to

44,000 tonnes of microplastics are introduced into cropland in North America per annum, while approximately 63,000 tonnes to 430,000 tonnes enter the cropland in Europe. Microplastics in Australia are projected into the agroecosystems in biosolid form at a rate of 2,800 to 19,000 tonnes per year (Zhang et al., 2020a). As mentioned earlier, the number of microplastics that collect in the sludge of wastewater treatment plants are highly depended on the population served. The majority shape of microplastics found in the sludge from wastewater treatment plants were flakes and fibres, with the percentage of 60.4% and 35%, respectively. Flake-shaped microplastics are mainly formed due to the degradation of larger plastic debris into smaller-sized microplastics, while fibre-shaped microplastics are obtained from the washing process of synthetic textiles. According to Zhang et al. (2020a), there are no microbeads used in personal care and cosmetics products that were detected in this study. This might be due to the habits of the consumers in the selected regions which have different consuming habits or awareness in resisting the usage of personal care and cosmetics products with the presence of microbeads. Most of the microplastics detected were polypropylene and polyethylene, which are the most used polymers in the plastic manufacturing and production industries (Zhang et al., 2020a). According to Alavian Petroody et al. (2021), the number of microplastics in the final sludge could be further reduced by undergoing a series of sludge-processing steps. However, the development of suitable guidelines for proper disposal of final sludge is necessary to avoid the discharge of microplastics trapped in the final sludge into the environment. This is because the sludge disposed from wastewater treatment plants will be further used as fertilizers in farmland and may lead to a pathway for the microplastics to enter the environment (Zhang et al., 2020b).

1.2.3 Microplastics as Vector of Pollutants

By referring to Huang et al. (2021), microplastics with smaller sizes have a greater surface area and higher hydrophobic surface, which will lead to the enhancement effect on the absorption mechanism between microplastics and other hydrophobic organic pollutants. In general, there are various types of adsorption mechanisms that could attract the hydrophobic pollutants onto the microplastics, which are hydrophobic bonding, van der Waals force, halogen bonding, and π-π interactions. According to Fu et al. (2021), different types of adsorption mechanisms may be applied for different types of hydrophobic, organic pollutants to adsorb onto the microplastics. On the other hand, the heavy metals in the terrestrial or aquatic environment can also adsorb onto the microplastics. By referring to Cao et al. (2021), the types of mechanisms that are involved in the interaction of microplastics and heavy metals are electrostatic interactions, surface completion, precipitation, and biofilms. Besides, various chemical additives have been added into plastic products during the manufacturing process to improve the performance properties of plastics products. During the degradation process of plastics products, the plastic products experience a fragmentation reaction into smaller microplastics, and the chemical additives store inside the microplastics. The fragmented microplastics further degrade into a smaller size of microplastics during the long period of the degradation process. The continuous degradation process will further cause the leaching of the chemical additives into aquatic environment. The microplastics adsorbed with hydrophobic organic pollutants and heavy metals would also be released into the aquatic environment through the desorption process. The ingestion of these contaminated microplastics by aquatic organisms may further cause the occurrence of leaching and desorption inside the body of the organism in the aquatic environments. This phenomenon has further increased the potential health risk of the ingested organisms (Huang et al., 2021).

1.2.4 Trophic Transfer of Microplastic along Aquatic Food Chains

Globally, there are approximately 5 trillion microplastics, with more than 0.25 million tonnes floating on the ocean surface (Madhav et al., 2020). These microplastics can be transferred indirectly, through the trophic, along the food chains, when the high-trophic-level predator consumes the low-trophic-level prey, which is microplastic-containing. Since the aquatic organisms such as fish, shellfish, etc., are one of the common foods of human daily life, the microplastics can be passed into humans' bodies through trophic transfer along the aquatic food chains. Basically, aquatic organisms with higher trophic levels are easily being affected by contaminants in the aquatic environment, and these contaminants will further bioaccumulate and biomagnify in the bodies of consumers (Nelms et al., 2019). The transferring of microplastics through food chains will occur along different paths and ways, where microplastics can be transferred from lower-trophic-level prey to higher-trophic-level predator, and the microplastics can bioaccumulate in the bodies of consumers.

The primary producers in aquatic ecosystems such as green algae, duckweed, seagrass, seaweed, and phytoplankton are the first trophic level in the nutrient system. The primary products can undergo the photosynthesis process to produce food for the primary consumers in the nutrient system and can also serve as the supplier of oxygen gas in aquatic ecosystems. The microplastics present in aquatic environments can attract to the surface primary producers through a variety of mechanisms, including electrostatic interactions, biofilms, and producer surface morphology. Microplastics will stain on the surface of the primary producers and cause the accumulation of microplastics in primary producers. Primary consumers such as herbivorous aquatic organisms will ingest the primary producers with microplastics stained on the surfaces of the primary producers, thus creating migration pathways for the microplastics in in the food chains system (Huang et al., 2021). Several studies have been conducted to investigate the transferring pathways of microplastics from primary producers to primary consumers. According to Goss et al. (2018), the seagrass species that they collected were turtle grass (*Thalassiatestudium*) and 16 seagrass blades, and these collected seagrasses were separated into four categories, in accordance with the coverage of epibiont on the surface of seagrass blades. The types of microplastics found in these seagrass blades were microfibres, microbeads, and plastic debris, which account for 81%, 16%, and 3%, respectively. There was no significant relationship between the coverage of epibiont on the surface of seagrass leaves and the number of microplastics found on the surface. However, the epibiont and biofilms stained on the surface of seagrass leaves are assumed to capture the microplastics in the aquatic environment. More amounts of epibiont and biofilms coverage could lead to more microplastics being captured and affixed to the seagrass blades. No significant results were obtained due to the limitation effect on the sampling size in collecting the seagrass blades. Besides, the parrotfish in the aquatic environment preferred to consume the seagrass with high epibiont coverage, as the obtained results showed that, the more epibiont coverage on the seagrass blades' surface, the higher fish gazing scars could be significantly found. In another study conducted by Goss et al. (2018), they investigated the adsorption effect of the microplastics on the seaweed surfaces and the transfer of microplastics on the seaweed surface to herbivore species. In this study, the seaweed species used is *Fucusvesiculosus*, while the species of herbivore used is periwinkle *Littorinalittorea*. From the results, the number of microbeads and plastic debris affixed to the surface of seaweed significantly increased with the increase of the particles' concentration in the seawater. However, the highest number of microfibres was found to be affixed to the seaweed's surface at the moderate particle concentration in seawater when compared to the seawater with low and high particle concentrations. This might be due to the fact that a moderate particle concentration of microfibres in seawater tend to float on the water's surface, while the microbeads and plastic debris tend to sink to the bottom. On the other hand, the feeding activity of the periwinkle *Littorinalittorea* was found to be less sensitive towards the appearance of microplastics in the food. Microplastics were found in the stomach and intestines of periwinkle, which indicates the ingestion of microplastics. However, the majority of microplastics were also found in the faecal matter of the periwinkle, which indicates the accumulation of the microplastics in the intestines of periwinkle for a period. As a result, the microplastics stained on the surface of seaweed could be ingested by the herbivores or higher trophic level organisms, and thus, enter the food chains. Since the periwinkle is a food source to organisms with higher trophic levels, the microplastics could enter the food chains by transferring to higher-trophic-level organisms.

In addition, the trophic transfer of microplastics from the primary consumers with trophic level 2 to secondary consumer with trophic level 3 was investigated by observing the trophic transfer of microplastics from mussel to crab (Farrell and Nelson, 2013; Santana et al., 2017; Watts et al., 2014). According to Farrell and Nelson (2013), they conducted an experiment on the trophic transfer of microplastics from the blue mussel *Mytilusedulis* to shore crab *Carcinusmaenaswas*. After one hour of feeding the crabs with polystyrene microspheres for a period of time, the stomach of the crabs was found to contain approximately 1,025 ± 556 microspheres. After two and four hours of feeding, the stomach of another two crabs were examined to have approximately 883 ± 589 and 1,007 ± 572 microspheres, respectively. Besides, the microspheres were also found to appear inside the gills, ovary, and hepatopancreas of the crabs. No microspheres could be found in the crabs' stomachs after 21 days; this might be due to the discharging of the ingested microplastics to outside of the body through the faecal matter. The ingestion of mussels containing microplastics caused microspheres to enter the tissues and haemolymph of crabs, and thus, these microspheres could bioaccumulate and biomagnify in the crabs' body (Farrell and Nelson, 2013). This will increase the potential risk of transferring the microplastics to crab predators, as it will take a longer time to remove the microplastics completely from the crabs. According to Watts et al. (2014), the absence of microplastics in the haemolymph of crabs indicates no translocation of microplastics from foregut to haemolymphs.

1.3 MICROPLASTICS AS CARRIERS OF POLLUTANTS

Currently, the microplastics generated from the activities of humans, industrial activities, and agriculture have been highly discharged into the terrestrial and aquatic environments. The discharge of microplastics into the aquatic environment has led to a detrimental impact on the aquatic environment and has threatened the health of consumers by entering the food chains through the trophic transfer method. On the other hand, other pollutants such as hydrophobic organic compounds, heavy metals, and plastic additives discharged from households and industrial areas have also been discharged into the aquatic environment. Microplastics will form complex products, which are higher toxicity to the consumers and environment by associating with these pollutants. The adsorption and desorption characteristic of microplastics have caused microplastics to have the ability to carry and transfer these pollutants together to other places. Therefore, the adsorption mechanisms between the microplastics and contaminants and the factors affecting the adsorption process are very important.

1.3.1 Microplastics and Non-Additive, Hydrophobic Compounds

The adsorption ability of microplastics and non-hydrophobic compounds could form and carry the concentrated hydrophobic compounds to the surrounding area and release them into the aquatic environment (Huang et al., 2021). According to a study conducted by Tang et al. (2021), the amount of antibiotics that is disposed into the aquatic environment in China in 2013 was approximately 53,800 tonnes. Hydrophobic compounds with high toxicity were discharged from the industrial and household areas into the aquatic environment. Hydrophobic compounds can stay and exist in the aquatic environment for a long time, which will cause pollution of the surrounding environment. According to the research of Fu et al. (2021), the combination of microplastics with hydrophobic compounds through adsorption lead to serious environmental issues. In a study of Besseling et al. (2013), they examined the presence of 19 polychlorinated biphenyls (PCBs), carcinogenic chemical compounds that were formerly used in industries in the aquatic environment of the North Sea. They found that a gram of microplastics was observed to contain an average of approximately 4 ng to 980 ng of polychlorinated biphenyls (PCBs) globally, while the amount of 169 ng to 324 ng of polychlorinated biphenyls were found in North Sea. The presence of microplastics will lead to the enhancement of the movement of hydrophobic compounds such as polychlorinated biphenyls (PCBs) to other places, as the smaller microplastics can travel easily to the other sites through wind, leachates, and water. This will further promote the spread of hydrophobic compounds and microplastics as environmental pollutants into the aquatic environments globally. The spread of microplastics with hydrophobic compounds into environments globally provided a route for hydrophobic compounds with high toxicity to enter food chains. Hydrophobic compounds deposited on microplastics can be digested by low-trophic organisms, and thus, the ingested microplastics and hydrophobic compounds will stay inside the body of the low-trophic organisms. The low-trophic organisms with microplastics and hydrophobic compounds in their body will be further consumed by the higher-trophic organisms, and thus, the microplastics and hydrophobic compounds will enter the food chains and threaten the health of organisms (Fu et al., 2021). The combination effects of microplastics and hydrophobic compounds had been investigated to prove that the absorption of hydrophobic compounds on microplastics could lead to increase the detrimental impacts on the environment and the health of organisms.

According to Fu et al. (2021), they found that the mechanisms involved in the adsorption of microplastics and hydrophobic compounds (also known as organics pollutants) are hydrophobic interactions, electrostatic attraction, hydrogen bonding, Van der Waals forces, partition effect π-π interactions, and halogen bonding. The application of multiple mechanisms tends to attract organic compounds to adsorb on the microplastics during the adsorption process. The criteria of organic matters to achieve the adsorption of organic matters on microplastics through different types of interactions bonding and the examples of organic matter is summarized in Table 1.11.

TABLE 1.11
The Criteria of the Adsorption of Organic Matters on Microplastics through Various Types of Interactions Bonding and the Examples of Organic Matter

Types of interaction	Criteria	Examples
Hydrogen bonding	Hydrophilic organic compounds.	Antibiotics and microplastics polyamide
Hydrophobic	High octanol/water partition coefficient (Log K_{ow}), which means the organic compounds are more hydrophobic characteristics, high adsorption coefficient.	Perfluorooctane sulfonate and microplastics polyethylene; Bisphenol
Partition effect (van der Waals force)	The value of n in terms of adsorption isotherm is equal to 1, which indicates a strong linearity.	Antibiotics; Dibutyl phthalate (DBP)
Electrostatics attraction	The pH of microplastics obtained when the net charge of the surface of microplastics is equal to 0 needs to be less than the pH surrounding area; the pH surrounding area needs less than the acid dissociation constant (pKa) of organic matters if exceeds, repulsion occurs, and adsorption will be inhibited.	Sulfamethoxazole (a type of antibiotics); Tylosin (a type of antibiotics)
Halogen bonding	Microplastics contain the halogen atoms.	Bisphenol and microplastics polyvinyl chloride
π-π interactions	Aromatics organic compounds.	PAHs and microplastics polystyrene

According to research conducted by Besseling et al. (2013), polystyrene microplastics were selected and used in this research, as polystyrene microplastic is the most found in the world. They observed that the addition of microplastics had effectively diluted the organic content in the sludge from 1.73% to about 1.64%. This observation indicates that the appearance of microplastics could adsorb the organic matters and thus cause the decrement of organic matters in sludge. Besides, the bioaccumulation of polychlorinated biphenyls in the body of marine organisms would decrease when the concentration of microplastics in sediment increased, while the bioaccumulation increased when the concentration of microplastics in sediment were decreased. This did not indicate that the risks of persistent organic pollutants and microplastics for marine organisms had been reduced, but the accumulation of microplastics in marine environments might also use the same method to transfer the organic matters from sediment into the body of marine organisms. Thus, the presence of microplastics will affect the bioaccumulation of organic matters, but a detailed conclusion cannot be drawn, as more experiments are needed to prove that the increase of microplastics will cause the increase of PCBs in marine organisms.

Xu et al. (2018) had investigated the adsorption effect between sulfamethoxazole and polyethylene microplastics. Sulfamethoxazole is a common, ionizable, and polar antibiotic commonly found in estuaries with an average concentration of 43.8 ng/L. The type of interaction between the sulfamethoxazole and polyethylene is van der Waals forces. In this study, the Log Kow obtained was approximately 0.79, which is low in value and indicated lack of hydrophobic interaction mechanisms in this adsorption process. Besides, the occurrence of electrostatic repulsion at pH 6.8 was observed to the highest, which is due to the fact that most of the sulfamethoxazole would form anions (negative charge) during the adsorption process. The negative charge on the surface of microplastics will repulse the anions, and thus, inhibits the adsorption process. However, the pH value was observed to pose a less significant effect on the result, which proved that the electrostatic attraction is not the primary mechanism in the adsorption process of sulfamethoxazole onto microplastics. Sulfamethoxazole contributed a greater affinity for polyethylene than other dissolved organic compounds, which showed that sulfamethoxazole could be adsorbed onto polyethylene through van der Waals interaction, and polyethylene could become the vector for sulfamethoxazole to be widely spread in the aquatic environment (Xu et al., 2018).

Guo and Wang (2019) also conducted an investigation on the adsorption of three types of antibiotics (sulfamethazine, sulfamethoxazole, and cephalosporin C) on aged microplastics (polystyrene and polyethylene) in freshwater and seawater. The rougher surface of aged polystyrene in comparison to the surface of polyethylene is mainly attributed to the fact that polystyrene is an amorphous polymer which can be degraded faster than the polyethylene, a semi-crystalline polymer. The degradation process will cause the surface of microplastics to become rougher and irregular. In freshwater, the adsorption of sulfamethazine onto microplastics was conducted for a duration of eight hours, while the adsorption process of sulfamethoxazole and cephalosporin C required a duration of 16 hours to achieve equilibrium adsorption with microplastics. These three types of antibiotics could form strong linear in the adsorption isotherm, which showed the appearance of van der Waals forces. These antibiotics had posed approximately 0.0293 L/g to 0.0383 L/g of adsorption capacity, which is higher than the adsorption capacity of primary microplastics. This had also proved that the aged microplastics could improve the adsorption affinity to organic matters due to rougher surfaces and the formation of carbonyl groups during the degradation process. In seawater, the adsorption of antibiotics onto aged microplastics could only be achieved when using the antibiotic cephalosporin C and the required time to reach equilibrium was 24 hours, and this mechanism was surface complexation. The adsorption process was challenging achieve in between antibiotics and microplastics because the pH of seawater was 7.6, and the pKa for sulfamethoxazole and sulfamethazine were 7.2 and 7.4, respectively, which led to the presence of electrostatic repulsion. This showed that electrostatic attraction is the required mechanism for the adsorption of antibiotics and microplastics. In this study, the three main mechanisms for the adsorption of microplastics and antibiotics were electrostatics attraction, van der Waals, and hydrophobic interaction (Guo and Wang, 2019).

1.3.2 Microplastics and Heavy Metals

Microplastics are known as pollutants carriers because microplastics will associate with the heavy metals in wastewater systems and then release the carried heavy metals into the environment through the adsorption and desorption process. Microplastics can be easily carried by wind and water due to the smaller size and light weight of microplastics. Microplastics have adsorbed the heavy metals discharged from industrial areas and then will be transported together with microplastics from one area to another area (Liu et al., 2022). According to Cao et al. (2021), approximately 40% of the aquatic environment and 235 million hectares of farm land over the globe have been polluted by the presence of heavy metals. The two primary sources of heavy metals are the wastes discharged from industrial activities and the usage of additives in plastics industries (Xiang et al., 2021; Campanale et al., 2020). According to Xiang et al. (2021), various industries such as mining, smelting, paper making, manufacturing, electroplating, and machinery are contributing to the release of heavy metals into the environment through sewage wastewater systems. The release of heavy metals from industries, households, and agriculture into the terrestrial and aquatic environments will cause serious pollution and threaten the health of organisms. The release of heavy metals into environments is a serious matter, as the heavy metal compounds will cause damage to DNA and nuclear proteins, which lead to the abnormal growth of cells and thus cancer, and thus destroy the nervous systems of the organisms (Cao et al., 2021). The combination of microplastics and heavy metals has highly increased the toxicity of microplastics to the environment and the health of organisms. The combined microplastics with heavy metals has also caused heavy metals to be able to enter easily into food chains because microplastics with smaller sizes can be easily ingested by organisms (Huang et al., 2021). Hence, the combined toxicity of microplastics and heavy metals has been investigated to evaluate the possible risks of microplastics associated with heavy metals to the environments and the health of organisms.

Kim et al. (2017) investigated the toxicity of two different types of microplastics with heavy metals on *Daphnia magna*. The microplastics that were used in this research were polystyrene with a functional group of –COOH and polystyrene without a functional group. On the other hand, the heavy metal used in this research was nickel (Ni). Based on the results of this research, the toxicity effect on *Daphnia magna* was examined and arranged in the order of Ni + polystyrene with –COOH > Ni + polystyrene > Ni. This result proved that the toxicity would increase when combining microplastics and heavy metals. The presence of the functional group in microplastics were observed to pose significant influence on the toxicity of heavy metals to organisms as the functional group might affect the adsorption capacity of heavy metals onto microplastics. Moreover, a study conducted by Lin et al. (2020) investigated the impact of microplastics polyacrylonitrile (PAN), copper (II) (Cu2+), and a combination of PAN and Cu2+ on green algae *Chlorella pyrenoidosa*. The combination of PAN and Cu2+ was found to cause the amount of chlorophyll-a in green algae to lower when compared with PAN or Cu2+. Overall, the increasing of the concentration of PAN, Cu2+ and a combination of PAN and Cu2+ would cause the level of chlorophyll-a, chlorophyll-b, and total chlorophyll to be significantly decreased (Lin et al., 2020). However, some studies showed that the combination of microplastics and heavy metals would not cause the combined effect on the health of organisms. They claimed that the effects may only increase slightly or reduce the toxicity of heavy metals to the health of organisms. Davarpanah and Guilhermino (2015) investigated the impacts of polyethylene microplastics, copper, and a combination of microplastics and copper (Cu) on the microalgae *Tetraselmis chuii*. Their results showed that the microplastics would not affect the growth of microalgae, but the Cu would significantly decrease the growth of microalgae. They also found that, when exposed to a mixture of microplastics and Cu, the presence of microplastics would not affect the toxicity curve of Cu. Besides, in a study conducted by Wen et al. (2018), they investigated the effects of microplastics and the combination of cadmium (Cd) and microplastics on the accumulation of Cd, innate immunity, and the antioxidant defence of discus fish. They found that the increasing number of polystyrene microplastics would decrease the accumulation of Cd. They also observed that the growth and survival of discus fish were not affected by the exposure to microplastics polystyrene and Cd, while the combination of polystyrene microplastics and Cd will cause severe oxidative damage and stimulate the innate immune response of discus fish (Wen et al., 2018).

According to Cao et al. (2021), the adsorption process between microplastics and heavy metals usually happens in the liquid medium. The mechanisms involved in the direct interaction adsorption process of microplastics and heavy metals in a liquid medium are electrostatic interaction, surface complexation, and precipitation. They also observed that the electrostatic interaction between microplastics and heavy metals can be enhanced by increasing the polarity of the microplastics, such as adding the charged compounds or additives into the microplastics, which could charge the surface of microplastics. For the surface complexation, the biofilm containing different functional groups can improve the adsorption capacity of microplastics and heavy metals. However, the adsorption process between microplastics and heavy metals is mainly affected by various factors such as the types of microplastics, physical and chemical properties of microplastics, size of microplastics, ageing of microplastics, types of heavy metals, amount of heavy metals to reach the criteria required for the occurrence of the adsorption process, chemical properties of heavy metals, and conditions of the surrounding area. Hence, the adsorption mechanism and factors influencing the adsorption mechanisms were analysed to have a better understanding of the adsorption process between microplastics and heavy metals.

In these few years, many studies have been conducted to investigate the adsorption mechanism between microplastics and heavy metals and the factors that influence the adsorption process. Zou et al. (2020) conducted a study to investigate the adsorption process of three different heavy metal ions, which are Cd2+, Cu2+, Pb2+, and four different types of microplastic polymers, which were polyvinyl chloride (PVC), chlorinated polyethene (CPE), high crystallinity polyethene (HPE), and low-crystallinity polyethene (LPE). They found that the strength of the adsorption affinity of microplastics to the heavy metals Cd, Cu, and Pb increased from LPE, HPE, PVC, and then CPE. Although the CPE microplastics had an irregular surface due to the amorphous characteristics when compared with other microplastics with higher crystallinity characteristics, the CPE has the strongest adsorption affinity among LPE, HPE, and PVC. This observation showed that the strength of adsorption affinity does not depend on the crystallinity of microplastic polymer but on the chemical structure and electronegativity of microplastics. Among these three heavy metals, Pb2+ has the highest adsorption affinity compared to Cu2+ and Cd2+ due to several reasons such as hydrated radius of ions, electronegativity, softness, and hydration energy. The hydrated radius of Pb2+, Cu2+, and Cd2+ are 0.401 nm, 0.419 nm, and 0.426 nm, respectively. The smaller the hydrated radius could lead to lower the hydration energy and induce electrostatic interactions with the microplastics. Because of this, Pb2+ had the highest adsorption affinity to microplastics among these three heavy metals. This also showed that the electrostatic interaction is one of the adsorption mechanisms between heavy metals and microplastics. Lin et al. (2021) also found that the presence of hexabromocyclododecane (HBCD) in microplastics would improve the heavy metal lead (II) (Pb2+) and malachite green (MG) to adsorb onto the microplastics polystyrene (PS). The adsorption capacity of the Pb2+ was found to be greater than MG, where the Pb2+ was 3.33 μmol.g-1 and MG was 1.87 μmol.g-1. The result showed that Pb2+ had a greater adsorption affinity than MG on the microplastics. When the adsorption process of Pb2+ and MG on the microplastics happened, the energy of C-Br bonding in HBCD would decrease from 285.4 eV to 285.2 eV, which could induce the potential of attracting electrons, and the Br would be acted on as the electron-withdrawing group in the adsorption process. The Br in HBCD played an important role in promoting the adsorption process of Pb2+ and MG on the microplastics through an electrostatic interaction mechanism. According to a study conducted by Tang et al. (2021) on the adsorption of heavy metals such as copper (II) (Cu2+), nickel (II) (Ni2+), and zinc (II) (Zn2+) on the microplastics nylon, the adsorption capacity of Cu2+ was observed to be the highest (16.712 μmol/g), followed by Zn2+ (12.726 μmol/g), and lastly, Ni2+ (10.565 μmol/g). For the hydrated radius of ions, Zn2+ had the largest size (0.430 nm), followed by Cu2+ with size of 0.419 nm, and lastly, Ni2+ with smaller in size of 0.404 nm. In this research, the adsorption process is mainly dependent on the maximum adsorption capacity of different heavy metals (Cu2+, Ni2+, and Zn2+)

TABLE 1.12
The Results on the Adsorption Process between the Different Types of Microplastics such as PS, PP, and PA with Heavy Metals of Sr^{2+}

	PS	PP	PA
Specific Surface Area (m^2/g)	0.584	0.314	0.480
Maximum Adsorption Capacity of Sr^{2+} (μg/g)	51.4	52.4	31.8
Interaction Energy with Sr^{2+} (kcal/mol)	−6.418	−13.05	−5.638
Total Interaction Energy with Sr^{2+} (kcal/mol)	2,024.99	1,816.27	−1,457.68

Source: Data compiled from Guo et al. (2020).

and the surface complexation. The adsorption affinity of the heavy metals on the microplastics nylon is arranged in the order of Cu2+ > Zn2+ > Ni2+. According to research conducted by Guo et al. (2020) to investigate the interaction effect between microplastics and heavy metal Sr2+, he types of microplastics used in this research are polystyrene (PS), polypropylene (PP), and polyamide (PA). The result showed that the microplastics of PS and PP have almost similar adsorption affinity, as summarized in Table 1.12. They also found that the adsorption affinity of heavy metals is higher on the microplastics of PA. This observation also indicates that the adsorption affinity is highly depended on the maximum adsorption capacity but independent from the specific surface area. As a conclusion, the mechanism involved in this research is electrostatic interactions.

Guan et al. (2020) investigated the effects of biofilm, and complexation with a functional group in biofilm can improve the interaction of microplastics and heavy metals through the surface complexation mechanism. The adsorption of these heavy metals on the biofilm covered microplastics, microplastics polystyrene, suspended particles, and surficial sediment. The maximum adsorption capacity of heavy metals on these solid particles collected from Nanhu Lake and Shitoukoumen Reservoir is summarized in Table 1.13. From Table 1.13, the adsorption capacity of Cu (II) was observed to be more significant than other heavy metals. Comparing the solid particles, the adsorption affinity of heavy metals on the suspended particles was found to be the highest, then followed by surficial sediment, biofilm covered microplastics and microplastics, which showed that the presence of biofilm could improve the adsorption affinity of heavy metals on microplastics. Chemical adsorption had applied in the biofilm-covered microplastics, suspended particles, and surficial sediment, while physical adsorption was used in microplastics.

1.3.3 Microplastics and Plastic Additives

The degradation process of plastics to microplastics, mainly influenced by various factors such as UV radiation from sunlight, movement of water, pH of the environment, and salinity, will cause the additives inside the plastics to be released into the environment. The additives that are most used in plastics industries are phthalates, benzotriazoles, bisphenol A (BPA), and polybrominated diphenyl ethers (PBDEs). These plastics additives play an important role in the manufacturing of plastics to

TABLE 1.13
Maximum Adsorption Capacity of Heavy Metals on These Solid Particles Collected from Nanhu Lake and Shitoukoumen Reservoir

Source of Solid Particles	Solid Particles	Maximum Adsorption Capacity (μg/g)					
		Ag (I)	Cd (II)	Co (II)	Cu (II)	Ni (II)	Zn (II)
Nanhu Lake (Water comes from land runoff and sewage of city)	Biofilm covered microplastics	364.6	825.5	377	1048.6	338.5	610.6
	Microplastics polystyrene	304.4	508.9	197.7	643.1	148.6	474.9
	Suspended particles	432.5	1032.1	810.1	2732.6	1059	1304.3
	Surficial sediment	392.6	958.1	658.3	1897.1	792.2	838.8
Shitoukoumen Reservoir (Water to be provided water to city, agriculture, fish farms)	Biofilm covered microplastics	326.4	763.7	292.8	819.9	329.5	531.4
	Suspended particles	419	1159	769.3	2521.9	949.6	1251.1
	Surficial sediment	401.2	963.5	511.7	1700.9	742.1	786.1
Nanhu Lake (Water comes from land runoff and sewage of city)	Biofilm covered microplastics	364.6	825.5	377	1048.6	338.5	610.6
	Microplastics polystyrene	304.4	508.9	197.7	643.1	148.6	474.9
	Suspended particles	432.5	1032.1	810.1	2732.6	1059	1304.3
	Surficial sediment	392.6	958.1	658.3	1897.1	792.2	838.8

increase the properties of plastics required in certain application fields. Some additives are added to function as a stabilizer, colour pigments, flame retardants, and antioxidants to improve the properties of plastics. Plastic additives can be organic compounds or heavy metals. The fragmentation of plastics into microplastics had led to the leaching of harmful plastic additives from fragmented microplastics to the environment. This is one of the reasons microplastics are known as the carrier of pollutants. Microplastics become the pathway to spread plastic additives to surrounding areas. The leaching of plastic additives to environments has further caused environmental issues, as most of the plastic additives are harmful to the health of consumers (Huang et al., 2021). When small-sized microplastics have been ingested by organisms, the plastic additives will bioaccumulate in the body of organisms and affect the health of consumers. The high concentration of aged microplastics severely affected the photosynthesis rate and the growth rate of microalgae due to leaching of the plastic additives in the microplastics (Luo et al., 2020). Besides, the leaching of plastic additives is also found to depend on the condition of the environment and the properties of microplastics (Luo et al., 2019). Therefore, most studies were focused on the factors that caused the release of plastic additives from microplastics into the environment.

Luo et al. (2019) investigated the factors that affected the leaching of the plastic additives, which act as a fluorescent function from polyurethane microplastics. They found that a high concentration of fluorescent additives had been leached and detected in basic water, followed by saline water, seawater, West Lake, rivers, and wetlands in the duration of 12 hours to 24 hours. Besides, the results also showed that plastic additives were preferably leached from microplastics under conditions of a high pH. However, the leaching concentration of fluorescent additives was observed to be high at high pH initially and was almost similar after 48 hours in different pH conditions. This indicates that the leaching of plastic additives depended on the pH of the condition and the time. Based on a study conducted by Yan et al. (2021), the characteristics of microplastics and pH solution were investigated to evaluate the factors that affect the leaching of plastic additives and dibutyl phthalate (DnBP) plasticizer from microplastics polyvinyl chloride (PVC). The release of DnBP was observed to be decreased after one day because equilibrium had been reached, and the DnBP might be reabsorbed back into microplastics when the outside concentration of DnBP is higher than the concentration inside the microplastics. Besides, they also found that a higher concentration of DnBP was released from microplastics with a smaller size. This is because microplastics of smaller sizes have greater total surface areas, and the diffusion pathway will be reduced and become shorter. As a result, the additives could easily leach out from smaller microplastics. Then, the pH of the solution was less affected by the leaching of DnBP. The presence of fulvic acid, a dissolved organic matter component in nature, would also enhance the release of DnBP from microplastics because of the solubilization effect of DOM on hydrophobic organic matters. Therefore, smaller microplastics had a higher potential to release more concentration amounts of additives, and the condition of the environment also affected the leaching process of plastic additives.

1.4 DETECTION OF MICROPLASTICS IN DRINKING WATER SOURCES

Microplastics are detected in many drinking water sources, such as bottled water, tap water, groundwater, surface water, and wastewater. The water samples are collected from different sources of drinking water. After that, the collected samples underwent different processing methods to avoid contamination of the collected water samples and increase the accuracy of the result obtained. However, the detection of microplastics is hard to do with the naked eye, as the size of microplastics is too small to be detected by the naked eye. Therefore, the presence of microplastics in the drinking water sources must be detected by scientific methods. The water samples are collected, and microplastics are extracted from the water samples. The extracted microplastics are analysed with technologies such as microscopes and FTIR (Smith et al., 2018).

1.4.1 Detection of Microplastics in Bottled Water

In the study conducted by Oßmann et al. (2018), 22 samples were collected from bottled water made from polyethylene terephthalate (PET), whereas 10 samples were collected from glass bottled water. Firstly, 250 mL of collected water was filtered from the initial sample through an aluminium-coated polycarbonate membrane filter to obtain a number of analysable particles. Then, the foam was destroyed by rinsing the filter funnel with 5 mL of 50% ethanol solution and ultrapure water. Finally, the filters were flattened so that the microscope could focus on the flat surface of the filter for analysis. The filters were stored in closed petri dishes to prevent pollution of the collected samples (Oßmann et al., 2018). Micro-Raman spectroscopy was used to identify and detect microplastics present in bottled water in the studies conducted by Oßmann et al. (2018) and Schymanski et al. (2017). In the study by Oßmann et al. (2018), micro-Raman spectroscopy was used together with an XploRa Plus system, and it was equipped with a cooled, charge-coupled device detector and lasers. Five randomly chosen spots with an area of 1 mm^2 were analysed for each water sample (Imhof et al., 2016; Oßmann et al., 2018). However, all the particles larger than or equal to 1 μm were examined with Particle Finder Module software, and their size and coordinates were recorded. The Raman spectra obtained were compared with the spectra in the database to identify the type of plastic (Oßmann et al., 2017, 2018).

Zhou et al. (2020b) also investigated to detect the presence of microplastics in various brands of bottled water samples. In this research, the bottled water samples were obtained from 23 different bottled water brands packaged in PET plastic bottles. A 0.4-μm gold-plated polycarbonate membrane was used to filter every sample using a vacuum pump operated at 1 bar. First, every bottle was rinsed with 50 mL of Milli-Q water, and the samples were processed three times. Then, the filter was removed carefully to prevent scratching on the surface of the filter and stored in a petri dish that was dried at room temperature for the analysis (Zhou et al., 2020a). Moreover, in the study conducted by Schymanski et al. (2017), the water samples in this research were collected from different bottled water: 22 samples from the PET bottles, three samples from beverages bottles, and nine samples from glass bottles. First, the bottom of the filter funnel in the filtering apparatus was highly polished to ensure the filter membrane was tightly embedded. Then, all the bottled water was vacuum filtered through the filter, and the container and filter funnel were rinsed thoroughly with Milli-Q water to ensure no particles were left. After that, the filter was stored in a petri dish before analysis (Schymanski et al., 2017).

Mason et al. (2018) conducted an investigation of the detection of microplastics in bottled water by selecting 10 bottles of 500 mL to 600 mL bottled water, six bottles of 750 mL bottled water, and four bottles of 2000 mL bottled water. Nile Red solution was added to every bottle, and the samples were placed inside a laminar flow fume chamber. All the bottled water was incubated for 30 minutes and vacuum filtered using a glass fibre filter (Mason et al., 2018). Makhdoumi et al. (2021) investigated the presence of microplastics in bottled by selecting 11 brands of PET bottled water from various supermarkets in Iran. Firstly, they added 50 mL of 20 μg/mL of Rose Bengal solution to every sample. The water samples that contained Rose Bengal solution were then incubated for 30 minutes, and the water was filtered through a glass fibre filter with a vacuum. Finally, a petri dish was used to store the filter membranes for further analysis (Makhdoumi et al., 2021).

The abundance of microplastics in bottled water obtained from five studies is shown in Table 1.14. PET, PP and PE are primarily found in all the samples collected from bottled water (Oßmann et al., 2018; Schymanski et al., 2017; Zhou et al., 2020a). PE and PP microplastics were found in the glass bottles in the study conducted by Oßmann et al. (2018), while PET, PE, and PP microplastics were found in the glass

TABLE 1.14
Abundance of Microplastics in Bottled Water Samples

Bottled water samples	Abundance of microplastics	References
12 reusable PET bottled water	4889.00 ± 5432.00 microplastics/L	Oßmann et al., 2018
10 single-use PET bottled water	2649.00 ± 2857.00 microplastics/L	
10 glass bottles	6292.00 ± 10521.00 microplastics/L	
11 PET bottled water	8.50 ± 10.20 microplastics/L (average)	Makhdoumi et al., 2021
10 bottles of 500–600 mL	325.00 microplastics/L	Mason et al., 2018
6 bottles of 750 mL	(>100 μm)	
4 bottles of 2000 mL	315.00 microplastics/L (6.5 – 100 μm)	
12 reusable PET bottled water	118.00 ± 88.00	Schymanski et al., 2017
10 single-use PET bottled water	microplastics/L	
9 glass bottles	14.00 ± 14.00	
3 beverages bottles	microplastics/L	
	50.00 ± 52.00 microplastics/L	
	11.00 ± 8.00 microplastics/L	
23 PET bottled water	2.00–23.00 microplastics/bottle	Zhou et al., 2020b

bottles analysed by Schymanski et al. (2017). In addition, PE, PET, and PP were detected in the beverage's carton obtained by Schymanski et al. (2017). According to Schymanski et al. (2017), the potential sources of microplastics in the plastic bottled water are the bottles and caps. PET is found in a high percentage in water samples because it is the manufacturing material of plastic bottles (Oßmann et al., 2018). PP and PE plastics are mainly used in manufacturing the caps, and thus, both can be found in the water samples (Mason et al., 2018). Furthermore, the number of microplastics detected in reusable plastics bottles is higher than single-use plastics bottles due to the frequent stress applied on the bottles during the washing and reutilization process (Oßmann et al., 2018; Schymanski et al., 2017). Reusable bottles with frequent washing action could cause the inner surface of the plastic bottles to be torn off during the reutilization process (Oßmann et al., 2018). According to the study conducted by Singh (2021), the number of microplastics detected in bottled water increases with the number of open-close cycles of the bottle cap. Another study by Singh (2021) also reported that 16 open-close cycles were detected with the highest number of microplastics when compared to a lower number of open-close cycles. An elevation in the number of microplastics shows that the bottle's cap can be torn off due to the abrasion effect.

In addition, the microplastics present in the glass bottle might be attributed to the abrasion effect of the bottles between the glass bottle and plastic caps during the sealing process in the factory. PE is mainly found in beverage cartons because it is commonly used as the packaging material of beverage cartons (Schymanski et al., 2017). Thus, the high probability of plastics being torn from the inner surface of the beverage's carton due to the abrasion effect and forming microplastic particles in the water bottled, as detected in the study.

1.4.2 Detection of Microplastics in Tap Water

The presence of microplastics, also detected in tap water, is one of the drinking water sources in our daily life. This is the reason various studies were conducted to investigate the detection of microplastics in tap water. In the study carried out by Chanpiwat and Damrongsiri (2021), tap water samples were collected at the canal systems in four drinking-water treatment plants in Thailand. They sampled 100 L of water samples from the four different drinking-water treatment plants and then filtered using a stainless-steel container and 50 µm of wire mesh. SDS solution was used to rinse the particles on the mesh into a 1 L glass bottle. After that, the non-plastic particles were removed through density separation with saturated sodium chloride solution (NaCl) by adding saturated NaCl into the funnel, and the settlement in the separating funnel was disposed of after 24 hours. After 24 hours, 200 to 250 mL of the mixture's supernatant was vacuum pumped via the filter paper. Next, the microplastics were leached from the filter paper by soaking them in hydrogen peroxide solution (H_2O_2) and sonication at 40 kHz for 10 minutes. Lastly, H_2O_2 solution containing microplastics was filtered through filter paper and stored in a petri dish after a drying process (Chanpiwat and Damrongsiri, 2021). On the other hand, research of Zhang et al. (2019) was conducted by collecting water samples (tap water) from seven residential areas of Qingdao in China. The tap water samples were collected three times using 4.5 L glass bottles after the tap water was run for 10 minutes. After that, the collected tap water was filtered using 0.45 µm of nitrocellulose membrane to filter the microplastics in the water samples through a vacuum and then further cleaned with ultrapure water. The nitrocellulose membranes containing filtered microplastics were stored between watch glasses for analysis (Zhang et al., 2019). In research conducted by Kosuth et al. (2018), they collected and analysed tap water samples collected in 14 different countries, while Lam et al. (2020) collected tap water in Hong Kong for further analysis. On the other hand, Abdulmalik Ali (2019) and Pratesi et al. (2020) conducted an investigation into the tap water samples collected at the University of North Dakota and Brasilia, respectively, while another study was conducted by Tong et al. (2020); they were collecting tap water samples from different places in China. The collection method of the tap water samples for all research conducted was mainly the same. First, tap waters were opened and kept flowing for 1 minute, and 500 mL or 1 L of high-density polyethylene (HDPE) bottles were filled to the point of overflowing to collect the water samples (Abdulmalik Ali, 2019; Kosuth et al., 2018; Lam et al., 2020; Pratesi et al., 2020; Tong et al., 2020). The tap water collection was done three times, where the first and second collections were used to rinse the bottle and the last collection was processed for analysis (Abdulmalik Ali, 2019; Kosuth et al., 2018; Lam et al., 2020; Pratesi et al., 2020).

By comparing all the research mentioned previously, the staining of the particles was observed to be different for all of the research. Rose Bengal solution was used to stain the particles in the studies of Kosuth et al. (2018) and Lam et al. (2020), while Nile Red solution was used by Abdulmalik Ali (2019), Pratesi et al. (2020) and Tong et al. (2020). In the study conducted by Kosuth et al. (2018), a Whatman cellulose filter with 2.5 μm was used to filter the water samples through a vacuum. Then, deionized water was used to rinse the sample bottles. A second filtration was performed to investigate the possible penetration of contaminants (Kosuth et al., 2018). 2 mL of Rose Bengal solution was applied to stain the particles (Kosuth et al., 2018; Lam et al., 2020). Besides, the Rose Bengal solution on the filter was rinsed with deionized water, and the filter was allowed to dry (Lam et al., 2020). In the study carried out by Abdulmalik Ali (2019), 10 mL of Nile Red solution was injected into all the water samples, and then they were incubated for a duration of 30 minutes. The glass microfibre filter was used to perform vacuum filtration, and the sample bottles were rinsed with distilled water. A second filtration was conducted to collect the penetrated particles in the first filtration. Finally, a sterile petri dish was used to store the filter papers and dried in the desiccator for 48 hours (Abdulmalik Ali, 2019). In a study by Pratesi et al. (2020), zinc chloride (ZnCl_2) solution was added to the water samples and allowed to settle. 0.2 mL of Nile Red solution was added to the supernatant collected from the settling process. The mixture was moved to an Eppendorf tube and left for 10 minutes before moving to a Sedgewick-Rafter counting cell chamber (Pratesi et al., 2020). Finally, in a study by Tong et al. (2020), 200 mL of Nile Red solution was added to the black polycarbonate membrane that had undergone vacuum filtration before with the water samples. The membrane was allowed to dry for 30 minutes at ambient temperature (Tong et al., 2020).

The abundance of microplastics in tap water obtained from seven studies is shown in Table 1.15. The potential sources of the contamination of microplastics are the abrasion of synthetic clothing, population density, anthropogenic activities, and mechanical abrasion of pipes (Kosuth et al., 2018; Lam et al., 2020; Zhang et al., 2019). In a study by Kosuth et al. (2018), the highest average concentration of microplastics was 9,240 microplastic/m^3 found in the United States. A higher concentration of microplastics is found in the water samples from more developed countries. This may be due to the water source and the human population density that causes the difference

TABLE 1.15
Abundance of Microplastics in Tap Water Samples

Location	Abundance of microplastics	References
University of North Dakota	66.00–472.00 microplastics/L	Abdulmalik Ali, 2019
Eastern and Western canal systems in Thailand	0.40–2.40 microplastics/L (Eastern) 0.40–2.10 microplastics/L (Western)	Chanpiwat and Damrongsiri, 2021
Cuba, Ecuador, England, France, Germany, India, Indonesia, Ireland, Italy, Lebanon, Slovakia, Switzerland, Uganda, and the United States	0.00–61.00 microplastics/L	Kosuth et al., 2018
110 urban sites in Hong Kong	0.00–8.61 microplastics/L	Pratesi et al., 2020
North and South wings of Brazil	0.19 ± 0.11 microplastics/L (South) 0.44 ± 0.32 microplastics/L(North)	
38 different places in China	0.00–1247.00 microplastics/L	Tong et al., 2020
7 residential districts of Qingdao in China	0.30–1.60 microplastics/L	Zhang et al., 2019

in the concentration of microplastics found in the tap water samples (Kosuth et al., 2018). However, microplastics' concentration in the tap water from private residences was higher than in public spaces because water filtration systems are used in public spaces. The filtration system is PP-prefilter-mesh equipped in the carbon filter media, which removes the microplastics (Abdulmalik Ali, 2019). The fact can be proven by the high amount of microplastics detected in the low-income residential area based on a survey conducted by students from the University of Brasilia. More microplastics were detected because there is no water treatment facility for 50% of the residents. Thus, the tap water contains microplastics, as it is not treated before being consumed (Pratesi et al., 2020). According to Chanpiwat and Damrongsiri (2021), microplastics in the tap water might be due to plastic debris discarded near the rivers because they disintegrate into microplastics through degradation processes. According to Lam et al. (2020), the potential sources of microplastics in tap water may be due to anthropogenic activities near the water treatment system and mechanical abrasion of the pipes used in water distribution. The higher amount of microplastics detected in the north wing may also be due to the plumbing used in the north wing being PVC compared to galvanized steel tubes used in the south wing. The abrasion effect can tear off the plastic from the plumbing when water flows inside the plumbing, which results in a high amount of microplastics detected (Pratesi et al., 2020). The main types of microplastics identified were PET, PVC, PE, PP, polystyrene (PS), polyamide (PA), rayon, and acrylonitrile butadiene styrene (Chanpiwat and Damrongsiri, 2021; Pratesi et al., 2020; Tong et al., 2020; Zhang et al., 2019). This is because the plastic materials of PVC, PP, and PE are used as pipes in the piping system for the drinking-water treatment plant and households. Furthermore, most of the pipe fittings are made mainly from PA. Therefore, the plastics can be leaked into the tap water due to the abrasion effect and the formation of microplastics (Tong et al., 2020). Furthermore, rayon, a synthetic textile fibre, and PET are also used in the manufacturing of textiles industries and clothes. Therefore, the detected fibres also show that the synthetic textiles are a potential source of rayon and PET in the tap water samples because fibres are produced from the abrasion effect of clothes (Browne et al., 2011; Kuczenski and Geyer, 2010; Zhang et al., 2019).

1.4.3 Detection of Microplastics in Groundwater

Groundwater is one of the drinking water sources in our daily life. Groundwater samples were collected from various sources such as Huangshui River and Dagu River in Shangdong in China to determine the presence of microplastics in groundwater in research conducted by Su et al. (2021). A diaphragm vacuum pump is used to filter the collected water samples through a fibre membrane with 0.45 μm and then washed with distilled water to rinse the microplastics trapped on the membrane, then it was kept in a petri dish. The organic matter was oxidized by 10 mL of 30% H_2O_2 solution. The petri dish was stored at a temperature of 60 °C for 1 hour. After that, 0.45 μm of polycarbonate membrane was used to filter and collect the microplastics. The staining process of microplastics was done by dripping two to three drops of Nile Red solution onto the polycarbonate membrane. The membrane was then incubated in a dark environment for 0.5 hours (Su et al., 2021). Samandra et al. (2021) conducted an investigation by collecting the groundwater samples from Bacchus Marsh in Australia. The samples were collected using a bailer connected to a braided PA rope from the mid-point of seven monitoring bores three times. 15-μm-pore-sized polycarbonate filter paper was used to filter the groundwater samples (Samandra et al., 2021). The organic matter in the water samples was dissolved by using 0.04 L of 30% H_2O_2 solution at the temperature of 60 °C for a duration of 12 hours to 24 hours. Then, the water samples were filtered again using 15 μm of polycarbonate filter paper, and density separation was performed using 35 mL of saturated calcium chloride $CaCl_2$ solution (Liu et al., 2020a; Samandra et al., 2021). The microplastics trapped on the 5 μm of polycarbonate filter paper were moved to an infrared reflective slide for analysis purposes (Samandra et al., 2021). The sampling collection and processing methods in Panno et al. (2019) and Bharath et al. (2021) were the same. Eleven groundwater samples were collected from shallow wells and springs in the Driftless Area, and six samples were collected from springs in Salem Plateau in a study conducted by Panno et al. (2019). 2 L of HDPE bottles were used to collect the water samples (Panno et al., 2019). However, Bharath et al. (2021) collected samples at 20 different locations in two solid-waste disposal sites at Kodungaiyur and Perungudi in India. As a result, 20 samples were collected using 1 L glass bottles with Teflon-lined bottle caps at a borewell water depth of 3 m to 30.48 m at two sampling sites (Bharath et al., 2021). In addition, a 0.45-μm-pore-sized Whatman filter was used in both studies to filter the water samples. The filter was dried at a temperature of 75 °C for 24 hours and at room temperature in the studies conducted by Panno et al. (2019) and Bharath et al. (2021), respectively (Bharath et al., 2021; McNeish et al., 2018; Panno et al., 2019).

Table 1.16 shows the abundance of microplastics in the groundwater samples collected in four studies. The possible sources of microplastics in groundwater originated from atmospheric fallout and drainage of septic systems. Besides, the cervices, conduits, and springs are also pathways for microplastics to enter the groundwater (Panno et al., 2019). The leachate from the sampling sites in Bharath et al. (2021) can also leach from marshland, as the sampling sites were located near marshland. Furthermore, the leachate is also leached due to the landfill sites' poor management, which contaminates the groundwater (Bharath et al., 2021; Chatterjee, 2010; Su et al., 2019). The types of microplastics detected in the groundwater samples were PA, PET, PVC, PP, PE, PS and nylon (Bharath et al., 2021; Panno et al.,

TABLE 1.16
Abundance of Microplastics in Groundwater Samples

Location	Abundance of microplastics	References
Chennai (India)	2.00–80.00 microplastics/L	Bharath et al., 2021
Salem Plateau and Driftless Area (United States)	6.40 microplastics/L (median) 15.20 microplastics/L (maximum)	Panno et al., 2019
Bacchus Marsh (Australia)	16.00–97.00 microplastics/L	Samandra et al., 2021
Huangshui River and Dagu River (China)	3352.00 microplastics items	Su et al., 2021

2019; Samandra et al., 2021; Su et al., 2021). The presence of PA may be because of the contamination due to improper sample collection (Su et al., 2021). The type of plastics detected can also depend on the sampling location. For example, PVC, PE, PP, and PET were found in an industrial site because the collection site was near a meat processing plant (Samandra et al., 2021). This is because the packaging materials for meat and the transport trays of meat are made of these few materials (Cenci-Goga et al., 2020; Heinz and Hautzinger, 2007; Júnior et al., 2020; Maga et al., 2019; McMillin, 2017; Samandra et al., 2021). Moreover, PE was primarily found in the residential site, and it may be due to the domestic cleaning process of clothes and the application of compost (Boucher and Friot, 2017; Samandra et al., 2021). On the other hand, PE and PU were found in the agricultural site, as the plastic mulch and PU-coated fertilizers were used because PU-coated fertilizers ensure greater control of the release of nitrogen and phosphorus (Bortoletto-Santos et al., 2020; Liao et al., 2020; Samandra et al., 2021). The presence of nylon, PE, PET, and PVC in the groundwater samples collected by Bharath et al. (2021) might be due to the disposal of cosmetic products and clothes from industries. The sampling sites are waste-dumping sites that contain various types of plastics wastes (Bharath et al., 2021; Qiu et al., 2020).

1.4.4 Detection of Microplastics in Surface Water

Various studies were conducted to investigate the detection of microplastics in surface water. In a study conducted by Wang et al. (2020), 2.5 L of stainless-steel sampler was used to collect water samples from reservoirs from a depth of 0.3 m and river water samples by wading from the centre of the river. The water samples were collected three times and stored in a 1-L glass bottle. The water samples were vacuum filtered through a 0.45-µm Whatman filter (Wang et al., 2020). A 5-L-of-water bucket was used by Zhang et al. (2020a) to collect the surface-water samples. On the other hand, the water samples collected were filtered through a 50-µm sieve and the residue trapped on the sieve was rinsed into a 1-L bottle with distilled water (Zhang et al., 2020b). Di and Wang (2017) used a Teflon pump to collect 25 L of water at 1 m from the water surface for two replicates. A 48-µm stainless-steel sieve was used to filter the water collected, and the water samples were stored in a 5% formalin solution (Di and Wang, 2017). Then, the samples were subjected to a wet peroxide oxidation (WPO) process to remove organic matter, and the detailed process was described in the later section (Di and Wang, 2017; Wang et al., 2020; Zhang et al., 2020b). Besides, a plankton net was applied by Dris et al. (2017) and Kameda et al. (2021) to collect surface water samples. The sizes of the mesh of the plankton net were 10 µm (Kameda et al., 2021) and 80 µm (Dris et al., 2017). The mixture was treated with SDS solution before the WPO process (Dris et al., 2017). Meanwhile, the water collected by Kameda et al. (2021) was filtered through a 5-µm hydrophilic PTFE filter (Kameda et al., 2021). On the other hand, the water samples collected by Dris et al. (2017) and Kameda et al. (2021) were subjected to the WPO process. A neuston net was used by Baldwin et al. (2016) and Mataji et al. (2019). The mesh sizes of the neuston net were 50 µm (Mataji et al., 2019) and 333 µm (Baldwin et al., 2016). Firstly, the collected water samples were transferred to a glass jar and preserved in isopropyl alcohol solution. Then, filtration was undergone by the water samples collected, which were subjected to a filtration process, and lastly, the WPO process (Baldwin et al., 2016; Mataji et al., 2019). Finally, a manta trawl was used by four studies to collect surface water samples. First, the manta trawl was rinsed with deionized water or surface water before collecting samples by towing. The mesh sizes of the manta trawl used in these studies were 300 μm (Mani et al., 2015) and 333 μm (Anderson et al., 2017; Eriksen et al., 2013; Mason et al., 2018). Then, the manta trawl was either towed along the side of the boat or on the surface of the lake (Eriksen et al., 2013; Mani et al., 2015; Mason et al., 2016). Finally, a few reagents such as 70% ethanol solution (Anderson et al., 2017), 100% NaCl solution (Mani et al., 2015) and 70% isopropyl alcohol solution (Mason et al., 2016) were used to preserve the samples. A filtration process was undergone by some of the water samples collected by the manta trawl before undergoing the WPO process. The water samples collected by Anderson et al. (2017) and Mason et al. (2016) underwent the filtration process through the sieves prior to the WPO process.

Table 1.17 shows the abundance of microplastics detected in the surface water samples detected in 11 studies. According to the results obtained from 11 studies, the abundance of microplastics depends on anthropogenic activities due to the population densities because a higher abundance of microplastics is detected at the sampling location with a higher population (Eriksen et al., 2013; Wang et al., 2020). Furthermore, the abundance of microplastics detected is higher at tourist spots, city centres, or WWTPs (Di and Wang, 2017; Eriksen et al., 2013; Mani et al., 2015; Wang et al., 2020; Zhang

TABLE 1.17
Abundance of Microplastics in Bottled Water Samples

Location	Abundance of microplastics	References
Lake Winnipeg (Canada)	53,000.00–748,000.00 microplastics/km^2	Anderson et al., 2017
Great Lake tributaries	0.05–32.00 microplastics/m^3	Baldwin et al., 2016
Yangtze River from Chongqing to Yichang (China)	1,597.00–12,611.00 microplastics/m^3	Di and Wang, 2017
River Seine (Greater Paris)	38.20–101.60 microplastics/m^3	Dris et al., 2017
Laurentian Great Lakes (United States)	450.00–450,000.00 microplastics/km^2	Eriksen et al., 2013
Tsurumi River (Japan)	298.00 ± 105.00 – 1240.00 ± 295.00 microplastics/m^3	Kameda et al., 2021
Rhine River (Germany)	1.01 microplastics/m^3	Mani et al., 2015
Lake Michigan (United States)	1,400.00–100,000.00 microplastics/km^2	Mason et al., 2016
Caspian Sea	34,491.00 ± 18,827.00 microplastics/km^2	Mataji et al., 2019
Manas River Basin	14,000.00 ± 2,000.00 microplastics/m^3(average)	Wang et al., 2020
Guangdong (China)	3,000.00–19,000.00 microplastics/m^3	Zhang et al., 2020a

et al., 2020b). In addition, the abundance of microplastics collected near tourist spots is high because plastic wastes are disposed of, such as plastic bottles (Wang et al., 2020). The highest concentration of microplastics was found in the sample collected near two WWTPs. In contrast, a lower concentration of microplastics was detected at the sampling location near the rural area with less population (Di and Wang, 2017). The abundance of microplastics detected in the wet season is higher than that detected in the dry season. Higher concentrations of microplastics are found in the water samples after rain (Zhang et al., 2020b). This is attributed to the wash away of microplastics into the surface water by the high water-flow rate (Suteja et al., 2021). However, the abundance of microplastics collected during the pre-monsoon season (dry season) can also be higher than in the post-monsoon season (wet season). The flushing mechanism occurs during the post-monsoon season due to the water's elevation in flow rate and volume, which causes the microplastics to be flushed away from the surface water (Zhao et al., 2019; Zhang et al., 2020b).

Firstly, the detected microplastics in the surface water can be formed by the degradation of large plastics when the WWTPs and industry are absent near the sampling location (Auta et al., 2017; Browne et al., 2007; Cole et al., 2014; Kameda et al., 2021; Wang et al., 2020). Next, the potential source of the contamination is atmospheric fallout, as atmospheric deposition can carry the microplastics to the sampling location (Anderson et al., 2017; Baldwin et al., 2016; Zhang et al., 2020b). Then, wind can transport the microplastics to the surface water because the microplastics collected during heavy-wind events are higher than during low-wind events (Kameda et al., 2021). Surface runoff can also carry microplastics into the surface water, especially during rain (Zhang et al., 2020b). For instance, surface runoff can transport the fibres (Baldwin et al., 2016; Kameda et al., 2021). The effluent of WWTPs also transports the microplastics into the surface water (Mani et al., 2015; Zhang et al., 2020b). The absence of WWTPs also leads to an increment in the abundance of microplastics in surface water due to the lack of wastewater treatment before discharging into the surface water (Mataji et al., 2019). For example, the absence of the treatment of the effluent discharged from the textile industry causes synthetic fibres detected in the surface water to have 92.00% (Lahens et al., 2018). A few types of plastics were found in the water samples, such as PE, PP, and PS (Di and Wang, 2017; Eriksen et al., 2013; Kameda et al., 2021; Mason et al., 2016; Mataji et al., 2019; Wang et al., 2020). PE originates from domestic plastics products, such as household utensils, the caps of plastics bottles, and plastic bags (Wang et al., 2020). PP is also used as packaging materials of food products, and thus, PP can be transported into the surface water through household wastes (Mataji et al., 2019). PP is also released from utilizing fishing tools such as fishing gear (Di and Wang, 2017). Personal care products' microbeads can also be manufactured from PE and PP. Thus, personal care products are one of the potential sources of PE and PP in surface water (Eriksen et al., 2013; Fendall and Sewell, 2009; Mason et al., 2016). Furthermore, the source of PS is domestic waste because it is used as materials for the packaging of food products and cosmetic materials (Di and Wang, 2017; Mataji, Taleshi and Balimoghaddas, 2019; Wang et al., 2020). Fibres, fragments, microbeads, and foams were found as the microplastics detected in the water samples. In a study by Baldwin et al. (2016), the microbeads found in the water samples were estimated to come from personal care products, and domestic sewage carried them into the sampling location (Baldwin et al., 2016; Fendall and Sewell, 2009). Atmospheric deposition may also carry the detected fibres (Baldwin et al., 2016; Wang et al., 2020). Fibres are the shape of microplastics detected dominantly in the water samples. The fibres can originate from the washing process of garments (Kameda et al., 2021; Wang et al., 2020). Next, fragments are mainly formed from the degradation of large plastic items, such as photodegradation (Di and Wang, 2017; Derraik, 2002; Mason et al., 2016; Mataji et al., 2019).

1.4.5 Detection of Microplastics in Wastewater

As discussed earlier, the microplastics discharged from household and industrial areas entered wastewater treatment plants through sewage systems. Various studies have been conducted to evaluate the detection of microplastics in wastewater. In the research of Kazour et al. (2019), they used a pump to collect wastewater samples and filtered the collected wastewater through a few sieves, which have different sizes of mesh in different studies, such as 500 μm, 200 μm, 80 μm, 20 μm (Kazour et al., 2019). Besides, various studies used metal buckets (Blair et al., 2019; Murphy et al., 2016) and glass bottles (Leslie et al., 2017) to collect the wastewater samples from WWTPs. The samples from WWTPs collected from these two methods were filtered through sieves (Blair et al., 2019; Leslie et al., 2017; Murphy et al., 2016). Furthermore, a plastic container was used to collect the wastewater samples. The plastic containers were rinsed with deionized water and dried with air before collecting the wastewater samples (Michielssen et al., 2016). The collected samples from some studies were digested using a WPO process (Blair et al., 2019; Simon et al., 2018). Density separation was done on the samples in Kazour et al. (2019) and Leslie et al. (2017) to separate solids from liquid. On the other hand, a steel scuttle (Franco et al., 2021) and autosampler (Horton et al., 2020; Simon et al., 2018) were also used to collect the wastewater samples. The wastewater samples collected by Franco et al. (2021) were filtered through a few sieves with mesh sizes of 100 μm, 355 μm, and 1000 μm. On the other hand, in a study by Horton et al. (2020), influent of WWTPs was collected behind a coarse screen, whereas effluent was collected with autosamplers (Horton et al., 2020). Wastewater samples collected by Simon et al. (2018) were filtered through a series of meshes with 500 μm, 1 mm, and 2 mm in the vibratory sieve shaker in wet conditions. Simon et al. (2018) treated the wastewater samples with SDS solution and incubated 200 mL of samples with enzymes before the WPO process. A WPO process (Franco et al., 2021; Horton et al., 2020; Simon et al., 2018) was also undergone by some studies.

The abundance of microplastics detected in the wastewater samples collected by eight studies were summarized in Table 1.18. The removal percentage of microplastics in the primary treatment of wastewater treatment plants range from 60.0% to 98.4%. Up to 95.0% of the microplastics can be eliminated after the secondary treatment (Blair et al., 2019; Kazour et al., 2019; Murphy et al., 2016). Tertiary treatment of wastewater treatment further removed 4.0% of microplastics (Blair et al., 2019; Kazour et al., 2019). The overall elimination percentage of the microplastics in WWTPs can reach 99.2% (Franco et al., 2021; Horton et al., 2020; Kazour et al., 2019; Michielssen et al., 2016; Simon et al., 2018). Although the elimination

TABLE 1.18
Abundance of Microplastics in Wastewater Samples

Location	Abundance of microplastics, microplastic/*L*	References
Scotland	Wastewater: 1,308 microplastics particles	Blair et al., 2019
Cádiz (Spain)	Urban WWTP influent: 645.00	Franco et al., 2021
	Urban WWTP effluent: 16.00	
	Industrial WWTP influent: 1567.00	
	Industrial WWTP effluent: 131.00	
United Kingdom	WWTP Effluent: 2.00–54.00 microp	Horton et al., 2020
Le Havre (France)	WWTP influent: 244.00	Kazour et al., 2019
	WWTP effluent: 2.84	
Netherlands and Germany	WWTP influent: 73.00 (average)	Leslie et al., 2017
	WWTP effluent: 65.00 (average)	
Detroit (United States)	WWTP influent: 133.00 ± 35.60	Michielssen et al., 2016
	WWTP effluent: 5.90	
Glasgow (Scotland)	WWTP influent: 15.70 ± 5.20	Murphy et al., 2016
	WWTP effluent: 0.25 ± 0.04	
Danish (Denmark)	WWTP influent: 7,216.00	Simon et al., 2018
	WWTP effluent: 54.00	

percentage of microplastics is high, microplastics are still present in the effluent and flow into the marine environment (Kazour et al., 2019; Murphy et al., 2016). Tiny microplastics are mainly detected in the effluent of wastewater treatment plants because their small sizes are difficult to trap in the treatment process (Blair et al., 2019; Franco et al., 2021; Kazour et al., 2019; Murphy et al., 2016). The higher concentration of the microplastics in the effluent might be due to the fragmentation of large plastics or the sonication process during the sample processing steps (Franco et al., 2021; Kazour et al., 2019; Simon et al., 2018). The sonication process can cause large plastics particles to become brittle and fragment (Andrady, 2011; Simon et al., 2018).

Fibres, fragments, films, and microbeads were found abundantly in the collected wastewater samples (Blair et al., 2019; Kazour et al., 2019; Leslie et al., 2017; Murphy et al., 2016). Many fibres are found at locations that have higher populations because fibres originate from synthetic textiles, and they are mainly removed during the pre-treatment process (Blair et al., 2019; Franco et al., 2021; Kazour et al., 2019). According to Blair et al. (2019), garments made from PA and polyester release many fibres (Michielssen et al., 2016). Therefore, PA and polyester discovered in the samples can be due to their release from synthetic textiles (Michielssen et al., 2016). However, PA and polyester were not found in the analysis because they may settle due to their high densities. Fragments are mainly removed at the primary and tertiary stages of wastewater, whereas the pre-treatment stage mainly removes films (Blair et al., 2019). Fragments are produced from the degradation of large plastics items and cleaning products (Franco et al., 2021; Kazour et al., 2019; Michielssen et al., 2016). Microbeads are mainly removed at the pre-treatment and primary stages of the wastewater treatment process (Blair et al., 2019; Franco et al., 2021; Michielssen et al., 2016; Murphy et al., 2016). However, the presence of microbeads in WWTPs is low because the legislation of most countries supports bans on the manufacturing of microbeads in personal care products. Consequently, the emission of microbeads to the marine environment is lower due to lower production, and hence, lower amount of microbeads are detected in wastewater samples (Franco et al., 2021; Pico and Barcelo, 2019). A high abundance of PP, PE, PVC, PA, PS, PET, and polyester were found as the types of microplastics collected (Blair et al., 2019; Horton et al., 2020; Murphy et al., 2016; Simon et al., 2018). PP and PE are used to manufacture the packaging materials and microbeads (Kazour et al., 2019; Murphy et al., 2016). In addition, PET, PA, and polyester are used to produce synthetic fibres, and they are mainly released during the cleaning process of garments (Franco et al., 2021; Kazour et al., 2019). PS is mainly used to manufacture the packaging of food products such as food containers (Kazour et al., 2019).

1.4.6 Comparison of the Detection of Microplastics

The detection of microplastic in different water sources are compared and discussed. All the types of microplastics found in different water sources are summarized in Table 1.18. In the water source of bottled water, the microplastics detected in the collected bottled water were identified to consist of PET, PP, and PE plastics. The presence of microplastics PET, PP, and PE mainly originated from the bottles and caps due to the abrasion of inner wall of bottles, as summarized in Table 1.19. Basically, the water stored in reusable bottles was detected to consist of higher microplastics than single use bottles. This might be attributed to the stress put on the inner wall surface of the bottles during the washing and utilization process. The tap water collected by various studies detected the presence of microplastics in tap water;

TABLE 1.19
The Types of Microplastics Present in Various Types of Water Sources

Water sources	Types of microplastics or form	Sources of microplastics
Bottled water	• PET, PP, PE	• Released from the bottles and caps.
Tap water	• PET, PVC, PE, PP, PS, PA, rayon, and acrylonitrile butadiene styrene • Mostly in fragment and microfibres forms	• The types of microplastics found depend on the human population density. • Abrasion of synthetic clothing, anthropogenic activities, and mechanical abrasion of pipes.
Ground water	• Mainly detected are PA, PET, PVC, PP, PE, PS, nylon. • PU (agriculture site)	• Atmospheric fallout and drainage of waste. • Leachate from landfill sites with microplastics contamination.
Surface water	• PE, PP, PS • In forms of fragments, microbeads, microfibre	• Depended on the size of the population. • Atmospheric fallout. • Surface runoff.
Wastewater	• PP, PE, PVC, PA, PS, PET, and polyester • Microfibres, fragments, films, and microbeads	• Depended on the size of the population.

these microplastics originated from PET, PVC, PE, PP, PS, PA, rayon, and acrylonitrile butadiene styrene. The potential sources of the microplastics contamination in tap water are contributed by the abrasion effect of synthetic clothing, the population density of the sampling areas, anthropogenic activities in the area, and the mechanical abrasion of pipes. The concentration of microplastics detected in tap water are highly dependent on the human population density in the sampling area. The presence of PVC, PP, PE, and PA in tap water is also attributed to the usage of pipe and fittings made of PVC, PP, PE, or PA in piping systems which are transport the drinking water from water treatment plants into household areas. Plastic of PVC, PP, and PE are mainly used to fabricate the pipes used in drinking water systems, while PA is commonly used to produce the fittings used in piping systems. On the other hand, PET is also attributed to the washing of textile fibres. For the source of groundwater, the types of microplastics found in groundwater is mainly dependent on the sampling location. In various research, the presence of PA microplastics is attributed to improper sampling during the sample-collection process. PVC, PE, PP, and PET were mainly detected in industrial sites. On the other hand, PE microplastic was primarily found in residential sites, while the presence of PU was mainly found in agricultural sites. The main sources contributing to the presence of microplastics in groundwater are atmospheric fallout and the drainage of wastewater containing microplastics. Besides, the leachate leaked from landfill sites due to poor management also contributes to the contamination of groundwater with microplastics.

Microplastics detected in the source of surface water were mainly PE, PP, and PS. The microplastics of PE, PP, and PS are discharged as domestic waste from food and cosmetics packaging. PE and PP are commonly used as microbeads in personal care cosmetics products. Furthermore, the microplastics detected in surface water were observed to be in the form of microfibres, fragments, and microbeads. The fragment form of microplastics mainly originated from the degradation of large plastics, while microbeads originated from personal care and cosmetics products. Microfibres are basically attributed to textile industrials or the washing of clothes and textiles products. The presence of microplastics found in surface water is mainly dependent on the human additives. In other words, higher population in an area could lead to detection of a higher abundance of microplastic amounts. Basically, most of the microplastics detected in surface water are attributed to the degradation of large particles and the release of microbeads from personal care and cosmetics products.

Another potential source of microplastics in surface water is surface runoff, especially heavy rain, which could transport microplastics from area to area. For another source of microplastics in wastewater, PP, PE, PVC, PA, PS, PET, and polyester were found in the collected wastewater. The presence of PS microplastics mainly originated from the manufacturing of food packaging products such as food containers. By referring to Blair et al. (2019), the overall elimination of microplastics can reach up to 99.2%. However, wastewater still has some remaining, tiny microplastics which are too tiny in size and unable to be eliminated in various treatment stages of the wastewater treatment process. Microplastics were mainly detected in the effluent of wastewater treatment plants due to the fact that the small sizes of microplastics can easily flow through the filtration system during various treatment stages (Blair et al., 2019). The microplastics found in the effluent of collected wastewater were mainly in the form of microfibres, fragments, films, and microbeads. The presence of microfibres in wastewater is observed to be influenced by the sampling locations. The locations with higher populations were found to detect higher microfibres in the collected wastewater. This is mainly due to the higher population areas usually accompanied by the higher washing frequency of synthetic textiles in the areas. Furthermore, clothes and garments made of PA and polyester may release high amounts of microfibres due to the abrasion effect during the washing process. Besides, the higher concentration of microplastics in the effluent of wastewater in comparison to the other sources such as surface water, groundwater, etc., might be attributed to the high amount of microplastics originating from the fragmentation of large plastic waste. The sonication process of wastewater during the sample-processing step also highly contributed to the higher amount of microplastics from the fragmentation process (Franco et al., 2021).

1.5 IMPACTS OF MICROPLASTICS ON AQUATIC ORGANISMS AND HUMANS

Most of the microplastics from the various sources can be transferred to the aquatic environment through stream runoff, leading to microplastic pollution issues in the aquatic environment. Microplastics with smaller in particle sizes can be easily ingested by low trophic-level aquatic organisms. Microplastics ingested by low trophics can further accumulate in the bodies of the low trophic-level aquatic organisms and then can be further transferred to higher trophic-level aquatic organisms and humans. The threatening impacts of microplastics in the food chains on the health of the aquatic organisms and humans will be discussed later.

1.5.1 Impacts of Microplastics on Aquatic Organisms

According to Huang et al. (2021), the primary producers are the base of the aquatic food chains, and they will provide the aquatic ecosystem with oxygen gas and food sources through the photosynthesis process. Phytoplankton and aquatic plants are the main contributors of the primary producers in the aquatic ecosystem (Agrilife, 2013). Due to the important role of primary producers in the aquatic ecosystem, especially phytoplankton, the impacts of microplastics on primary producers need to be considered to clarify the potential risks and effects of microplastics on the sustainability of the primary producers and aquatic ecosystem. Thus, further proper action can be taken before the microplastics issue becomes more serious (Huang et al., 2021). Several studies have been conducted to investigate the effects of microplastics on the growth and photosynthesis process of microalgae. Based on the research of Liu et al. (2022), the toxicity of the polystyrene microplastics of various sizes (0.1, 0.5, 1, and 2 μm), polystyrene microplastics with surfaces charged of amino groups (0.1 μm) and the addition of humic acid to the microalgae *Scenedesmus obliquus* have been investigated. The exposure of different sizes of polystyrene microplastics was found to have an insignificant impact on the growth inhibition rate of microalgae. Similar results were also obtained from another laboratory study, where the size of microplastics did not affect *Tetraselmis chuii* microalgae growth rate (Davarpanah and Guilhermino, 2015). However, the growth inhibition rate of the microalgae increased when exposed to microplastics with the amino group surface charged compared to the 0.1 μm of polystyrene microplastics without the surface charged. This might be due to the fact that the surface charge on the microplastics enhanced the adsorption of microplastic onto the cell wall of microalgae. Besides, the bigger size of polystyrene microplastics showed a greater effect on the photosynthesis process of microalgae because the big size microplastics acted as a barrier, which made it difficult for the chlorophyll in the cells of microalgae to receive light energy. Microplastics caused the destruction of the photosynthesis process of microalgae, which would lead to irreversible growth inhibition. Then, the presence of humic acid would adhere to the surface of polystyrene microplastics and reduce the number of microplastics adsorbed into the cells of microalgae. This would reduce the toxicity of microplastics to microalgae, as fewer microplastics accumulated in the microalgae but are only available for the smaller-size microplastics. The effect of bigger microplastics that affected the photosynthesis process of microalgae could not be solved by adding humic acid (Liu et al., 2020b). Moreover, another laboratory study investigated the impacts of microplastics on marine microalgae. The microalgae species was *Skeletonema costatum*, and the microplastics used were polyvinyl chloride with a size of 1 μm. The results showed that the size of the microplastics would affect the toxicity of microplastics to microalgae, as the polyvinyl chloride microplastics caused the growth inhibition of microalgae but not for polyvinyl chloride bulk. When exposed to a higher concentration of polyvinyl chloride microplastics, the efficiency of the photosynthesis process and chlorophyll content of microalgae was lower but increased again after a long exposure time (Zhang et al., 2017). According to Mao et al. (2018), the toxicity of microplastics to microalgae was examined along the growth period of the microalgae with a cultivation period of 30 days. The microalga involved in the experiment was *Chlorella pyrenoidosa*, a freshwater microalga, and polystyrene microplastics with a size of 0.1 and 1 μm were used. During the lag to logarithmic phase, the toxicity of microplastics to microalgae was observed, as the growth rate and efficiency of the photosynthesis process showed a negative effect. At the same time, the cell membrane of microalgae was also damaged, which indicated that the microplastics would lead to physical damage to the microalgae cells. However, during the logarithmic to stationary phase, the growth rate and efficiency of the photosynthesis process of microalgae recovered to a normal state as well as cell structure. This might be due to the several actions taken by the cell microalgae to recover to a normal state. The actions taken included thickening of the cell membrane, homo-aggregation, and hetero-aggregation. The thickening of the cell membrane could help avoid the entering of microplastics and prevent further damage to the cell membrane. The microalgae would then undergo a homo-aggregation process to protect the cells from the stress of microplastics. At the same time, the hetero-aggregation process could let the microplastics settle down to the bottom, and the microalgae would suspend in the aquatic environment, which would reduce the contact between the microalgae and microplastics. The polyethylene microplastics was observed to negatively affect the growth rate of duckweed and the chlorophyll content in the cells of duckweed. However, the solid-state of polyethylene microplastics would reduce the length of the duckweed roots because the polyethylene microplastics adsorbed on the surface of roots would become obstacles and affect the root growth. Besides, polyethylene microplastics with a rough surface would cause damage to roots, resulting in a decrease in the root cell-viability and root length. The smooth surface of polyethylene microplastics did not show a significant effect on the root cell-viability (Kalčíková et al., 2017). The growth rate and photosynthesis processes of the phytoplankton or aquatic plants was found to be affected by some types of microplastics, depending on various factors such as the different species of the primary producers may have a different structure of cells, types, sizes, and characteristics of microplastics.

Furthermore, as a primary consumer, the aquatic invertebrates play an important role to balance the aquatic ecosystem. The food source of aquatic invertebrates usually is the primary producer, and the aquatic invertebrates will act as food sources and be consumed by aquatic carnivores. The feeding characteristics and the food sources of aquatic invertebrates have increased the contact between microplastics and aquatic invertebrates such as amphipods, mussels, and oysters (Huang et al., 2021). According to Moos et al. (2012), the effect of high-density polyethylene microplastics without polymer additives on blue mussel *Mytilus edulis* has been investigated. The blue mussel was exposed to microplastics for several hours in two pathways, which are respiratory pathways (the intake of microplastics by blue mussels, microplastics found in gills) and digestive pathway (microplastics were found in the stomach and intestine). After three hours of exposure, the microplastics found in the stomach had been transferred to the digestive gland and gathered in the lysosomal. Then, after six hours of exposure, the granulocytomas formation on the connective tissue indicated a great inflammatory response of blue mussel cellular to the microplastics. An increase in the degree of instability of the lysosomal membrane was observed with increasing exposure time. Another study also examined the impacts of microplastics on blue mussels *Mytilus edulis* by using biodegradable polylactic acid and high-density polyethylene. By comparing the amount of the byssal threads produced, a great reduction was obtained when the blue mussels were exposed to high-density polyethylene, and the tenacity of blue mussels decreased by about 50% compared with the blue mussels without being exposed to microplastics. The reduction of the number of byssal threads produced might cause consequences such as weaker the resistance to predators, weaker attachment of blue mussels, and reducing the production yield of blue mussels. When exposed to polylactic acid and high-density polyethylene microplastics, the haemolymph proteome obtained was different from the blue mussels without being exposed to microplastics. The altered proteins due to the appearance of microplastics in the blue mussels might have a negative effect on the immune system, development of structure, detoxification, and metabolism process, as proteins are the main component in these systems or biological processes (Green et al., 2019). The Pacific oyster *Crassostrea gigas* was exposed to polystyrene microplastics with a size of 2 and 6 μm in the research of Sussarellu et al. (2016). The oyster preferred to ingest polystyrene microplastics with a size of 6 μm, and no accumulation of microplastics occurred in the stomach of the oyster, as the microplastics excreted through faeces, which indicated the oyster could excrete the microplastics easily due to the usage of smooth-surface microplastics. However, the actual microplastics could be in the shape of fibres and with a rough surface. Besides, the number of oocytes, the diameter of the oocyte, and the velocity of sperm decreased significantly with a reduction percentage of 38%, 5% and 23%, respectively, when exposed to microplastics (Sussarellu et al., 2016). However, several laboratory studies showed no significant impacts of microplastics on aquatic invertebrates such as exposure of the Mediterranean mussel, *Mytilus galloprovincialis*, to polystyrene microplastics with a spherical shape. The results showed that microplastics were observed in the lumen of the gut and excreted out of the body through faeces and the absence of microplastics in the gills and digestive gland. When exposed to microplastics with a smooth surface, the microplastics ingested could be filtered out through the excretion process (Gonçalves et al., 2019). Besides, another laboratory study investigated the impacts of polyethylene terephthalate microplastics on amphipod *Gammarus pulex*, a type of invertebrate that lives in freshwater, and found that the polyethylene terephthalate microplastics would not pose a significant negative effect on the amphipod. This might be due to the influence of various parameters such as exposure time, the concentration of microplastics exposed, characteristics of microplastics like shapes, surface conditions and types, the presence of additives, and the morphology of microplastics (Weber et al., 2018). Besides, the impacts of microplastics on the aquatic invertebrates were found to depend on the types of microplastics used and the characteristics of the microplastics.

Fish have a high probability of ingesting microplastics from their prey that contain microplastics due to the trophic transfer of microplastics along food chains. Besides, fish also can ingest microplastics from the surrounding environment or during the respiratory process, causing microplastics to enter the fish's body. This caused the accumulation of microplastics in the fish's gastrointestinal tract and gills (Huang et al., 2021). Huang et al. (2021) found the negative effects of polystyrene microplastics on the gut of the zebrafish *Danio rerio*. They found that the appearance of microplastics in the gut of the zebrafish would cause dramatic changes in tissue's metabolic profile and the gut microbiome, which might lead to oxidative stress and inflammation in the zebrafish's gut. Besides, microplastics would disturb the gut metabolism of zebrafish by resulting in a significant change in the inhibition of lipid metabolism and amino acid metabolism in zebrafish (Qiao et al., 2019). Polyethylene, polypropylene, polyamides, polystyrene, and polyvinyl chloride microplastics were selected to investigate the impacts of microplastics on the zebrafish *Danio rerio*. Qiao et al. (2019) found the rupture of the villus and the division of enterocytes indicated the damage of microplastics to the intestines of zebrafish. No significant results showed that microplastics would cause damage to the tissue of livers, kidneys, and gills of zebrafish (Lei et al., 2018). Moreover, another laboratory study showed the impacts of polystyrene microplastics on the juvenile, intertidal fish *Girella laevifrons*. The observation obtained was the rising of microplastic accumulation in the intestine of the fish when the exposure time increased. The microplastics accumulated might cause damage and the malfunction of tissue and cause the excretion system to be unable to properly remove the microplastics. Besides, the physical abrasion between the intestinal epithelium and microplastics might lead to cell death and injury of the intestinal epithelial cells. Then, the occurrence of more serious hyperemia and leukocyte infiltration was observed when exposed to a high concentration of microplastics. This indicated that the physical abrasion would cause an immune response by increasing the flushing of intestinal blood vessels, and thus, an influx of white blood cells into the damaged tissue. According to Chagnon et al. (2018), the yellow tuna *Thunnus albacares* that consumed microplastics-containing prey was observed to egest microplastics rapidly out of its body. Microplastics have a smaller size than the pylorus of fish and also could easily discharge from the body of the tuna. Furthermore, another study found that the potential for the accumulation of microplastics in the fish's gastrointestinal tract is low, but the majority of the microplastics contained chemical additives, which would cause adverse effects on fish when these chemical compounds leached from the ingested microplastics into the body of the fish. Besides, the chemical contaminants within microplastics might be adsorbed and bioaccumulated in fish, increasing the fish's health risks and further transferring through trophic transfer. Microplastics also will cause the starvation or satiation of planktonic fish, as fish need more time to digest the microplastics contained in food. Microplastics with a rough surface and sharp edges penetrate the intestinal wall of fish, causing physical injury and inflammation. Microplastics in the gut of fish could also translocate to the liver of fish, leading to metabolic changes, inflammation, and the aggregation of lipids in the liver. The ingestion of microplastics by fish causes adverse effects on fish health, such as damage to intestines, toxicity, growth rate reduction, physical damage to organs, and altered lipid metabolism (Jovanović, 2017). The impact of microplastics on the immune system of fish was observed to cause abnormal functioning of the intestinal immune cells, caused by the increasing population of pathogenic bacteria. Among various kinds of immune cells, the M1 macrophage was greatest affected by polystyrene microplastics. The disturbance in the initiation process of the immune system was detected, as most of the genes involved in the phagosomes and immune system regulation were suppressed in M1 macrophages. The inhibition of the immune system would increase the potential health risks in zebrafish because of an increase in the prevalence of pathogenic bacteria in the intestines of zebrafish (Gu et al., 2020). The impacts of microplastics on the tissue and reproductive system of marine medaka *Oryzias melastigma* had been investigated Bu Gu et al. (2020). They observed that microplastics could accumulate in the gills, intestine, and liver of marine medaka, where the number of microplastics detected in the liver was about 5% higher than the amount in the gut. The majority of microplastics ingested or inhaled into the body of marine medaka could be expelled by the digestive system, but the minority would migrate into the liver through the circulatory system. This study also demonstrated that microplastics would interrupt the hypothalamic-pituitary-gonadal axis and steroidogenesis pathway, which caused the imbalance of sex hormones. The consequences of the imbalance of sex hormones were the reduction of vitellogenin and choriogenin, which play an important role in the oogenesis process. This would lead to the development of the ovary being affected and delayed. These proved that microplastics would influence the reproduction of fish and might affect the whole population (Wang et al., 2019). In an investigation of the accumulation and toxic effects of polystyrene microplastics in zebrafish after an exposure time of seven days, microplastics with a size of 20 μm were detected in the gills and gut of zebrafish, while 5 μm of microplastics were identified in the gills, gut, and liver of zebrafish. This might be due to the smaller size of microplastics that could be able to transfer to the liver through the circulatory system. Microplastics would cause the inflammation of liver cells with symptoms such as vacuolation, infiltration, and necrosis. Besides, the alteration of fatty acids and triglycerides lipid metabolites, which are related to lipid and energy metabolisms, would cause the disturbance of lipid and energy metabolisms. This was verified by observing lipid droplets in fish liver (Lu et al., 2016). Microplastics have the possibility to bioaccumulate in the body of fish, and smaller sizes of microplastics are able to enter into the blood vessels, and further, be transported to the liver of fish. Microplastics also bring a lot of negative effects to fish, such as tissue damage, disruption of the immune system in the intestine of fish, imbalance of sex hormones, reproductive issues, and disruption of lipid metabolism. These health issues of fish may affect the fish populations, so further investigation into the impacts of microplastics on fish needs to be continued and explored.

According to Nelms et al. (2019), due to the ingestion of microplastics, the potential health impacts on top predators or marine mammals, including weakened immune systems or increased susceptibility to diseases, might only emerge when microplastics have passed through the whole body of top predators or marine mammals. According to Fossi et al. (2012), additives or chemical pollutants such as polycyclic aromatic hydrocarbon, polychlorinated biphenyl, and bisphenol A might be able to leach from microplastics after the ingestion of microplastics and exposure directly to the fin whale. Among these chemical pollutants or additives in the microplastics, some were recognized as disruptive endocrine compounds because they could interfere with the endogenous hormone-synthesis process and would cause toxicological effects on the fin whale. However, research about the impacts of microplastics on these organisms is still limited and challenging, as these organisms usually have a greater body size, making it more difficult to obtain accurate results about the toxicological effects of microplastics.

1.5.2 Impacts of Microplastics on Humans

According to Huang et al. (2021), microplastics in the aquatic environment can be transferred along food chains through trophic transfer between predator-prey relationships. Humans can be exposed to microplastics by eating seafood, an indispensable diet in daily life. Seafood consumption contributed to protein consumption and animal protein consumption with about 6.7% and 17%, respectively, with an overall seafood intake of at least 20 kg/capita/year up to year 2015 (Smith et al., 2018). An estimate of approximately 112 to 842 microplastics/g in fish can be consumed by a human in one year. The amount of microplastics ingested by humans from fish depends on the number of microplastics contained in fish and the number of fish consumed in each country. The amount of microplastics ingested by humans from fish is about 518 microplastics/year/person in Brazil, while it is approximately 3,078 microplastics/year/person in Portugal (Barboza et al., 2020). Therefore, it is important to investigate the risks of microplastics in the food chains to human health, as humans are exposed to these microplastics through the consumption of seafood.

Several studies have been conducted on the cytotoxicity of the microplastics on human cells and the potential risks of microplastics to human health. In order to determine the cytotoxicity of microplastics to human cells, several parameters such as the cell viability, cellular uptake of microplastics, reactive oxygen species (ROS) level, oxidative stress, mitochondrial membrane potential, cell apoptosis, and membrane integrity need to be considered (Zhang et al., 2021; Schirinzi et al., 2017; Wu et al., 2019). Table 1.20 shows the target cells used and the types, size, and concentration of microplastics used in different research study, while Table 1.21 shows the results obtained in each experimental study. From the results obtained, the polystyrene microplastics and polyethylene microplastics with a size range of 0.1 to 16 μm and with a concentration of 0.05 to 200 μg/mL had no significant change in the cell viability, which indicated that these microplastics were non-toxic to the human cells (Zhang et al., 2021; Schirinzi et al., 2017; Wu et al., 2019). After that, the cellular uptake of microplastics was always greater for microplastics with a smaller size (Zhang et al., 2021; Wu et al., 2019). More microplastics will be accumulated inside

TABLE 1.20
Target Cells Used and Types, Size, and Concentration of Microplastics Used

Details	Sources		
	(Zhang et al., 2021)	(Schirinzi et al., 2017)	(Wu et al., 2019)
Target cells	Human small intestinal epithelial cell line (HIEC-6); Human colonic epithelial cell line (CCD841CoN)	Cerebral human cells (T98G); Epithelial human cells (HeLa)	Human colorectal adenocarcinoma cells (Caco-2)
Types of microplastics	Polystyrene microplastics	Polyethylene microplastics, polystyrene microplastics	Polystyrene microplastics
Size of microplastics (μm)	0.1, 0.5, 1, 5	3–16	0.1, 5
Concentration of microplastics (μg/mL)	12.5, 25, 50, 100	0.05, 0.1, 1, 10	1, 10, 40, 80, 200

TABLE 1.21
Results of Cytotoxicity of Microplastics to Human Cells

Parameters	Sources		
	(Zhang et al., 2021)	(Schirinzi et al., 2017)	(Wu et al., 2019)
Cell viability	No significant change	No significant change	No significant change
Cellular uptake of microplastics	• Exposure time increased, more uptake of microplastics in cells. • The 0.1 μm of polystyrene microplastics had greater uptake than 5μm.	–	• The 0.1 μm of polystyrene microplastics had greater uptake than 5 μm.

cells when the cellular uptake of microplastics is greater, which means that smaller microplastics have more cytotoxicity to human cells than microplastics with larger particles. However, the cellular uptake of microplastics also depended on various factors, such as the types of cells and the functions of cells (Zhang et al., 2021). Caco-2 cells had weaker cellular uptake of polystyrene microplastics with a size of 1 μm than 4 μm. This might be due to the absorption mechanism where the 1 μm microplastics were taken by cells through phagocytosis and 4 μm microplastics were taken by cells through phagocytosis and pinocytosis (Stock et al., 2019).

Various studies found that a greater ROS effect was generated by microplastics of a smaller size and a high concentration of microplastics (Zhang et al., 2021; Schirinzi et al., 2017; Wu et al., 2019). The increasing of ROS levels indicates the occurrence of oxidative stress, as there is an imbalance between antioxidants and ROS production (Zhang et al., 2019). At first, the entering of microplastics will cause the reduction of ROS level because the appearance of microplastics leads to excessive reactive oxygen radicals, and the antioxidant will immediately eliminate the reactive oxygen radicals. After the extension of exposure time, the ROS level may go back to a normal level or an excess of the normal level. When ROS exceeds the normal level, the antioxidant is in insufficient status and unstable for the reactive oxygen radicals. This will lead to intracellular oxidative stress and damage to cells, so the cytotoxicity of the microplastics to human cells can be determined through an oxidative stress experiment (Zhang et al., 2021). The possible diseases that will be formed due to oxidative stress include cancer, obesity, cardiovascular diseases, and diabetes (Zhang et al., 2019). Furthermore, polystyrene microplastics with a bigger size (5 μm) were observed to have a greater effect on mitochondrial membrane potential (Zhang et al., 2021; Wu et al., 2019). The function of mitochondria in the cells is to produce energy stored in the form of adenosine triphosphate (ATP) for energy metabolism. Polystyrene microplastics can cause the reduction of the mitochondrial membrane potential, which consequently leads to membrane depolarization, where the larger size of microplastics showed a greater decrease in the mitochondrial membrane potential level. Larger microplastics will adhere to the cell membrane and damage the mitochondrial electron transport and thus reduce the mitochondrial membrane potential level and cause cell apoptosis (Zhang et al., 2021). According to Wu et al. (2019), the mitochondrial depolarization issue would cause the inhibition of the activity of ATP-binding cassette transporters, which will directly enhance the accumulation of toxic pollutants in human cells. This indicates that a larger size of microplastics has higher cytotoxicity to human cells compared with a smaller size of microplastics.

1.6 CONCLUSIONS

In conclusion, severe issues of microplastic pollution have gradually been caused by the rapid development of plastic in manufacturing consumer products. Primary and secondary microplastics are present in sources of drinking water. Primary microplastics such as microbeads used in personal care and cosmetics products and microplastics in paint are commercially manufactured in tiny sizes (Fendall and Sewell, 2009). Besides, primary microplastics also exist in paint products and abrasive blasting agents due to the additional characteristics or benefits offered by microplastics

(Sundt et al., 2014; Verschoor et al., 2016). Secondary microplastics such as microfibres and microplastics from tyres' erosion result from degradation. They are mostly produced from photodegradation, and the rate of photodegradation is higher on land (GESAMP, 2015; Pegram and Andrady, 1989).

Microplastics originating in urban areas are mainly contributed by manufacturing and production activities in the cities. These microplastics in cities can be transported into different drinking water sources, including surface water and the ocean, through different processes. The transportation processes of microplastics include atmospheric fallout, surface runoff, effluent from wastewater treatment plants, and leaching from soil (Qiu et al., 2020). Microplastics are proven to be carried by the atmospheric fallout to the marine environment in research conducted by Dris et al. (2015) and Dris et al. (2016). Furthermore, surface and stormwater runoff can also carry microplastics from the land to surface water because they can be washed by heavy rainfall (Qiu et al., 2020). The transportation of microplastics by stormwater runoff is proven by studies conducted by Liu et al. (2019b) and Liu et al. (2019a) at stormwater retention ponds as the stormwater is treated in retention ponds (Liu et al., 2019b). The effluents from wastewater treatment plants are also transportation methods for microplastics to the marine environment, since wastewater is typically treated in a wastewater-treatment plant before being directed to drinking-water sources. Microplastics are also found to be transferred and carried by leaching from soil into groundwater, such as landfill leachates, as proved in the research of He et al. (2019). Discrete devices such as water buckets and net-based devices such as manta trawls can be used to collect the water samples to determine the presence of microplastics in drinking water sources. The sources of the microplastics in the marine environment is also discussed, such as wastewater treatment plants in tourist areas and urban areas. Eventually, many harmful impacts can be introduced by the exposure to the additives leached from microplastics into drinking water sources, such as BPA, phthalates, arsenic, and BFR. In general, these four additives lead to damage in the male and female reproductive system, including sperm count reduction and alteration in male and female reproductive hormones. Furthermore, the alteration in the development of the brain and behaviour is affected by exposure to BPA, arsenic, and BFR. Apart from that, diabetes and obesity can be induced by exposure to all stated additives except BFR. BFR exposure also leads to the alteration in thyroid hormones' homeostasis, including a reduction in circulating thyroxine. The respiratory system is also caused by exposure to arsenic, such as lung cancer, pulmonary oedema, and haemorrhagic bronchitis. Ultimately, skin diseases, including spotted hyperpigmentation, skin lesions, and skin cancer, are also affected by arsenic exposure. Hence, severe health impacts can be caused by exposure to the additives leached from microplastics contained in drinking water.

REFERENCES

Abdulmalik Ali, M. G., 2019. Presence and characterization of microplastics in drinking (tap/bottled) water and soft drinks. Masters' Degree. University of North Dakota.

Agrilife, 2013. Aquatic ecology and the food web. Aquatic Ecology, [online] Available at: <http://agrilife.org/fisheries2/files/2013/10/Aquatic-Ecology-And-The-Food-Web.pdf [Accessed 01 March 2022].

Alavian Petroody, S. S., Hashemi, S. H. and van Gestel, C. A. M., 2021. Transport and accumulation of microplastics through wastewater treatment sludge processes. Chemosphere, [e-journal] 278, p. 130471. https://doi.org/10.1016/j.chemosphere.2021.130471

Alvim, B. C., Mendoza-Roca, J. A. and Bes-Piá, A., 2020. Wastewater treatment plant as microplastics release source – Quantification and identification techniques. Journal of Environmental Management, [e-journal] 255, p. 109739. https://doi.org/10.1016/j.jenvman.2019.109739

Anderson, P. J., Warrack, S., Langen, V., Challis, J. K., Hanson, M. L. and Rennie, M. D., 2017. Microplastic contamination in Lake Winnipeg, Canada. Environmental Pollution, [e-journal] 225, pp. 223–231. https://doi.org/10.1016/j.envpol.2017.02.072

Andrady, A. L., 2011. Microplastics in the marine environment. Marine Pollution Bulletin, [e-journal] 62(8), pp. 1596–1605. https://doi.org/10.1016/j.marpolbul.2011.05.030

Auta, H. S., Emenike, C. U. and Fauziah, S. H., 2017. Distribution and importance of microplastics in the marine environment: A review of the sources, fate, effects, and potential solutions. Environment International, [e-journal] 102, pp. 165–176. https://doi.org/10.1016/j.envint.2017.02.013

Baldwin, A. K., Corsi, S. R. and Mason, S. A., 2016. Plastic debris in 29 Great Lakes Tributaries: Relations to watershed attributes and hydrology. Environmental Science and Technology, [e-journal] 50(19), pp. 10377–10385. https://doi.org/10.1021/acs.est.6b02917.

Barboza, L. G. A., Lopes, C., Oliveira, P., Bessa, F., Otero, V., Henriques, B., Raimundo, J., Caetano, M., Vale, C. and Guilhermino, L., 2020. Microplastics in wild fish from North East Atlantic Ocean and its potential for causing neurotoxic effects, lipid oxidative damage, and human health risks associated with ingestion exposure. Science of the Total Environment, [e-journal] 717, p. 134625. https://doi.org/10.1016/j.scitotenv.2019.134625

Baresel, C., Westling, K., Samuelsson, O., Andersson, S., Royen, H., Andersson, S. and Dahlén, N., 2017. Membrane bioreactor processes to meet todays and future municipal sewage treatment requirements? International Journal of Water and Wastewater Treatment, [e-journal] 3(2). http://dx.doi.org/10.16966/2381-5299.140.

Besseling, E., Wegner, A., Foekema, E. M., Heuvel-Greve, M. J. and Koelmans, A. A., 2013. Effects of microplastic on fitness and PCB bioaccumulation by the lugworm Arenicola marina (L.). Environmental Science and Technology, [e-journal] 47, pp. 593–600. https://doi.org/10.1021/es302763x

Bhada-Tata, P. and Hoornweg, D., 2012. What a waste?: A global review of solid waste management. [online] Available at: <https://documents1.worldbank.org/curated/en/302341468126264791/pdf/68135-REVISED-What-a-Waste-2012-Final-updated.pdf>

Bharath K, M., Natesan, U., Vaikunth, R., Oraveen Kumar, R., Ruthra, R. and Srinivasalu, S., 2021. Spatial distribution of microplastic concentration around landfill sites and its potential risk on groundwater. Chemosphere, [e-journal] 277. https://doi.org/10.1016/j.chemosphere.2021.130263

Blair, R. M., Waldron, S. and Gauchotte-Lindsay, C., 2019. Average daily flow of microplastics through a tertiary wastewater treatment plant over a ten-month period. Water Research, [e-journal] 163. https://doi.org/10.1016/j.watres.2019.114909

Bläsing, M. and Amelung, W., 2017. Plastics in soil: Analytical methods and possible sources. Science of the Total Environment, [e-journal] 612, pp. 422–435. https://doi.org/10.1016/j.scitotenv.2017.08.086

Bortoletto-Santos, R., Plotegher, F., Majaron, V. F., da Silva, M. G., Polito, W. L., Ribeiro, C., 2020. Polyurethane nanocomposites can increase the release control in granulated fertilizers by controlling nutrient diffusion. Applied Clay Science, [e-journal] 199. https://doi.org/10.1016/j.clay.2020.105874

Boucher, J. and Friot, D., 2017. Primary microplastics in the oceans: A global evaluation of sources. [online]. Available at: <https://portals.iucn.org/library/node/466; [Accessed 27 June 2021].

Briggs, H., 2021. Plastic pollution: Take-out food is littering the oceans. BBC News. [online] 11 June. Available at: <www.bbc.com/news/science-environment-57436143> [Accessed 25 June 2021].

Boucher, J. and Friot, D., 2017. Primary microplastics in the oceans: A global evaluation of sources. [online]. Available at: <https://portals.iucn.org/library/node/466> [Accessed 27 June 2021].

Browne, M. A., Crump, P., Niven, S. J., Teuten, E., Tonkin, A., Galloway, T. and Thompson, R., 2011. Accumulation of microplastic on shorelines worldwide: Sources and sinks. Environmental Science & Technology, [e-journal] 42(21) pp. 9175–9179. https://doi.org/10.1021/es201811s

Browne, M. A., Galloway, T. and Thompson, R., 2007. Microplastics: An emerging contaminant of potential concern? Integrated Environmental Assessment and Management, [e-journal] 3(4), pp. 559–561. https://doi.org/10.1002/ieam.5630030412.

Browne, M. A., Galloway, T. and Thompson, R., 2010. Spatial patterns of plastic debris along estuarine shorelines. Environmental Science and Technology, [e-journal] 44(9), pp. 3404–3409. https://doi.org/10.1021/es903784e

Bui, X. T., Vo, T. D. H., Nguyen, P. T., Nguyen, V. T., Dao, T. S. and Nguyen, P. D., 2020. Microplastics pollution in wastewater: Characteristics, occurrence and removal technologies. Environmental Technology and Innovation, [e-journal] 19. https://doi.org/10.1016/j.eti.2020.101013

Cai, L., Wang, J., Peng, J., Tan, Z., Zhan, Z., Tan, X. and Chen, Q., 2017. Characteristic of microplastics in the atmospheric fallout from Dongguan city, China: Preliminary research and first evidence. Environmental Science and Pollution Research, [e-journal] 24, pp. 24928–24935. https://doi.org/10.1007/s11356-017-0116-x.

Campanale, C., Massarelli, C., Savino, I., Locaputo, V. and Uricchio, V. F., 2020. A detailed review study on potential effects of microplastics and additives of concern on human health. International Journal of Environmental Research and Public Health, [e-journal] 17(4). https://doi.org/10.3390/ijerph17041212.

Cao, Y., Zhao, M., Ma, X., Song, Y., Zuo, S., Li, H. and Deng, W., 2021. A critical review on the interactions of microplastics with heavy metals: Mechanism and their combined effect on organisms and humans. Science of the Total Environment, [e-journal] 788, p. 147620. https://doi.org/10.1016/j.scitotenv.2021.147620

Cenci-Goga, B. T., Iulietto, M. F., Sechi, P., Borgogni, E., Karama, M. and Grispoldi, L., 2020. New trends in meat packaging. Microbiology Research, [e-journal] 11(2), pp. 56–67. https://doi.org/10.3390/microbiolres11020010.

Chagnon, C., Thiel, M., Antunes, J., Lia, J., Sobral, P. and Christian, N., 2018. Plastic ingestion and trophic transfer between Easter Island flying fish (Cheilopogon rapanouiensis) and yellowfin tuna (Thunnus albacares) from Rapa Nui (Easter Island). Environmental Pollution, [e-journal] 243, pp. 127–133. https://doi.org/10.1016/j.envpol.2018.08.042

Chanpiwat, P. and Damrongsiri, S., 2021. Abundance and characteristics of microplastics in freshwater and treated tap water in Bangkok, Thailand. Environmental Monitoring and Assessment, [e-journal] 193(5). https://doi.org/10.1007/s10661-021-09012-2.

Chatterjee, R., 2010. Municipal solid waste management in Kohima city: India. Iranian Journal of Environmental Health Science and Engineering, [e-journal] 7(2), pp. 173–180. Available at: <www.bioline.org.br/pdf?se10020> [Accessed 30 September 2021].

Chen, G., Feng, Q. and Wang, J., 2019. Mini-review of microplastics in the atmosphere and their risks to humans. Science of the Total Environment, [e-journal] 703, p. 135504. https://doi.org/10.1016/j.scitotenv.2019.135504.

Chen, R., Qi, M., Zhang, G. and Yi, C., 2018. Comparative experiments on polymer degradation technique of produced water of polymer flooding oilfield. IOP Conference Series: Earth and Environmental Science, [e-journal] 113. https://doi.org/10.1088/1755-1315/113/1/012208

Clarke, B. O. and Smith, S. R., 2010. Review of 'emerging' organic contaminants in biosolids and assessment of international research priorities for the agricultural use of biosolids. Environment International, [e-journal] 37(1), pp. 226–247. https://doi.org/10.1016/j.envint.2010.06.004.

Cole, M., Webb, H., Lindeque P. K., Fileman, E. S., Halsband, C. and Galloway, T. S., 2014. Isolation of microplastics in biota-rich seawater samples and marine organisms. Scientific Reports, [e-journal] 4. https://doi.org/10.1038/srep04528.

Davarpanah, E. and Guilhermino, L., 2015. Single and combined effects of microplastics and copper on the population growth of the marine microalgae Tetraselmis chuii. Estuarine, Coastal and Shelf Science, [e-journal] 167, pp. 269–275. https://doi.org/10.1016/j.ecss.2015.07.023

De Falco, F., Di Pace, E., Cocca, M. and Avella, M., 2019. The contribution of washing processes of synthetic clothes to microplastic pollution. Scientific Reports, [e-journal] 9(6633). https://doi.org/10.1038/s41598-019-43023-x.

Derraik, J. G. B., 2002. The pollution of the marine environment by plastic debris: A review. Marine Pollution Bulletin, [e-journal] 44(9), pp. 842–852. https://doi.org/10.1016/S0025-326X(02)00220-5

Di, M. and Wang, J., 2017. Microplastics in surface waters and sediments of the Three Gorges Reservoir, China. Science of the Total Environment, [e-journal] 616–617, pp. 1620–1627. https://doi.org/10.1016/j.scitotenv.2017.10.150.

Dris, R., Gasperi, J., Rocher, V., Saad, M., Renault, N. and Tassin, B., 2015. Microplastic contamination in an urban area: A case study in Greater Paris. Environmental Chemistry, [e-journal] 12(5), pp. 592–599. https://doi.org/10.1071/EN14167.

Dris, R., Gasperi, J., Rocher, V. and Tassin, B., 2017. Synthetic and non-synthetic anthropogenic fibers in a river under the impact of Paris Megacity: Sampling methodological aspects and flux estimations. Science of the Total Environment, [e-journal] 618, pp. 157–164. https://doi.org/10.1016/j.scitotenv.2017.11.009.

Dris, R., Gasperi, J., Saad, M., Mirande, C. and Tassin, B., 2016. Synthetic fibers in atmospheric fallout: A source of microplastics in the environment? Marine Pollution Bulletin, [e-journal] 104(1–2), pp. 290–293. https://doi.org/10.1016/j.marpolbul.2016.01.006.

Eriksen, M., Mason, S., Wilson, S., Box, C., Zellers, A., Edwards, W., Farley, H. and Amato, S., 2013. Microplastic pollution in the surface waters of the Laurentian Great Lakes. Marine Pollution Bulletin, [e-journal] 77(1–2), pp. 177–182. https://doi.org/10.1016/j.marpolbul.2013.10.007.

European Chemicals Agency., 2019. Annex XV restriction report: Microplastics. [online] Available at: <https://echa.europa.eu/documents/10162/05bd96e3-b969-0a7c-c6d0-441182893720> [Accessed 30 June 2021].

Farrell, P. and Nelson, K., 2013. Trophic level transfer of microplastic: Mytilus edulis (L.) to Carcinus maenas (L.). Environmental Pollution, [e-journal] 177, pp. 1–3. https://doi.org/10.1016/j.envpol.2013.01.046

Fendall, L. S. and Sewell, M. A., 2009. Contributing to marine pollution by washing your face: Microplastics in facial cleansers. Marine Pollution Bulletin, [e-journal] 58(8), pp. 1225–1228. https://doi.org/10.1016/j.marpolbul.2009.04.025.

Fossi, M. C., Panti, C., Guerranti, C., Coppola, D., Giannetti, M., Marsili, L. and Minutoli, R., 2012. Are baleen whales exposed to the threat of microplastics? A case study of the Mediterranean fin whale (Balaenoptera physalus). Marine Pollution Bulletin, [e-journal] 64, pp. 2374–2379. https://doi.org/10.1016/j.marpolbul.2012.08.013

Franco, A. A., Arellao, J. M., Albendín, G., Rodríguez-Barroso, R., Quiroga, J. M. and Coello, M. D., 2021. Microplastic pollution in wastewater treatment plants in the city of Cádiz: Abundance, removal efficiency and presence in receiving water body. Science of the Total Environment, [e-journal] 776. https://doi.org/10.1016/j.scitotenv.2021.145795.

Freeman, S., Booth, A. M., Sabbah, I., Tiller, R., Dierking, J., Klun, K., Rotter, A., Ben-David, E., Javidpour, J. and Angel, D. L., 2020. Between source and sea: The role of wastewater treatment in reducing marine microplastics. Journal of Environmental Management, [e-journal] 266. https://doi.org/10.1016/j.jenvman.2020.110642.

Fu, L., Li, J., Wang, G., Luan, Y. and Dai, W., 2021. Adsorption behavior of organic pollutants on microplastics. Ecotoxicology and Environmental Safety, [e-journal] 217, p. 112207. https://doi.org/10.1016/j.ecoenv.2021.112207

GESAMP., 2015. Sources, Fates and Effects of Microplastics in the Marine Environment: A Global Assessment. Exeter: Polestar Wheatons.

Gigault, H., Halle, A. T., Baudrimont, M., Pascal, P. -Y., Gauffre, F., Phi, T. -L., Hadri, H. E., Grassl, F. and Reynaud, S., 2018. Current opinion: What is a nanoplastic? Environmental Pollution, [e-journal] 235, pp. 1030–1034. https://doi.org/10.1016/j.envpol.2018.01.024

Golwala, H., Zhang, X., Iskander, S. M. and Smith, A.L., 2021. Solid waste: An overlooked source of microplastics to the environment. Science of the Total Environment, [e-journal] 769, p. 144581. https://doi.org/10.1016/j.scitotenv.2020.144581

Gonçalves, C., Martins, M., Sobral, P., Costa, P. M. and Costa, M. H., 2019. An assessment of the ability to ingest and excrete microplastics by filter-feeders: A case study with the Mediterranean mussel. Environmental Pollution, [e-journal] 245, pp. 600–606. https://doi.org/10.1016/j.envpol.2018.11.038

Goss, H., Jaskiel, J. and Rotjan, R., 2018. Thalassia testudinum as a potential vector for incorporating microplastics into benthic marine food webs. Marine Pollution Bulletin, [e-journal] 135, pp. 1085–1089. https://doi.org/10.1016/j.marpolbul.2018.08.024

Green, D. S., Colgan, T. J., Thompson, R. C. and Carolan, J. C., 2019. Exposure to microplastics reduces attachment strength and alters the haemolymph proteome of blue mussels (Mytilus edulis). Environmental Pollution, [e-journal] 246, pp. 423–434. https://doi.org/10.1016/j.envpol.2018.12.017

Gu, W., Liu, S., Chen, L., Liu, Y., Gu, C., Ren, H. Q. and Wu, B., 2020. Single-cell RNA sequencing reveals size-dependent effects of polystyrene microplastics on immune and secretory cell populations from zebrafish intestines. Environmental Science and Technology, [e-journal] 54, pp. 3417–3427. https://doi.org/10.1021/acs.est.9b06386

Guan, J., Qi, K., Wang, J., Wang, W., Wang, Z., Lu, N. and Qu, J., 2020. Microplastics as an emerging anthropogenic vector of trace metals in freshwater: Significance of biofilms and comparison with natural substrates. Water Research, [e-journal] 184, p. 116205. https://doi.org/10.1016/j.watres.2020.116205

Guo, X., Liu, Y. and Wang, J., 2020. Equilibrium, kinetics and molecular dynamic modeling of Sr2+ sorption onto microplastics. Journal of Hazardous Materials, [e-journal] 400, p. 123324. https://doi.org/10.1016/j.jhazmat.2020.123324

Guo, X. and Wang, J., 2019. Sorption of antibiotics onto aged microplastics in freshwater and seawater. Marine Pollution Bulletin, [e-journal] 149, p. 110511. https://doi.org/10.1016/j.marpolbul.2019.110511

Harrison, J. P., Hoellein, T. J., Sapp, M., Tagg, A. S., Nam, Y. J. and Ojeda, J. J., 2018. Microplastic-associated biofilms: A comparison of freshwater and marine environments. In: Wagner, M. and Lambert, S. eds. Freshwater Microplastics. Cham: Springer, 181–201.

Hayes, D. G., 2021. Enhanced end-of-life performance for biodegradable plastic mulch films through improving standards and addressing research gaps. Current Opinion in Chemical Engineering, [e-journal] 33, p. 100695. https://doi.org/10.1016/j.coche.2021.100695

He, P., Chen, L. Shao, L., Zhang, H. and Lu¨, F., 2019. Municipal solid waste (MSW) landfill: A source of microplastics?: Evidence of microplastics in landfill leachate. Water Research, [e-journal] 15, pp. 38–45. https://doi.org/10.1016/j.watres.2019.04.060

Heinz, G. and Hautzinger, P., 2007. Meat Processing Technology for Small- to Medium- Scale Producers. Bangkok: FAO.

Hidayaturrahman, H. and Lee, T. G., 2019. A study on characteristics of microplastic in wastewater of South Korea: Identification, quantification, and fate of microplastics during treatment process. Marine Pollution Bulletin, [e-journal] 146, pp. 696–702. https://doi.org/10.1016/j.marpolbul.2019.06.071.

Hillel, D., 2005. Water harvesting. Encyclopedia of Soils in the Environment, [e-journal] 2005, pp. 264–270. https://doi.org/10.1016/B0-12-348530-4/00306-4

Hopewell, J., Dvorak, R. and Kosior, E., 2009. Plastics recycling: Challenges and opportunities. Philosophical Transactions of the Royal Society B: Biological Sciences, [e-journal] 364(1526), pp. 2115–2126. https://doi.org/10.1098/rstb.2008.0311.

Horton, A. A., Cross, R. K., Read, D. S., Jürgens, M. D., Ball, H. L., Svendsen, C., Vollertsen, J. and Johnson, A. C., 2020. Semi automated analysis of microplastics in complex wastewater samples. Environmental Pollution, [e-journal] 268. https://doi.org/10.1016/j.envpol.2020.115841.

Horton, A. A. and Dixon, S. J., 2017. Microplastics: An introduction to environmental transport processes. WIREs Water, [e-journal] 5(2). https://doi.org/10.1002/wat2.1268.

Horton, R. E., 1993. The role of infiltration in the hydrologic cycle. Eos, Transactions American Geophysical Union, [e-journal] 14(1), pp. 446–460. https://doi.org/10.1029/TR014i001p00446.

Huang, W., Song, B., Liang, J., Niu, Q., Zeng, G., Shen, M., Deng, J., Luo, Y., Wen, X. and Zhang, Y., 2021. Microplastics and associated contaminants in the aquatic environment: A review on their ecotoxicological effects, trophic transfer, and potential impacts to human health. Journal of Hazardous Materials, [e-journal] 405, p. 124187. https://doi.org/10.1016/j.jhazmat.2020.124187

Huang, Y., Liu, Q., Jia, W., Yan, C. and Wang, J., 2020. Agricultural plastic mulching as a source of microplastics in the terrestrial environment. Environmental Pollution, [e-journal] 260, p. 114096. https://doi.org/10.1016/j.envpol.2020.114096

Hunt, P. A., Koehler, K. E., Susiarjo, M., Hodges, C. A., Ilagan, A., Voigt, R. C., Thomas, S., Thomas, B. F. and Hassold, T. J., 2003. Bisphenol a exposure causes meiotic aneuploidy in the female mouse. Current Biology, [e-journal] 13(7), pp. 546–553. https://doi.org/10.1016/S0960-9822(03)00189-1

Imhof, H. K., Laforsch, C., Wiesheu, A. C., Schmid, J., Anger, P. M., Niessner, R. and Ivleva, N. P., 2016. Pigments and plastic in limnetic ecosystems: A qualitative and quantitative study on microparticles of different size classes. Water Research, [e-journal] 98, pp. 64–74. https://doi.org/10.1016/j.watres.2016.03.015

International Union for Conservation of Nature (2021). Marine Plastic Pollution. Available at https://www.iucn.org/resources/issues-brief/marine-plastic-pollution. Assessed on 6 Feb 2024.

Isari, E. A., Papaioannou, P., Kalavrouziotis, I. K. and Karapangioti, K., 2021. Microplastics in agricultural soils : A case study in cultivation of watermelons and canning tomatoes. Water, [e-journal] 13, p. 2168. https://doi.org/10.3390/w13162168

Issac, M. N. and Kandasubramanian, B., 2021. Effect of microplatics in water and aquatic systems. Environmental Science and Pollution Research, [e-journal] 28, pp. 19544–19562. https://doi.org/10.1007/s11356-021-13184-2

Iyare, P. U., Ouki, S. K. and Bond, T., 2020. Microplastics removal in wastewater treatment plants: A critical review. Environmental Science: Water Research and Technology, [e-journal] 6(10), pp. 2664–2675. https://doi.org/10.1039/D0EW00397B.

Jambeck, J. R., Geyer, R., Wilcox, C., Siegler, T. R., Perryman, M., Andrady, A., Narayan, R. and Law, K. L., 2015. Plastics waste inputs from land into the ocean. Science, [e-journal] 347(6223), pp. 768–771. https://doi.org/10.1126/science.1260352.

Jovanović, B., 2017. Ingestion of microplastics by fish and its potential consequences from a physical perspective. Integrated Environmental Assessment and Management, [e-journal] 13, pp. 510–515. https://doi.org/10.1002/ieam.1913

Júnior, L. M., Cristianini, M. and Anjos, C. A. R., 2020. Packaging aspects for processing and quality of foods treated by pulsed light. Journal of Food Processing and Preservation, [e-journal] 44(11). https://doi.org/10.1111/jfpp.14902.

Kalčíková, G., Alič, B., Skalar, T., Bundschuh, M. and Gotvajn, A. Z., 2017. Wastewater treatment plant effluents as source of cosmetic polyethylene microbeads to freshwater. Chemosphere, [e-journal] 188, pp. 25–31. https://doi.org/10.1016/j.chemosphere.2017.08.131.

Kameda, Y., Yamada, N. and Fujita, E., 2021. Source- and polymer-specific size distributions of fine microplastics in surface water in an urban river. Environmental Pollution, [e-journal] 284. https://doi.org/10.1016/j.envpol.2021.117516.

Kasirajan, S. and Ngouajio, M., 2012. Polyethylene and biodegradable mulches for agricultural applications: A review. Agronomy for Sustainable Development, [e-journal] 32, pp. 501–529. https://doi.org/10.1007/s13593-011-0068-3

Kazour, M., Terki, S., Rabhi, K., Jemaa, S., Khalaf, G. and Amara, R., 2019. Sources of microplastics pollution in the marine environment: Importance of wastewater treatment plant and coastal landfill. Marine Pollution Bulletin, [e-journal] 146, pp. 608–618. https://doi.org/10.1016/j.marpolbul.2019.06.066.

Kim, D., Chae, Y. and An, Y. J., 2017. Mixture toxicity of nickel and microplastics with different functional groups on Daphnia magna. Environmental Science and Technology, [e-journal] 51, pp. 12852–12858. https://doi.org/10.1021/acs.est.7b03732

Kirkby, M. J. and Chorley, R. J., 1967. Throughflow, overland flow and erosion. International Association of Scientific Hydrology. Bulletin, [e-journal] 12(3), pp. 5–21. http://dx.doi.org/10.1080/02626666709493533.

Kjeldsen, P., Barlaz, M. A., Rooker, A. P., Baun, A., Ledin, A. and Christensen, T. H., 2002. Present and long-term composition of MSW landfill leachate: A review. Critical Reviews in Environmental Science and Technology, [e-journal] 32(4), pp. 297–336. https://doi.org/10.1080/10643380290813462.

Klein, M. and Fischer, E. K., 2019. Microplastic abundance in atmospheric deposition within the Metropolitan area of Hamburg, Germany. Science of the Total Environment, [e-journal] 685, pp. 96–103. https://doi.org/10.1016/j.scitotenv.2019.05.405.

Kole, P. J., Löhr, A. J., Van Belleghem, F. and Ragas, A. M. J., 2017. Wear and tear of tyres: A stealthy source of microplastics in the environment. International Journal of Environmental Research and Public Health, [e-journal] 14(10). https://doi.org/10.3390/ijerph14101265.

Kosuth, M., Mason, S. A. and Wattenberg, E. V., 20180 Anthropogenic contamination of tap water, beer, and sea salt. PLoS One, [e-journal] 13(4). https://doi.org/10.1371/journal.pone.0194970.

Krystynik, P., Strunakova, K., Syc, M. and Kluson, P., 2021. Notes on common misconceptions in microplastics removal from water. Applied Science, [e-journal] 11(13). https://doi.org/10.3390/app11135833.

Kuczenski, B. and Geyer, R., 2010. Material flow analysis of polyethylene terephthalate in the US, 1996–2007. Resources, Conservation and Recycling, [e-journal] 54(12), pp. 1161–1169. https://doi.org/10.1016/j.resconrec.2010.03.013.

Lahens, L., Strady, E., Kieu-Le, T. C., Dris, R., Boukerma, K., Rinnert, E., Gasperi, J. and Tassin, B., 2018. Macroplastic and microplastic contamination assessment of a tropical river (Saigon River, Vietnam) transversed by a developing megacity. Environmental Pollution, [e-journal] 236, pp. 661–671. https://doi.org/10.1016/j.envpol.2018.02.005.

Lares, M., Ncibi, M. C., Sillanpa ̈a ̈, M. and Sillanpa ̈a ̈, M., 2018. Occurrence, identification and removal of microplastic particles and fibers in conventional activated sludge process and advanced MBR technology. Water Research, [e-journal] 133, pp. 236–246. https://doi.org/10.1016/j.watres.2018.01.049.

Laskar, N. and Kumar, U., 2019. Plastics and microplastics: A threat to environment. Environmental Technology & Innovation, [e-journal] 14. https://doi.org/10.1016/j.eti.2019.100352.

Lassen, C., Foss, H. S., Kerstin, M., Nanna, H. B., Pernille, R. J., Gissel, T. and Anna, B., 2015. Microplastics occurrence, effects and sources of releases to the environment in Denmark. Danish Environmental Protection Agency, [online] Available at: <https://backend.orbit.dtu.dk/ws/portalfiles/portal/118180844/Lassen_et_al._2015.pdf> [Accessed 10 July 2021].

Lei, L., Wu, S., Lu, S., Liu, M., Song, Y., Fu, Z., Shi, H., Raley-Susman, K. M. and He, D., 2018. Microplastic particles cause intestinal damage and other adverse effects in zebrafish Danio rerio and nematode Caenorhabditis elegans. Science of the Total Environment, [e-journal] 619–620, pp. 1–8. https://doi.org/10.1016/j.scitotenv.2017.11.103

Li, R., Liu, Y., Sheng, Y., Xiang, Q., Zhou, Y. and Cizdziel, J. V., 2020. Effect of prothioconazole on the degradation of microplastics derived from mulching plastic film: Apparent change and interaction with heavy metals in soil. Environmental Pollution, [e-journal] 260, p. 113988. https://doi.org/10.1016/j.envpol.2020.113988

Liao, Y., Cao, B., Liu, L., Wu, X., Guo, S., Mi, C., Li, K. and Wang, M., 2020. Structure and properties of bio-based polyurethane coatings for controlled-release fertilizer. Journal of Applied Polymer Science, [e-journal] 138(15), 50179. https://doi.org/10.1002/app.50179

Lin, L., Tang, S., Wang, X., Sun, X. and Yu, A., 2021. Hexabromocyclododecane alters malachite green and lead(II) adsorption behaviors onto polystyrene microplastics: Interaction mechanism and competitive effect. Chemosphere, [e-journal] 265, p. 129079. https://doi.org/10.1016/j.chemosphere.2020.129079

Lin, W., Su, F., Lin, M., Jin, M., Li, Y., Ding, K., Chen, Q., Qian, Q. and Sun, X., 2020. Effect of microplastics PAN polymer and/or Cu2+ pollution on the growth of Chlorella pyrenoidosa. Environmental Pollution, [e-journal] 265, p. 114985. https://doi.org/10.1016/j.envpol.2020.114985

Liu, F., Olesen, K. B., Borregaard, A. R. and Vollertsen, J., 2019a. Microplastics in urban and highway stormwater retention ponds. Science of the Total Environment, [e-journal] 671, pp. 992–1000. https://doi.org/10.1016/j.scitotenv.2019.03.416.

Liu, F., Vianello, A. and Vollertsen, J., 2019b. Retention of microplastics in sediments of urban and highway stormwater retention ponds. Environmental Pollution, [e-journal] 255, p. 113335. https://doi.org/10.1016/j.envpol.2019.113335.

Liu, K., Wang, X., Fang, T., Xu, P., Zhu, L. and Li, D., 2019c. Source and potential risk assessment of suspended atmospheric microplastics in Shanghai. Science of the Total Environment, [e-journal] 675, pp. 462–471. https://doi.org/10.1016/j.scitotenv.2019.04.110.

Liu, M., Lu, S., Chen, Y., Cao, C., Bigalke, M. and He, D., 2020a. Analytical methods for microplastics in environments: Current advances and challenges. In: He, D. and Luo, Y. eds. Microplastics in Terrestrial Environments. Cham: Springer, 3–24.

Liu, W., Zhang, J., Liu, H., Guo, X., Zhang, X., Yao, X., Cao, Z. and Zhang, T., 2020b. A review of the removal of microplastics in global wastewater treatment plants: Characteristics and mechanisms. Environment International, [e-journal] 146, 106277. https://doi.org/10.1016/j.envint.2020.106277

Liu, Y., Zhang, K., Xu, S., Yan, M., Tao, D., Chen, L., Wei, Y., Wu, C., Liu, G. and Lam, P. K. S., 2022. Heavy metals in the "plastisphere" of marine microplastics: Adsorption mechanisms and composite risk. Gondwana Research, [e-journal] 108, pp. 171–180. https://doi.org/10.1016/j.gr.2021.06.017

Lu, Y., Zhang, Y., Deng, Y., Jiang, W., Zhao, Y., Geng, J., Ding, L. and Ren, H., 2016. Uptake and accumulation of polystyrene microplastics in zebrafish (Danio rerio) and toxic effects in liver. Environmental Science and Technology, [e-journal] 50, pp. 4054–4060. https://doi.org/10.1021/acs.est.6b00183

Luo, H., Xiang, Y., He, D., Li, Y., Zhao, Y., Wang, S. and Pan, X., 2019. Leaching behavior of fluorescent additives from microplastics and the toxicity of leachate to Chlorella vulgaris. Science of the Total Environment, [e-journal] 678, pp. 1–9. https://doi.org/10.1016/j.scitotenv.2019.04.401

Luo, H., Zhao, Y., Li, Y., Xiang, Y., He, D. and Pan, X., 2020. Aging of microplastics affects their surface properties, thermal decomposition, additives leaching and interactions in simulated fluids. Science of the Total Environment, [e-journal] 714, p. 136862. https://doi.org/10.1016/j.scitotenv.2020.136862

Lutz, N., Fogarty, J. and Rate, A., 2021. Accumulation and potential for transport of microplastics in stormwater drains into marine environments, Perth region, Western Australia. Marine Pollution Bulletin, [e-journal] 168. https://doi.org/10.1016/j.marpolbul.2021.112362.

Madhav, V. N., Gopinath, K. P., Krishnan, A., Rajendran, N. and Krishnan, A., 2020. A critical review on various trophic transfer routes of microplastics in the context of the Indian coastal ecosystem. Watershed Ecology and the Environment, [e-journal] 2, pp. 25–41. https://doi.org/10.1016/j.wsee.2020.08.001

Maga, D., Hiebel, M. and Aryan, V., 2019. A comparative life cycle assessment of meat trays made of various packaging materials. Sustainability, [e-journal] 11(19), 5324. https://doi.org/10.3390/su11195324

Magnusson, K. and Noren, F., 2014. Screening of Microplastic Particles in and Down-stream a Wastewater Treatment Plant (Number C 55). Stockholm: National Environmental Monitoring Commissioned by The Swedish Environmental Protection Agency.

Makhdoumi, P., Amin, A. A., Karimi, H., Pirsaheb, M., Kim, H. and Hossini, H., 2021. Occurrence of microplastic particles in the most popular Iranian bottled mineral water brands and an assessment of human exposure. Journal of Water Process Engineering, [e-journal] 39, 101708. https://doi.org/10.1016/j.jwpe.2020.101708

Mani, T., Hauk, A., Walter, U. and Burkhardt-Holm, P., 2015. Microplastics profile along the Rhine River. Scientific Reports, [e-journal] 5. https://doi.org/10.1038/srep17988

Mao, Y., Ai, H., Chen, Y., Zhang, Z., Zeng, P., Kang, L., Li, W., Gu, W., He, Q. and Li, H., 2018. Phytoplankton response to polystyrene microplastics: Perspective from an entire growth period. Chemosphere, [e-journal] 208, pp. 59–68. https://doi.org/10.1016/j.chemosphere.2018.05.170

Mason, S. A., Kammin, L., Eriksen, M., Aleid, G., Wilson, S., Box, C., Williamson, N. and Riley, A., 2016. Pelagic plastic pollution within the surface waters of Lake Michigan, USA. Journal of Great Lakes Research, [e-journal] 42(4), pp. 753–759. https://doi.org/10.1016/j.jglr.2016.05.009

Mason, S. A., Welch, V. G. and Neratko, J., 2018. Synthetic polymer contamination in bottled water. Frontiers in Chemistry, [e-journal] 6, 407. https://doi.org/10.3389/fchem.2018.00407

Mataji, A., Taleshi, M. S. and Balimoghaddas, E., 2019. Distribution and characterization of microplastics in surface waters and the Southern Caspian Sea Coasts sediments. Archives of Environmental Contamination and Toxicology, [e-journal] 78, pp. 86–93. https://doi.org/10.1007/s00244-019-00700-2

McMillin, K. W., 2017. Advancements in meat packaging. Meat Science, [e-journal] 132, pp. 153–162. https://doi.org/10.1016/j.meatsci.2017.04.015

McNeish, R. E., Kim, L. H., Barrett, H. A., Mason, S. A., Kelly, J. J. and Hoellein, T. J., 2018. Microplastic in riverine fish is connected to species traits. Scientific Reports, [e-journal] 8, 11639. https://doi.org/10.1038/s41598-018-29980-9

Meng, F., Fan, T., Yang, X., Riksen, M., Xu, M. and Geissen, V., 2020. Effects of plastic mulching on the accumulation and distribution of macro and micro plastics in soils of two farming systems in Northwest China. PeerJ, [e-journal] 8, p. e10375. https://doi.org/10.7717/peerj.10375

Michielssen, M. R., Michielssen, E. R., Ni, J. and Duhaime, M. B., 2016. Fate of microplastics and other small anthropogenic litter (SAL) in wastewater treatment plants depends on unit processes employed. Environmental Science: Water Research and Technology, [e-journal] 2(6), pp. 1064–1073. https://doi.org/10.1039/c6ew00207b

Minelgaite, G., Nielsen, A. H., Pedersen, M. L. and Vollertsen, J., 2015. Photodegradation of three stormwater biocides. Urban Water Journal, [e-journal] 14(1), pp. 53–60. https://doi.org/10.1080/1573062X.2015.1076489.

Moos, V. N., Burkhardt-Holm, P. and Köhler, A., 2012. Uptake and effects of microplastics on cells and tissue of the blue mussel Mytilus edulis L. after an experimental exposure. Environmental Science and Technology, [e-journal] 46, pp. 11327–11335. https://doi.org/10.1021/es302332w

Murphy, F., Ewins, C., Carbonnier, F. and Quinn, B., 2016. Wastewater treatment works (WwTW) as a source of microplastics in the aquatic environment. Environmental Science and Technology, [e-journal] 50(11), pp. 5800–5808. https://doi.org/10.1021/acs.est.5b05416.

Napper, I. E., Bakir, A., Rowland, S. and Thompson, R. C., 2015. Characterisation, quantity and sorptive properties of microplastics extracted from cosmetics. Marine Pollution Bulletin, [e-journal] 99(1–2), pp. 178–185. https://doi.org/10.1016/j.marpolbul.2015.07.029.

Neetha, K., Varghese, L. M., Harshitha, M. R. and Jinsha, V. K., 2020. Microplastic detection in water using image processing. International Journal of Applied Engineering Research, [e-journal] 15(1). Available at: <www.ripublication.com/ijaerspl20/ijaerv15n1spl_12.pdf> [Accessed 29 June 2021].

Nelms, S. E., Barnett, J., Brownlow, A., Davison, N. J., Deaville, R., Galloway, T. S., Lindeque, P. K., Santillo, D. and Godley, B. J., 2019. Microplastics in marine mammals stranded around the British coast: Ubiquitous but transitory? Scientific Reports, [e-journal] 9, p. 1075. https://doi.org/10.1038/s41598-018-37428-3

Ngo, P. L., Pramanik, B. K., Shah, K. and Roychand, R., 2019. Pathway, classification and removal efficiency of microplastics in wastewater treatment plants. Environmental Pollution, [e-journal] 255(2). https://doi.org/10.1016/j.envpol.2019.113326.

Nizetto, L., Futter, M. and Langaas, S., 2016. Are agricultural soils dumps for microplastics of urban origin? Environmental Science and Technology, [e-journal] 50(20), pp. 10777–10779. https://doi.org/10.1021/acs.est.6b04140

Oßmann, B. E., Sarau, G., Holtmannspötter, H., Monika, P., Christiansen, S. H. and Dicke, W., 2018. Small-sized microplastics and pigmented particles in bottled mineral water. Water Research, [e-journal] 14, pp. 307–316. https://doi.org/10.1016/j.watres.2018.05.027

OECD., 2020. Workshop on microplastics from synthetic textiles: Knowledge, mitigation, and policy. [online] Available at: <www.oecd.org/water/Workshop_MP_Textile_Summary_Note_FINAL.pdf> [Accessed 30 June 2021].

Panno, S. V., Kelly, W. R., Scott, J., Zheng, W., McNeish, R. E., Holm, N., Hoellein, T. J. and Baranski, E. L., 2019. Microplastic contamination in karst groundwater systems. Groundwater, [e-journal] 57(2), pp. 189–196. https://doi.org/10.1111/gwat.12862

Parvin, F. and Tareq, S. M., 2021. Impact of landfill leachate contamination on surface and groundwater of Bangladesh: A systematic review and possible health risks assessment. Applied Water Science, [e-journal] 11, 100. https://doi.org/10.1007/s13201-021-01431-3

Pegram, J. E. and Andrady, A. L., 1989. Outdoor weathering of selected polymeric materials under marine exposure conditions. Polymer Degradation and Stability, [e-journal] 26(4), pp. 333–345. https://doi.org/10.1016/0141-3910(89)90112-2

Pico, Y. and Barcelo, D., 2019. Analysis and prevention of microplastics pollution in water: Current perspectives and future directions. ACS Omega, [e-journal] 4(4), pp. 6709–6719. https://doi.org/10.1021/acsomega.9b00222

Pilgrim, D. H., Huff, D. D. and Steele, T. D., 1978. A field evaluation of subsurface and surface runoff: II. runoff processes. Journal of Hydrology, [e-journal] 38(3–4), pp. 319–341. https://doi.org/10.1016/0022-1694(78)90077-X

Plastics Europe., 2020. Plastics: The facts 2020. [online] Available at: <www.plasticseurope.org/en/resources/publications/4312-plastics-facts-2020> [Accessed 27 June 2021].

Prata, J. C., 2017. Airborne microplastics: Consequences to human health? Environmental Pollution, [e-journal] 234, pp. 115–126. https://doi.org/10.1016/j.envpol.2017.11.043

Prata, J. C., 2018. Microplastics in wastewater: State of the knowledge on sources, fate and solutions. Marine Pollution Bulletin, [e-journal] 129(1), pp. 262–265. https://doi.org/10.1016/j.marpolbul.2018.02.046.

Pratesi, C. B., Almeida, M. A. A. L. S., Paz, G. S. C., Teitonio, M. H. R., dos Santos Cardia, F. M., Gandolfi, L., Pratesi, R. and Hecht, M., 2020. Presence and Quantification of Microplastic in Urban Tap Water: A Pre-Screening in Brasilia, Brazil. Sustainability, 13(11), 6404. [e-journal]. https://doi.org/10.21203/rs.3.rs-117255/v1

Qiao, R., Sheng, C., Lu, Y., Zhang, Y., Ren, H. and Lemos, B., 2019. Microplastics induce intestinal inflammation, oxidative stress, and disorders of metabolome and microbiome in zebrafish. Science of the Total Environment, [e-journal] 662, pp. 246–253. https://doi.org/10.1016/j.scitotenv.2019.01.245

Qiu, R., Song, Y., Zhang, X., Xie, B. and He, D., 2020. Microplastics in urban environments: Sources, pathways, and distribution. In: He, D. and Luo, Y. eds. Microplastics in Terrestrial Environments. Cham: Springer, 41–61.

Ramkissoon, S., 2020. Two million too many: Microplastic pollution in the ocean. [photograph] Available at: <www.envmedia.org/article/two-million-too-many/> [Accessed 12 July 2021].

Reisser, J., Shaw, J., Wilcox, C., Hardesty, B. D., Proietti, M., Thums, M. and Pattiaratchi, C., 2013. Marine plastic pollution in waters around Australia: Characteristics, concentrations, and pathways. PLoS One, [e-journal] 8(11). https://doi.org/10.1371/journal.pone.0080466.

Ren, S. Y., Kong, S. F. and Ni, H. G., 2021. Contribution of mulch film to microplastics in agricultural soil and surface water in China. Environmental Pollution, [e-journal] 291, p. 118227. https://doi.org/10.1016/j.envpol.2021.118227

Rezaei, M., Riksen, M. J. P. M., Sirjani, E., Sameni, A. and Geissen, V., 2019. Wind erosion as a driver for transport of light density microplastics. Science of the Total Environment, [e-journal] 669, pp. 273–281. https://doi.org/10.1016/j.scitotenv.2019.02.382

Rillig, M. C., Ziersch, L. and Hempel, S., 2017. Microplastic transport in soil by earthworms. Scientific Reports, [e-journal] 7, p. 1362. https://doi.org/10.1038/s41598-017-01594-7

Rochman, C. M., Kross, S. M., Armstring, J. B., Bogan, M. T., Darling, E. S., Green, S. J., Symth, A. R. and Verissimo, D., 2015. Scientific evidence supports a ban on microbeads. Environmental Science and Technology, [e-journal] 49(18), pp. 10759–10761. https://doi.org/10.1021/acs.est.5b03909.

Rolsky, C., Kelkar, V., Driver, E. and Halden, R. U., 2019. Municipal sewage sludge as a source of microplastics in the environment. Current Opinion in Environmental Science and Health, [e-journal] 14, pp. 16–22. https://doi.org/10.1016/j.coesh.2019.12.001

Romeo, T., Pietro, B., Peda, C., Consoli, P., Andaloro, F. and Fossi, M. C., 2015. First evidence of presence of plastic debris in stomach of large pelagic fish in the Mediterranean Sea. Marine Pollution Bulletin, [e-journal] 95, pp. 358–361. https://doi.org/10.1016/j.marpolbul.2015.04.048

Ryan, P. G., Moore, C. J., van Franeker, J. A. and Moloney, C. L., 2009. Monitoring the abundance of plastic debris in the marine environment. Philosophical Transactions of the Royal Society B: Biological Sciences, [e-journal] 364(1526), pp. 1992–2012. https://doi.org/10.1098/rstb.2008.0207.

Samandra, S., Johnston, J. M., Jaegar, J. E., Symons, B., Xie, S., Currell, M., Ellis, A. V. and Clarke, B. O., 2021. Microplastic contamination of an unconfined groundwater aquifer in Victoria, Australia. Science of the Total Environment, [e-journal] 802, p. 149727. https://doi.org/10.1016/j.scitotenv.2021.149727

Santana, M. F. M., Moreira, F. T. and Turra, A., 2017. Trophic transference of microplastics under a low exposure scenario: Insights on the likelihood of particle cascading along marine food-webs. Marine Pollution Bulletin, [e-journal] 121, pp. 154–159. https://doi.org/10.1016/j.marpolbul.2017.05.061

Schlesinger, W. H. and Bernhardt, E. S., 2013. Biogeochemistry: An Analysis of Global Change. 3rd ed. Beijing: Elsevier.

Schirinzi, G.F., Pérez-Pomeda, I., Sanchís, J., Rossini, C., Farré, M. and Barceló, D., 2017. Cytotoxic effects of commonly used nanomaterials and microplastics on cerebral and epithelial human cells. Environmental Research, 159, 579–587.

Schymanski, D., Goldbeck, C., Humpf, H. U. and Fürst, P., 2017. Analysis of microplastics in water by micro-Raman spectroscopy: Release of plastics particles from different packaging into mineral water. Water Research, [e-journal] 129, pp. 154–162. https://doi.org/10.1016/j.watres.2017.11.011

Sembiring, E., Fajar, M. and Handajani, M., 2021. Performance of rapid sand filter: Single media to remove microplastics. Water Supply, [e-journal] 21(5), pp. 2273–2284. https://doi.org/10.2166/ws.2021.060.

Shahnawaz, M., Sangale, M. K. and Ade, A. B., 2019. Microplastics. In: Bioremediation Technology for Plastic Waste. Singapore: Springer, 11–19.

Simon, M., van Alst, N. and Vollertsen, J., 2018. Quantification of microplastic mass and removal rates at wastewater treatment plants applying Focal Plane Array (FPA)-based Fourier Transform Infrared (FT-IR) imaging. Water Research, [e-journal] 142, pp. 1–9. https://doi.org/10.1016/j.watres.2018.05.019

Singh, B. and Sharma, N., 2007. Mechanistic implications of plastic degradation. Polymer Degradation and Stability, [e-journal] 93(3), pp. 561–584. https://doi.org/10.1016/j.polymdegradstab.2007.11.008.

Singh, T., 2021. Generation of microplastics from the opening and closing of disposable plastic water bottles. Journal of Water & Health, 29(3), 488–498.

Smith, M., Love, D. C., Rochman, C. M. and Neff, R. A., 2018. Microplastics in seafood and the implications for human health. Current Environmental Health Reports, [e-journal] 5, pp. 375–386. https://doi.org/10.1007/s40572-018-0206-z

Sol, D., Laca, A., Laca, A. and Díaz, M., 2020. Approaching the environmental problem of microplastics: Importance of WWTP treatments. Science of the Total Environment, [e-journal] 740. https://doi.org/10.1016/j.scitotenv.2020.140016.

Stock, V., Böhmert, L., Lisicki, E., Block, R., Cara-Carmona, J., Pack, L. K., Selb, R., Lichtenstein, D., Voss, L., Henderson, C. J., Zabinsky, E., Sieg, H., Braeuning, A. and Lampen, A., 2019. Uptake and effects of orally ingested polystyrene microplastic particles in vitro and in vivo. Archives of Toxicology, [e-journal] 93, pp. 1817–1833. https://doi.org/10.1007/s00204-019-02478-7

Su, S., Zhou, S. and Lin, G., 2021. Existence of microplastics in soil and groundwater in Jiaodong Peninsula. 2021 International Conference on Tourism, Economy and Environmental Sustainability (TEES 2021), 251, 02045. https://doi.org/10.1051/e3sconf/202125102045

Sundt, P., Schulze, P. E. and Syversen, F., 2014. Source of microplastics: Pollution to the marine environment. [online] Available at: <https://d3n8a8pro7vhmx.cloudfront.net/boomerangalliance/pages/507/attachments/original/1481155578/Norway_Sources_of_Microplastic_Pollution.pdf?1481155578> [Accessed 28 June 2021].

Sussarellu, R., Suquet, M., Thomas, Y., Lambert, C., Fabioux, C., Pernet, M. E. J., Goïc, N. Le, Quillien, V., Mingant, C., Epelboin, Y., Corporeau, C., Guyomarch, J., Robbens, J., Paul-Pont, I., Soudant, P. and Huvet, A., 2016. Oyster reproduction is affected by exposure to polystyrene microplastics. Proceedings of the National Academy of Sciences. https://doi.org/10.1073/pnas.151901911

Suteja, Y., Atmadipoera, A. S., Riani, E., Nurjaya, I. W., Nugroho, D. and Cordova, M. R., 2021. Spatial and temporal distribution of microplastic in surface water of tropical estuary: Case study in Benoa Bay, Bali, Indonesia. Marine Pollution Bulletin, [e-journal] 163, p. 111979. https://doi.org/10.1016/j.marpolbul.2021.111979

Talvitie, J., Mikola, A., Koistinen, A. and Seta¨la¨, O., 2017. Solutions to microplastic pollution: Removal of microplastics from wastewater effluent with advanced wastewater treatment technologies. Water Research, [e-journal] 123, pp. 401–407. https://doi.org/10.1016/j.watres.2017.07.005.

Tang, S., Lin, L., Wang, X., Yu, A. and Sun, X., 2021. Interfacial interactions between collected nylon microplastics and three divalent metal ions (Cu(II), Ni(II), Zn(II)) in aqueous solutions. Journal of Hazardous Materials, [e-journal] 403, p. 123548. https://doi.org/10.1016/j.jhazmat.2020.123548

Tong, H., Jiang, Q., Hu, X. and Zhong, X., 2020. Occurrence and identification of microplastics in tap water from China. Chemosphere, [e-journal] 252, p. 126493. https://doi.org/10.1016/j.chemosphere.2020.126493

Verschoor, A., de Poorter, L., Dröge, R., Kuenen, J. and de Valk, E., 2016. Emission of microplastics and potential mitigation measures: Abrasive cleaning agents, paints and tyre wear. National Institute for Public Health and the Environment. Available at: <https://rivm.openrepository.com/bitstream/handle/10029/617930/2016-0026.pdf?sequence=3> [Accessed 7 Feb 2024].

Vollertsen, J., Åstebøl, S. O., Coward, J. E., Fageraas, T., Nielsen, A. H. and Hvitved-Jacobsen, T., 2009. Performance and modelling of a highway wet detention pond designed for cold climate. Water Quality Research Journal, [e-journal] 44(3), pp. 253–262. https://doi.org/10.2166/wqrj.2009.027.

Wang, J., Li, Y., Lu, L., Zheng, M., Zhang, X., Tian, H., Wang, W. and Ru, S., 2019. Polystyrene microplastics cause tissue damages, sex-specific reproductive disruption and transgenerational effects in marine medaka (Oryzias melastigma). Environmental Pollution, [e-journal] 254, p. 113024. https://doi.org/10.1016/j.envpol.2019.113024

Wang, Z., Lin, T. and Chen, W., 2020. Occurrence and removal of microplastics in an advanced drinking water treatment plant (ADWTP). Science of the Total Environment, [e-journal] 700. https://doi.org/10.1016/j.scitotenv.2019.134520.

Weber, A., Scherer, C., Brennholt, N., Reifferscheid, G. and Wagner, M., 2018. PET microplastics do not negatively affect the survival, development, metabolism and feeding activity of the freshwater invertebrate Gammarus pulex. Environmental Pollution, [e-journal] 234, pp. 181–189. https://doi.org/10.1016/j.envpol.2017.11.014

Wen, B., Jin, S. R., Chen, Z. Z., Gao, J. Z., Liu, Y. N., Liu, J. H. and Feng, X. S., 2018. Single and combined effects of microplastics and cadmium on the cadmium accumulation, antioxidant defence and innate immunity of the discus fish (Symphysodon aequifasciatus). Environmental Pollution, [e-journal] 243, pp. 462–471. https://doi.org/10.1016/j.envpol.2018.09.029

White, C. P., DeBry, R. W. and Tytle, D. A., 2012. Microbial survey of a full-scale, biologically active filter for treatment of drinking water. Applied and Environmental Microbiology, [e-journal] 78, p. 17.

Wu, B., Wu, X., Liu, S., Wang, Z. and Chen, L., 2019. Size-dependent effects of polystyrene microplastics on cytotoxicity and efflux pump inhibition in human Caco-2cells. Chemosphere, 221, 333–341.

Xiang, Y., Jiang, L., Zhou, Y., Luo, Z., Zhi, D., Yang, J. and Lam, S. S., 2021. Microplastics and environmental pollutants: Key interaction and toxicology in aquatic and soil environments. Journal of Hazardous Materials, [e-journal] 422, p. 126843. https://doi.org/10.1016/j.jhazmat.2021.126843

Xu, B., Liu, F., Brookes, P. C. and Xu, J., 2018. The sorption kinetics and isotherms of sulfamethoxazole with polyethylene microplastics. Marine Pollution Bulletin, [e-journal] 131, pp. 191–196. https://doi.org/10.1016/j.marpolbul.2018.04.027

Xu, Z., Bai, X. and Ye, Z., 2021. Removal and generation of microplastics in wastewater treatment plants: A review. Journal of Cleaner Production, [e-journal] 291. https://doi.org/10.1016/j.jclepro.2021.125982

Yan, Y., Zhu, F., Zhu, C., Chen, Z., Liu, S., Wang, C. and Gu, C., 2021. Dibutyl phthalate release from polyvinyl chloride microplastics: Influence of plastic properties and environmental factors. Water Research, [e-journal] 204, p. 117597. https://doi.org/10.1016/j.watres.2021.117597

Zhang, G. S. and Liu, Y. F., 2018. The distribution of microplastics in soil aggregate fractions in southwestern China. Science of the Total Environment, [e-journal] 642, pp. 12–20. https://doi.org/10.1016/j.scitotenv.2018.06.004

Zhang, M., Li, J., Ding, H., Ding, J., Jiang, F., Ding, N. X. and Sun, C., 2019. Distribution characteristics and influencing factors of microplastics in urban tap water and water sources in Qingdao, China. Analytical Letters, [e-journal] 53(8), pp. 1312–1327. https://doi.org/10.1080/00032719.2019.1705476

Zhang, X., Chen, J. and Li, J., 2020a. The removal of microplastics in the wastewater treatment process and their potential impact on anaerobic digestion due to pollutants association. Chemosphere, [e-journal] 251. https://doi.org/10.1016/j.chemosphere.2020.126360

Zhang, C., Wang, S., Sun, D., Pan, Z., Zhou, A., Xie, S., Wang, J. and Zou, J., 2020b. Microplastic pollution in surface water from east coastal areas of Guangdong, South China and preliminary study on microplastics biomonitoring using two marine fish. Chemosphere, [e-journal] 256, p. 127202. https://doi.org/10.1016/j.chemosphere.2020.127202

Zhang, Y., Wang, S., Olga, V., Xue, Y., Lv, S., Diao, X., Zhang, Y., Han, Q. and Zhou, H., 2021. The potential effects of microplastic pollution on human digestive tract cells. Chemosphere, 291, 132714.

Zhang, X. L., Zhao, Y. Y., Zhang, X. T., Shi, X. P., Shi, X. Y. and Li, F. M., 2022. Re-used mulching of plastic film is more profitable and environmentally friendly than new mulching. Soil and Tillage Research, [e-journal] 216, p. 105256. https://doi.org/10.1016/j.still.2021.105256

Zhao, S., Wang, T., Zhu, L., Xu, P., Wang, X., Gao, L. and Li, D., 2019. Analysis of suspended microplastics in the Changjiang Estuary: Implications for riverine plastic load to the ocean. Water Research, [e-journal] 161, pp. 560–569. https://doi.org/10.1016/j.watres.2019.06.019

Zhou, X. J., Wang, J., Li, H. Y., Zhang, H. M., Jiang, H. and Zhang, D. L., 2020a. Microplastics pollution of bottled water in China. Journal of Water Process Engineering, [e-journal] 40, 101884. https://doi.org/10.1016/j.jwpe.2020.101884

Zhou, Y., Wang, J., Zou, M., Jia, Z., Zhou, S. and Li, Y., 2020b. Microplastics in soils: A review of methods, occurrence, fate, transport, ecological and environment risks. Science of the Total Environment, [e-journal] 748, p. 141368. https://doi.org/10.1016/j.scitotenv.2020.141368

Zou, J., Liu, X., Zhang, D. and Yuan, X., 2020. Adsorption of three bivalent metals by four chemical distinct microplastics. Chemosphere, [e-journal] 248, p. 126064. https://doi.org/10.1016/j.chemosphere.2020.126064

2 Policies on Plastic Use in Asian Countries

2.1 INTRODUCTION

Numerous Asian countries, such as Japan, India, China, Malaysia, Indonesia, Thailand, Singapore, and South Korea, play a substantial role in the accumulation of plastic waste. According to Chen et al. (2021), Asia is responsible for producing over 50% of the world's plastic, and a staggering 74% of globally exported plastic waste has ended up in Asia in recent years. This crisis demands immediate and effective action from governments and organizations, as the substantial increase in plastic production and waste generation leads to marine debris, soil pollution, climate change, CO_2 emissions, and other detrimental effects on the environment and human health. Numerous Asian countries have taken steps to address this issue by implementing various policies, and assessing the impact and outcomes of these initiatives is crucial in determining their effectiveness. Therefore, this chapter examines the policies adopted by Asian countries, evaluates the effects and outcomes of their implementation, and explores the challenges associated with these policies. This chapter provides a comprehensive review of the enforced policies in selected Asian countries, drawing from reliable sources. Each country is introduced briefly in terms of its plastic usage and waste generation, followed by an examination of the measures taken to combat plastic debris. The study primarily focuses on analysing these policies in detail. The common policies implemented among these eight nations include recycling enforcement, Extended Producer Responsibility (EPR) systems, tax incentives, plastic bans, bans on plastic waste importation, awareness campaigns for the public, and the establishment of milestones to gradually reduce plastic usage.

2.2 JAPAN

Japan is a prominent user of plastic packaging, which has led to the country being a significant generator of plastic waste. In 2017, Japan produced approximately 43 million tons of municipal solid waste (MSW), as reported by Akenji et al. (2020). Of this total, it is estimated that 9.4 million tonnes, equivalent to roughly 71.2 kg per capita, comprised plastic waste originating from both municipal and industrial sources. Recycling accounted for approximately 25% of the plastic waste, while around 57% was subjected to energy and heat recovery through incineration, a waste treatment method. The remaining 18% of the waste was disposed of in landfills. Although the amount of solid waste sent to landfills has been decreasing, Japan is currently facing a shortage of available landfill space. Furthermore, the incineration of solid waste releases a significant number of secondary pollutants, contributing to air pollution and climate change.

 DOI: 10.1201/9781003387862-2

a) Resource Circulating Strategy for Plastics

In response to global-scale challenges, Japan introduced the "Resource Circulation Strategy for Plastics" in May 2019 (Ministry of the Environment, 2019). This comprehensive strategy sets ambitious and effective goals for Japan to take a leading role in addressing global plastic issues. The strategy is based on the concept of "3R + Renewable," which builds upon Japan's pioneering adoption of the 3Rs (reduce, reuse, and recycle) and effective waste management practices for its domestic waste streams (Ministry of the Environment, 2019). By embracing this concept, Japan aims to promote sustainable economic development while addressing global concerns such as resource scarcity, waste management, marine plastic pollution, and climate change. As part of the strategy, Japan has established several significant milestones and initiatives, including the introduction of alternative biomass plastics (Aoki-Suzuki, 2016). The following milestones have been initiated under the resource circulation strategy:

i. By 2030, a 25% reduction in single-use plastics.
ii. By 2025, all containers, packaging, and products should be recyclable or reusable.
iii. By 2030, an increase in the recycling rate to 60% for containers and packaging.
iv. By 2035, achieving 100% effective reuse of used plastics, incorporating the concept of a circular economy.
v. By 2030, doubling the utilization of recycled materials.
vi. By 2030, producing approximately 2 million tonnes of biomass plastics.

These milestones aim to reduce the usage of single-use plastic packaging and products, thereby reducing the environmental impact. They also focus on improving the quality of recycled products, enhancing the practicality of bio-plastics as an alternative to non-renewable plastics (Aoki-Suzuki, 2016). This initiative has not only helped ASEAN nations in developing the 3Rs and effective waste management practices but has also supported infrastructure improvements (Aoki-Suzuki, 2016). Additionally, Japan has provided support to ASEAN nations in formulating their national action plans.

b) Plastics Smart Campaign

In October 2018, the Ministry of Environment in Japan, in collaboration with local individuals, NGOs, companies involved in plastic usage or production, and local research institutes, initiated the "Plastics Smart" campaign (Sawaji, 2019). The campaign aimed to raise awareness about plastic debris among the local residents and encourage companies and organizations to engage in the Plastics Smart Forum to generate alternative solutions for plastic pollution. Various initiatives were presented in online as part of the campaign, resulting in outcomes such as the "Picking up Litter as a Sport" contest, the mass production of "Wooden Straws," mannequins made of recycled paper, and more (Sawaji, 2019). One of the initiatives under the campaign is the "Spo-Gomi Contest" or "Sport-Gomi," which is a litter-picking competition organized by the Social Sports Initiative (Sawaji, 2019). The contest combines the idea of sports with cleaning up litter to make the participants enjoy the activity while properly disposing of waste. Each team, consisting of three to five people, is given one hour to pick up litter and separate it into different waste categories. The goal is to earn the highest overall score, determined by weight and type of litter. The contest is based on points, allowing people of all ages to participate and have a chance to win. Since its introduction in 2008, Spogomi tournaments have been hosted by local governments, businesses, schools, and various other individuals and organizations. Over 90,000 people have participated in more than 800 tournaments held in Japan. Furthermore, Aqura Home Co. Ltd., a Tokyo-based builder of wooden housing, successfully produced a large quantity of wooden straws using advanced technology (Kawaguchi, 2018). They roll wood ribbon carved to a thickness of 0.15 mm, making them the world's first industry to produce wooden straws in significant volume. The production sources include fallen trees from natural calamities, timber from forest thinning, and other domestic wood. The manufacturing process allows for adjustments in length, thickness, and wood type. The advantage of these wooden straws is the natural warmth provided by the grain in the wood. They have been trialled at hotel restaurants in Tokyo since January 2019. In the past, mannequins made of wax or paper were commonly used for clothing displays, but plastic became the norm in the 1960s. This led to the production of over 100,000 plastic mannequins annually in Japan, which were eventually discarded as industrial waste. However, Mode Kohgei Co. Ltd., a company in Fujimi City, Saitama Prefecture, has been using traditional methods to produce mannequins made of paper, which have gained attention and interest due to the increased awareness of the plastic crisis (Saori, 2020).

c) The Act on Promotion of Procurement of Eco-Friendly Goods and Services by the State and Other Entities

Green procurement is a method wherein purchasers actively seek products and services with lower environmental impacts throughout their entire lifecycle from suppliers who prioritize environmental consciousness (Ministry of the Environment, 2017). This approach encourages suppliers to produce goods with minimal environmental burdens and transform their overall operations to be more environmentally aware, while also promoting environmentally conscious behaviour among customers (Ministry of the Environment, 2017). Among Asian countries, Japan was the first to establish a well-structured framework for green public procurement. Since 1989, Japan has implemented policies and legislation to promote and implement green public procurement practices (UNEP, 2017). In April 2000, the Japanese government enacted a series of regulations to address the pressing issue of waste disposal facilities reaching their capacity due to population growth. The Green Procurement Act was one of the laws enacted to help alleviate this problem. The goals of this legislation were to promote and popularize eco-friendly products that reduce negative environmental impacts and establish a society that is less burdensome on the environment and more sustainable. To achieve this, the legislation called on the public sector, including the government, to promote the procurement of environmentally friendly items and disseminate information about such products. As early as January 2001, the government provided fundamental principles and a list of 101 approved procurement items, including plastic products, along with their specific requirements (UNEP, 2017). The Green Procurement Act incorporates three fundamental concepts to encourage the procurement of products and services that contribute to reducing environmental burdens. Firstly, it emphasizes the importance of purchasing from suppliers who consistently prioritize environmental concerns. Buyers should consider the environmental impacts of products, in addition to their price and quality. When procuring environmentally friendly items, it is essential to assess not only the environmental impacts of the goods but also the environmental management and information dissemination practices of manufacturers, distributors, and other third parties (Ministry of the Environment, 2017). Secondly, the act highlights the significance of considering the entire lifecycle of products and services. When making purchases, it is crucial to evaluate how to reduce the environmental impacts associated with resource extraction, manufacturing, and disposal. Local governments, which are often concerned about various environmental issues, may need to procure specific items and services to address these problems.

Lastly, the primary objective should be to reduce the overall volume of purchases. Encouraging the procurement of environmentally friendly items should not lead to an increase in the total procurement volume. Instead, the focus should be on reducing the volume of purchases. Proper and long-term utilization of purchased eco-friendly items is vital for continuously reducing environmental burdens (Ministry of the Environment, 2017). Therefore, this policy encourages the use of more environmentally friendly alternatives for single-use plastics, aligning with the three fundamental concepts, with the aim of addressing environmental issues.

d) The Packaging Recycling Act

The Containers and Packaging Recycling Act was initially implemented in 1995 to address the increasing need to reduce plastic waste volume and maximize recycling resources through proper sorting and collection. The primary objective of this act was to protect and preserve the environment in Japan and contribute to the sustainable development of the domestic economy by recycling waste containers and packaging. Additionally, the act aims to ensure compliance with sorting criteria established by designated business organizations, support container recycling and packaging, educate consumers, and collect and distribute relevant information (Ministry of the Environment, 2014a, 2014b).

Under this Act, "specified business entities" include manufacturers who use containers and packaging for shipping their products, retailers and wholesalers who use containers and packaging for selling merchandise, container manufacturers, and importers and exporters of plastic containers or packaging. These entities are responsible for recycling and packaging in accordance with the amount of plastic they produce or sell (Yamakawa, 2014). The act primarily targets waste packaging, particularly plastic packaging disposed of as municipal solid-waste (MSW). Plastic packaging encompasses all types of plastic containers and packaging for products, including plastic film and plastic components such as caps. If plastic waste consists of multiple materials, it is classified based on the material with the highest weight. For example, plastic bags with aluminium liners are classified as "other plastic packaging" if the weight of the plastic exceeds that of the aluminium (Yamakawa, 2014). Designated producers are obligated to appropriately recycle waste packaging collected in sorted form by municipalities from households, meeting the standards set for each packaging type. However, there are certain types of packaging that can be sold or provided to recyclers free of charge if they are collected in a sorted form, relieving companies from recycling fees. These items are not included in the list of mandatory recycling for manufacturers. As a result, PET bottles and plastic packaging are classified as materials that must be recycled by producers among the products covered by Municipal Sorted Collection Plans (Yamakawa, 2014).

e) Plastic Bag Charge

Since the 1960s, Japan's transition from coal to oil as an energy source has played a significant role in the widespread and heavy usage of plastic items seen today. This shift led to the integration of free plastic bags into the urban lifestyle, eventually becoming a necessity. To address this issue, Kumatori proposed an effective environmental policy for regulating the use of plastic bags, based on three main principles: command and control, market-based tools, and a voluntary approach (Kaminaga, 2018). Command and control involves implementing regulations that directly ban the usage or sale of plastic bags, controlling their import into the country. On the other hand, market-based tools employ economic strategies such as charging for plastic bags. This can be implemented at the enterprise level, allowing users who require plastic bags to purchase them as a commercial product. Additionally, the voluntary approach focuses on discontinuing the supply of free plastic bags to business customers through pre-arranged agreements (Kaminaga, 2018). Initially, there was no national prohibition policy in Japan, and waste management issues were handled independently by certain local governments in the 1970s and 1990s to address regional disputes related to waste disposal and treatment plants. For example, Suginami Ward in Tokyo, with a population of approximately 560,000, introduced a pilot project in 1998 to charge for plastic bags through a local agreement, although it has not been implemented to date (Kaminaga, 2018). Another successful attempt to reduce free plastic bags at supermarkets and stores took place in Toyama Prefecture in March 2008. Through a local government order, the use of plastic bags was significantly reduced, with an average of 95% of consumers bringing their own shopping bags (Kaminaga, 2018). This initiative was the first at the prefectural level to employ a market-based instrument (Kaminaga, 2018). On July 1, 2020, the Japanese government implemented a plastic bag fee policy for all retail stores across the country, encompassing the 47 prefectures, to address the growing environmental issue, particularly marine debris (Suzuki, 2021). This policy applies to nationwide retail outlets such as supermarkets, convenience stores, and department stores. The aim of this legislation is to promote a reduction in plastic waste that ends up in the ocean, causing severe environmental impacts (Suzuki, 2021). As part of the initiative, the Japan Environment Ministry called on consumers to use their own reusable bags, aiming to increase the number of customers who do not receive plastic bags during their purchases (Suzuki, 2021). Industries that use 50 metric tonnes or more of plastic per year are required to report their efforts in reducing plastic waste on an annual basis, under the "Act on the Promotion of Sorted Collection and Recycling of Containers and Packaging." Penalties, including the disclosure of company names and fines for non-compliance, are also included to address inadequate actions, such as the continued provision of free plastic bags (Suzuki, 2021).

2.2.1 Progress in Japan

Japan has formulated a comprehensive plan to introduce approximately 2 million tonnes of biomass-based plastic products by 2030 as part of its Resource Circulation strategy. Additionally, the Japanese government has set a goal to achieve zero greenhouse gas emissions by 2050. With these targets in mind, businesses in Japan are exploring opportunities to incorporate biomass plastics that align with environmental objectives while maintaining the high quality of plastic materials. In September 2020, ITOCHU announced a strategic partnership with Borealis and Borouge to explore ways of promoting the acceptance of Bio-Polypropylene in the Japanese market (Mesicek, 2021). Bio-Polypropylene packaging is produced using second-generation raw materials such as agricultural and food waste as well as waste cooking oils. ITOCHU began commercial production of Bio-PP packages in Japan in March 2020 and is currently working on expanding sales in other Asian countries (Mesicek, 2021). Life-cycle assessments have shown that Borealis' renewable PP contributes to higher greenhouse gas emissions reduction compared to virgin PP, without compromising product quality or performance (Mesicek, 2021).

Astellas Pharma Inc. has also taken a significant step by packaging pharmaceuticals in blister containers made from biomass-based polymers derived from plant-based resources. This marks the world's first use of biomass plastic for blister packaging in the pharmaceutical industry. The blister packages consist of biomass-based plastic, PE derived from sugarcane, which accounts for 50% of the raw material (Astellas, 2021). This eco-friendly packaging aligns with the concept of "carbon neutrality," which aims to balance greenhouse gas emissions and absorption (Astellas, 2021). Astellas has leveraged its extensive packaging technology expertise to develop biomass-based plastic sheets suitable for commercial production, ensuring both tablet protection and safe usage. Moreover, Japan has set a target to achieve a recycling rate of 60% under its Resource Circulation strategy (Klein, 2021). In recent years, Japan has surpassed this target with a recycling rate exceeding 85% (Klein, 2021). This success can be attributed to efficient solid-waste-management practices and advanced recycling technologies implemented in Japan. Given that households contribute significantly to solid-waste generation, waste sorting plays a crucial role in reducing costs and enhancing the efficiency of the recycling process. Japanese plastic recycling methods primarily include chemical recycling, thermal recycling (incineration for energy production), and material recycling. Thermal recycling is the most widely used method, accounting for approximately 58% of plastic waste treatment, followed by material recycling at 23%, and chemical recycling at 4%.

The sales volume of recycled plastics reached approximately 439 thousand metric tons in 2019, showing a significant increase from previous years (Klein, 2021).

Furthermore, the Nippon Foundation and 7-Eleven Japan Co. Ltd., in collaboration with the towns of Yokohama and Fujisawa in Kanagawa Prefecture, have initiated a PET plastic bottle recycling project. The aim of this project is to effectively utilize limited resources, promote recycling, and reduce ocean debris. Starting in August 2020 with 15 7-Eleven stores in Fujisawa, the project has expanded to include stores in Yokohama, with plans to install collection machines in 120 stores. The project employs a "bottle-to-bottle" recycling approach, transforming old PET bottles into new ones, and participants receive compensation based on the volume of PET bottles they deposit into the collection machine (The Nippon Foundation, 2020). Additionally, a plastic bag pricing policy has been implemented in nationwide retail outlets, including supermarkets, convenience stores, and department stores. Each retailer has the flexibility to determine the price of plastic bags, with many opting for rates that vary based on the bag's size, often costing less than 10 yen. Companies like 7-Eleven Japan Co., Family Mart Co., and Lawson Inc. have taken responsibility for the cost of shopping bags (Suzuki, 2021). However, some operators such as KFC Holdings Japan Ltd. and Yoshinoya Co. continue to provide free bioplastic bags to their customers (Suzuki, 2021). Efforts are also underway to replace plastic with alternative materials like paper and cloth. For instance, the renowned sportswear brand Adidas in Japan has started charging for paper bags as an alternative to plastics. Additionally, Isetan Mitsukoshi Holdings Ltd., a department store operator, has ceased providing free plastic shopping bags in their food departments since July and now charges customers for paper bags.

2.3 INDIA

India faces challenges in managing its growing plastic waste due to its widespread use in various essential items ranging from toothbrushes to debit cards. According to estimates from the Central Pollution Control Board, India generates approximately 26,000 tons of plastic waste per day. Unfortunately, a significant portion of this waste, exceeding 10,000 tons, remains uncollected (Aravind, 2019). As a result, uncollected plastic waste often finds its way into Indian seas, oceans, and accumulates on land.

a) Kerala Blanket Plastic Ban

Kerala, situated on the southwestern coast of India, has taken a significant step to combat plastic pollution. The state government of Kerala introduced a comprehensive ban on the use of plastics (Government of Kerala, 2019) after India's national attempt to adopt plastic waste management laws in 2016 fell short. As part of this initiative, the state government prohibited the use and sale of plastic carry bags with a thickness of less than 50 microns to promote better separation and recycling of plastic waste. However, this partial ban proved ineffective, as irresponsible plastic usage continued to cause environmental issues and posed health hazards to citizens. Consequently, the chief minister of Kerala decided to implement a complete

TABLE 2.1
The Type of Plastic Products Banned in Kerala (Government of Kerala, 2019)

Product	Type of material	Remarks
Carry bag	Plastic	Irrespective to the thickness.
Plastic sheets	Plastic	Cling film excluded.
Plates, cups, decorative materials	Thermocol and Styrofoam	–
Bags, flags, bunting	Non-woven and plastic	–
Water pouches and juice packets	Plastic	–
Bottle of drinking water	PET/PETE	Bottles with capacities less than 500 mL that are solely relevant to drinking water bottles. Branded juice PET bottles of all sizes, as well as drinking water PET bottles of 500 ml and larger, are subject to EPR regulations.
Garbage bags	Plastic	–
Flex materials	PVC	–
Packets	Plastic	EPR laws will apply to branded plastic juice packets.
Plates, cups, dishes, spoons, forks, straw stirrer	Plastic-coated paper	–
Plastic utensils- cups, plates, dishes, spoons, forks, straw, stirrer	Single-use plastic	Cups includes tumblers.
Packets	Plastic	The usage of plastic packets at retail outlets, including vendors/hawkers, for packing fruits and vegetables, as well as pre-cut fish and meat, pre-weighted cereals, pulses, sugar, flour, and so on, is forbidden.

ban on plastic usage, encouraging the adoption of more environmentally friendly alternatives (Government of Kerala, 2019). The table below provides a list of plastic products that have been completely banned in Kerala.

The Government of Kerala has issued an order that prohibits the manufacturing, transportation, storage, and distribution of certain plastic products categorized as single-use plastic or chlorinated plastic (Shrivastav, 2020). However, there are exemptions for plastics intended for individual use. For instance, plastic products specifically manufactured for export purposes as per export orders in the plastics industry, plastic products used for medical purposes such as medical equipment, and labelled plastic products made of compostable plastic are exempted. These exempted products must adhere to the standards outlined in the 2016 Plastic Waste Management and Disposal Standard – specifically, IS or ISO 17088:2008, titled "Compostable Plastics" (Shrivastav, 2020). On the other hand, the ban does not cover the use of plastics in branded items or products. Instead, the manufacturer or producer is required to comply with the Extended Producer Responsibility (EPR) plan. The EPR plan includes entities such as Kerala Beverage Company, Kerala Federation Kerakarshaka Sahakarana LTD, Kerala Milk Marketing Cooperation Federation, Kerala Water Authority, and other public sector companies that utilize plastic in their products. The generated waste must be managed in accordance with the EPR plan and transported to existing waste collection facilities. Failure to comply with this policy will result in penalties, including a fine of Rs. 10,000 for the first offense, Rs. 25,000 for the second offense, and Rs. 50,000 for subsequent offenses. In addition, non-compliance may lead to the cancellation of the company's license and the sealing of the company.

b) Tamilnadu Plastic Waste Management Rules

Tamil Nadu, a southern state in India with a population of 67.86 million, is recognized as the second-highest generator of plastic waste in the country. Consequently, on January 1, 2019, the Government of Tamil Nadu implemented a ban on throw-away and single-use plastics, without considering the thickness of the material (Preetha, 2020). This legislation prohibits the use of plastic bags that are included or form an integral part of the packaging used for sealing commodities at manufacturing or processing units (Preetha, 2020). To further enforce the ban, the Environment and Forest Department of Tamil Nadu eliminated the exception for plastics used in packaging within manufacturing and processing plants on June 5, 2020. The ban also extends to the food-processing industry, which previously utilized plastic for packaging edible products. While snacks were previously exempted, the lifting of the exclusion clause means that these companies can no longer rely on the use of plastic items. The Madras High Court, in response to a series of petitions, instructed the government to consider phasing out all types of plastics from the market, including those that were previously exempted. The court also noted that the Tamil Nadu Pollution Control Commission has requested exemptions for certain types of plastics. In accordance with government orders, the Central Plastics Engineering Technology Research Institute is studying alternatives such as compostable plastics that can serve as eco-friendly packaging materials as substitutes for plastics derived from fossil fuels. The list in Table 2.2 outlines the plastics that are banned and provides approved, eco-friendly alternatives.

c) Plastic Ban in Maharashtra

On March 23, 2018, the government of Maharashtra issued a notification aimed at curbing pollution by imposing restrictions on the use of plastic. The notification prohibits the production, usage, transportation, distribution, wholesale and retail sale, and storage of various types of plastic bags, including those with and without handles. Additionally, the import of disposable items made of plastic and thermoform, such as single-use cups, plates, glasses, forks, bowls, and containers used for food, is also prohibited (Environmental Department, 2018). These regulations apply to individuals, groups, government and non-government organizations, educational institutions, and any other entities using plastics in Maharashtra (Environmental Department, 2018). However, there are exceptions to these regulations for plastics

TABLE 2.2
The Banned Plastic Items and Eco-friendly Alternatives

List of banned plastic items	List of eco-friendly alternatives
• Plastic sheet/cling film used for food wrapping. • Plastic sheet used for spreading on dining table. • Thermocol plates made of plastic. • Paper plates coated with plastic. • Paper cups coated with plastic. • Plastic teacups, tumbler, cups. • Plastic carry bags in numerous sizes and thicknesses, as well as plastic lined and non-woven carry bags. • Water pouches/packets. • Plastic straws. • Plastic flags. • Plastic-coated carry bags.	• Plantain leaves, Areca nut plates. • Aluminium foil. • Paper rolls. • Lotus leaves. • Glass/metal tumblers. • Bamboo, wooden products. • Paper straw • Cloth/paper/jute bags. • Paper/cloth flags. • Ceramic wares. • Edible cutleries. • Earthen pots.

used in medical purposes and recyclable plastic bags used in plant nurseries, gardening, farming, and solid-waste disposal (Bhatia, 2018). Plastic bags or sheets used for these purposes must be clearly labelled as "Use exclusively for this specific purpose only." Manufacturers and sellers of biodegradable plastic carry bags must obtain a license from the Central Pollution Control Board before marketing or selling them (Bhatia, 2018). Plastic manufacturers are permitted to continue producing plastics, but only for export purposes. During the manufacturing stage, guidelines for recycling or reusing plastics must be printed on the cover or material of the plastic sheet or wrap. Plastic milk packaging bags should also be printed with a buy-back price of not less than Rs. 0.50 to establish a recycling buy-back system. Milk dairies, distributors, and dealers can buy back these used milk bags with pre-determined purchase prices printed on them, thereby setting up an effective collection system for recycling (Bhatia, 2018). Milk producers, retail dealers, and merchants must ensure the implementation of such purchase and recycling mechanisms within three months of the enforcement of the rule (Bhatia, 2018). Non-compliance with the regulations outlined in the policy may result in penalties, including a fine of $5,000 for the first offense, a fine of $10,000 for the second offense, and a three-month prison sentence, and a fine of $25,000 for the third offense, as stated by the government (Devashish, 2019). Table 2.3 provides a summary of permissible and prohibited plastic usage according to the policy.

2.3.1 Progress in India

The ban on plastic carry bags with a thickness of less than 30 microns in Kerala has unexpectedly benefited the local plastic sector. The Kerala Plastic Manufacturers Association reports that the restriction has led to an increase in the sale of higher-quality plastic bags and has also given a boost to the recycling sector. Previously,

TABLE 2.3
The Banned and Allowed Plastics

Banned	Allowed
• PET/PETE drinking water bottles having a liquid holding capacity of less than 200 mL. • Plastic with or without handle, single use plastic bags. • Plastic bags or non-woven bags. • Dishes, spoons, cups, plates, glasses, forks, bowls, and containers that are compost of single use thermocol or plastic. • Disposal dishes for kitchen utensils and straws used in hotels. Compostable plastic bags for plant nurseries and solid waste handling. • Aesthetic use of plastics and thermocol mineral water pouch.	• PET/PETE bottles with a liquid holding capacity of 200 ml or more must have a deposit and refund price or buy back price printed on them. • Bags made from cloth or paper. • Plastic household items. • Recyclable buy back plastics. • Manufacturing of plastic bags for export in special economic regions and export-oriented units. • Plastic material that contains at least 20% recyclable plastic and with a thickness larger than 50 microns that is used for wrapping the material during the manufacture stage or as an essential element of the manufacturing. • Virgin plastic milk bags with a minimum thickness of 50 microns and a printed purchase back price.

many businesses used to provide free plastic bags with an average thickness of 20 microns, but now they charge a nominal fee of about Rs. 1 to Rs. 2 for the thicker bags (The Economic Times, 2007). Stores that still provide free plastic bags now distribute them in smaller quantities. This rise in the sale of higher-priced plastic bags has resulted in increased income for the state's 150 plastic bag manufacturers. It has also revitalized the nearly 120 plastic recycling facilities in Kerala. Thinner plastic bags below 30 microns were not financially viable for recycling, so the recycling units had stopped collecting them, leading to environmental harm. As thicker bags are more expensive, people have started using them more prudently and selling them to recycling centres (The Economic Times, 2007). Previously, people in Kerala had the habit of using multiple carry bags when one would suffice, and they would discard them in public places due to a lack of proper disposal methods. These non-biodegradable bags clogged drainage systems and harmed soil quality, affecting the state's tourism potential. In addition to recycling plastic products, the state government plans to use plastic in road construction by mixing it with bitumen to create at least 100 kilometres of road in each region (The Economic Times, 2007).

The recycling units currently face a challenge in terms of collecting plastic items from households. To address this issue, the government has involved women's self-help groups in the collection, recycling, and sale of household plastics. Around 1,500 women have been provided with facilities and training for collecting and sorting materials from households, and a full-fledged recycling facility is being constructed for them (The Economic Times, 2007).

On the other hand, the Kerala Plastic Manufacturers Association has expressed dissatisfaction with the complete ban on plastics. They are not opposed to a ban but request some time to exit the plastic manufacturing industry, as their units were established with bank loans. They have urged the government to consider implementing the ban after six or eight months to allow manufacturers to sell their existing stock and transition to new businesses. Meanwhile, Tamil Nadu is making significant progress in eliminating plastic usage through various initiatives. The state has launched campaigns to ban single-use plastics, and efforts are being made to reach a wide audience through private channels and the participation of the Chief Minister (The Times of India, 2019). The "*Meendum Manjapai*" campaign, which promotes the use of traditional cloth bags as an alternative to plastic bags, has gained support and is seen as a successful outcome of the plastic ban. To enhance the effectiveness of the policy, the Tamil Nadu Pollution Control Board has held discussions with star hotels in the state to ensure their compliance with avoiding the use of throwaway plastics. Presentations have been made by hotel representatives on their current practices related to plastic-waste management and solid-waste management (The Hindu, 2019). The board has also encouraged star hotels to create short films showcasing their best practices in plastic-waste management to raise awareness among the public (The Hindu, 2019). Although there has been a significant decrease in plastic usage across Tamil Nadu over the past two years, the complete elimination of plastics has not yet been achieved. While many customers have switched to using their own cloth bags, small merchants still use prohibited plastics due to a lack of alternatives. The enforcement of the ban has led to the closure of illegal plastic

2.4 CHINA

China has witnessed a rapid growth in infrastructure and industrial development over the past two decades. However, this progress has come at a cost, leading to significant environmental challenges, including desertification, water depletion, land degradation, biodiversity loss, and pollution (Ogunmakinde, 2019). As the country with the largest population, China holds the title of the world's largest producer of plastic waste, responsible for one-fifth of global single-use plastic production. It accounts for approximately 31% of the world's plastic manufacturing output, making it the top plastic producer globally (Wong, 2021). In 2010, China generated 60 million tonnes of plastic waste, surpassing the United States, which produced 38 million tonnes. According to Hua'an Securities, the demand for biodegradable plastics is projected to reach 217 tonnes, with an economic value of approximately US$7.3 billion by 2025 (Joe, 2021).

a) Ban on Imported Plastic Waste

Since the 1980s, China has emerged as the leading global importer of trash. In 2012, China accounted for 56% of the world's exported plastic waste (Wen et al., 2021). The country alone imported nearly 9 million tonnes of plastic waste, of which around 70% was improperly disposed of or mishandled, resulting in various environmental problems (Wen et al., 2021). China has been grappling with pollution and environmental degradation, including persistent smog and soil contamination, as a consequence of rapid urbanization (Xinhua, 2017). To address this issue, the Chinese government implemented a new ban called the Solid Waste Import Management Reform Plan, also known as the "China ban," on July 27, 2017. This ban prohibits the importation of 24 types of solid waste across four broad categories, including plastic waste, paper waste, metal waste, and textile waste. Five government departments announced that the ban would take effect from December 31, 2017 (Zhang and Laney, 2017). In April 2018, the Ministry of Ecology and Environment added an additional 16 categories of solid waste, such as waste hardware and industrial waste plastics, to the list of banned items (Zhang and Laney, 2017). The plan outlines regulatory framework changes in relation to the import of foreign waste. Specifically, at the end of 2018, measures were revised to limit the ports where solid waste can be imported and to amend the Solid Waste Environmental Preventing and Control Law by the end of 2019 to increase penalties for smuggling foreign waste and illegal imports of solid waste. This sudden prohibition has led to both short- and long-term shifts in the global patterns of plastic waste flows as well as changes in waste management systems and processes for plastic waste disposal in many countries. The resulting environmental consequences necessitate significant attention and statistical evaluation from a global-sustainability perspective (Zhang and Laney, 2017).

b) Household Solid Waste (HSW) Sorting

In July 2019, China implemented a waste-classification strategy for households as a solution to waste-management issues. This policy introduced a stringent framework that requires households to separate garbage into four categories: dry, moist, recyclable, and hazardous, using color-coded containers (Wang and Jiang, 2020). Initially, the program was launched in eight cities in the early 2000s, but it was poorly managed and failed to achieve its objectives. The generation of 2 trillion tonnes of household solid waste in 202 cities across China in 2017 exemplifies the shortcomings of the earlier implementation (Wang and Jiang, 2020). In March 2017, China released a more ambitious national plan to address the growing mountains of domestic trash. This plan mandated 46 pilot cities, including Shanghai, to establish local regulations or laws for mandatory waste classification by the end of 2020 (Wang and Jiang, 2020), marking the beginning of the mandatory waste-sorting period in China.

Several legislations and rules support the establishment of the domestic and municipal household solid-waste-sorting policy in China. Article 26 of the country's 1982 Constitution, which was last revised in 2018, emphasizes the state's responsibility to safeguard and improve the living environment of its citizens, protect the natural environment, and regulate pollution and other public risks (Wang and Jiang, 2020). These constitutional provisions are complemented by significant national legislations such as the Law on the Prevention and Control of Environmental Pollution Caused by Solid Wastes and the Law on the Promotion of Circular Economy, as depicted in Table 2.4 (Wang and Jiang, 2020). Together, these laws form the fundamental legal basis for household-solid-waste management. Furthermore, the State Council and its agencies have the authority to establish administrative regulations and standards for enforcing national laws. Numerous national administrative rules and regulations, including the administrative measures for urban household solid-waste, published by the State Council or its agencies, provide overarching principles for household waste sorting (Wang and Jiang, 2020).

TABLE 2.4
The Constitution's Clause on Environmental Protection (Wang and Jiang, 2020)

Law	Effective date	Level of authority
Environmental Protection Law	Jan 1, 2015	Law
Circular Economy Promotion Law	October 26, 2018	Law
Law on the Prevention and Control of Environment Pollution Caused by Solid Wastes	November 7, 2016	Law
Regulation on the Administration of City Appearance and Environmental Sanitation	March 1, 2017	Administrative regulation
Administrative measures for urban HSW	May 4, 2015	Departmental rules

c) China's Five-Year Phase Out Plan

The National Development and Reform Commission of China has implemented a five-year ban on four specific types of single-use plastics, signalling the government's commitment to reducing plastic consumption. However, this is not the first attempt by China to address plastic usage. In 2008, China prohibited the distribution of free plastic bags in shops and banned the production of ultra-thin plastic bags. Unfortunately, this earlier plan proved ineffective and unsuccessful (Joe, 2021). As a result, the new five-year plan to phase out single-use plastics has been divided into two phases, aiming to gradually reduce usage, production, and waste generation. During phase one, non-degradable plastic bags were banned in shopping malls and supermarkets in major cities by the end of 2020. The goal is to expand this ban to all cities and towns by 2022 (Zhang and Laney, 2017). Supermarkets and shopping malls have also started offering customers recyclable bags for purchase, instead of providing one-time use plastic bags. Additionally, China has planned to replace plastic straws with paper straws or poly-lactic acid straws in restaurants, food stalls, and coffee shops nationwide. The complete ban on single-use straws was targeted for 2020, and single-use cutlery by 2022 (Joe, 2021). Furthermore, efforts are being made to reduce plastic usage in food delivery packaging by 30%, and hotels are encouraged to eliminate single-use plastic items by 2025 (Joe, 2021). In phase two, starting in 2022, all cities and municipalities are prohibited from using non-biodegradable plastic bags (Joe, 2021). However, vendors in the fresh products market have been given additional time until 2025 to completely stop using plastic. Moreover, the production and sale of one-time plastic bags with a thickness below 0.025mm are also advised to be completely halted (Joe, 2021). Furthermore, under the Environmental Pollution Prevention and Control Law, penalties ranging from 10,000 Yuan (US$1,400) to 100,000 Yuan (US$14,200) will be imposed on those who violate the five-year plan (Zhang and Laney, 2017). The plan also prohibits the use of PE agricultural mulching film with a thickness of less than 0.01mm, single-use moulds, and plastic cotton swabs in the production and sale of products.

2.4.1 Progress in China

In 2017, China imported approximately 5.7 billion kg of discarded plastic (Brady, 2021). However, following the implementation of China's importation ban policy, the government took strict measures to control and prohibit the import of waste materials, resulting in a 99.1% reduction in waste imports the following year (Brady, 2021). The ban was highly stringent and successfully achieved its primary objective of minimizing plastic-waste imports. However, it also exposed the vulnerability of the global reliance on a single importer, as it created a gap in the recyclables market. Before the ban, the United States, Japan, and seven European nations were the primary exporters of plastic waste to China, accounting for 77.9%, 87.6%, and 57.5%, respectively. After the ban, plastic waste started being redirected to five Southeast Asian countries: Thailand, Indonesia, Vietnam, Malaysia, and the Philippines. Prior to the prohibition, Figure 2.1 illustrates the worldwide trade flows of six types of plastic waste (Baseline Scenario). Among these, polyethylene (PE) accounted for 37% of the total plastic waste trade, with a volume of 11,404,697 tonnes. Polypropylene (PP), polyethylene terephthalate (PET), and other plastics comprised 23%, 14%, and 14%, respectively, while polystyrene (PS) and polyvinyl chloride (PVC) made up less than 8%. Notably, Hong Kong has a 46.2% market share in PE, making the movement of PE from Hong Kong to China significant in Figure 2.1 (Wen et al., 2021). In 2018, compared to the Baseline Scenario, the global trade flow of plastic waste decreased by 45.5%. Figure 2.2 depicts the worldwide plastic-waste trade flow after the ban (2018 Scenario). When comparing Figures 2.1 and 2.2, it can be observed that Hong Kong's export dominance has diminished. The changes in international trade patterns of plastic waste between countries highlight the complex shifts in trade flows. The exports of four countries, particularly Japan, the United States, Germany, and the United Kingdom, accounted for 46.1% of the global trade flow of plastic waste (Wen et al., 2021).

Large exporting nations were left scrambling for solutions to their waste problems, while smaller Southeast Asian countries experienced a significant surge in the

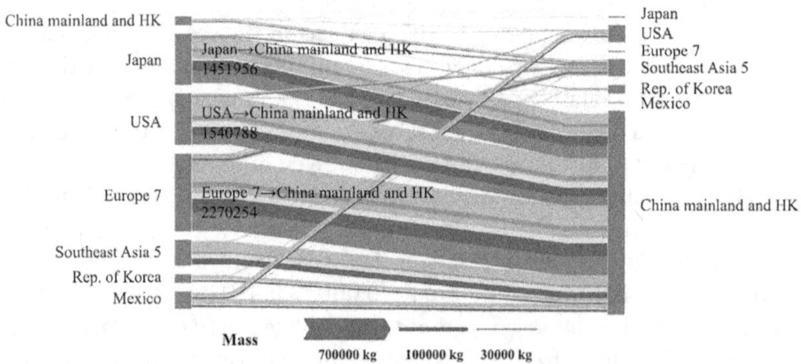

FIGURE 2.1 The trade flows of six types of plastic waste prior to the ban.

Source: Adapted from Wen et al., 2021, Nature Communications Open Access.

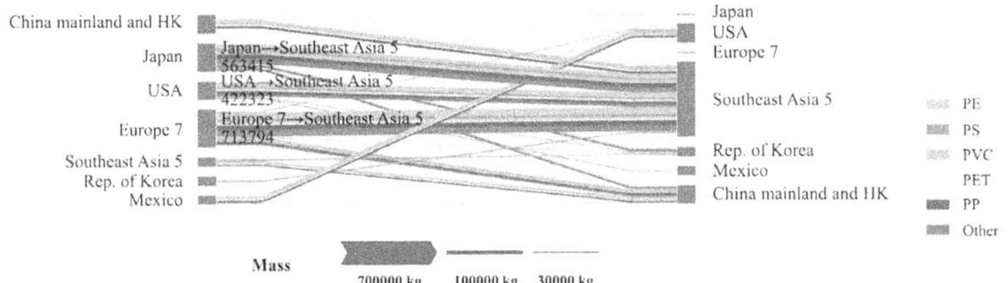

FIGURE 2.2 The trade flows of six types of plastic waste after the ban.

Source: Adapted from Wen et al., 2021, Nature Communications Open Access.

influx of garbage. In 2017, Thailand received 153 million kg of plastic waste, but by the end of 2018, just one year after China's ban came into effect, their scrap-plastic imports had more than quadrupled to half a billion kg. During the same period, scrap-plastic imports to Indonesia increased by 250%, and to Vietnam, by 200% (Brady, 2021). Malaysia also faced a substantial increase in waste imports following the prohibition. Due to its significant Chinese-speaking population, Chinese waste processing firms found it relatively easy to establish factories there. In early 2018, residents in a town near Kuala Lumpur noticed a significant rise in the number of plastic plants, leading to respiratory issues, as these facilities continuously released toxic chemicals into the air. By mid-2018, over 40 processing factories had been discovered in the town, most of them operating without permits. Continued pressure from government authorities resulted in the closure of 33 factories, although activists claim that many of them have simply relocated elsewhere in the country (Brady, 2021).

In response to the overwhelming flow of waste, Southeast Asian countries have begun taking precautionary measures to protect themselves from descending into chaos in their own waste-management systems. The involvement of civil society and environmental non-profit organizations in advocating for garbage import-restrictions in Southeast Asian countries has played a significant role in addressing this issue. By voicing concerns about the environment and public health, they have exerted political pressure on governments. As a result, Thailand announced a ban on foreign plastic waste imports by 2021 in October 2018. Vietnam has stopped granting new garbage import licenses and plans to cease collecting scrap plastic by 2025 (Neo, 2018). Malaysia temporarily suspended imports in 2018 and implemented measures to strengthen permission procedures. Recently, the Philippines returned hundreds of shipments of plastic waste to Canada after a Philippine court ruled that the import of 2,400 tonnes of Canadian rubbish was illegal (Neo, 2018). Since 2018, several Southeast Asian countries have banned plastic waste imports due to their inadequate waste management facilities' inability to properly sort, process, and recycle the overwhelming amount of waste. Furthermore, China has a large number of waste-plastic-recycling businesses, and extensive collection networks are present throughout the country. Programmes such as the National Sword were implemented during the transformation and upgrading period, particularly from 2016 to 2017.

By 2017, 87% of China's waste plastics recycling market had to comply with stringent environmental protection standards. Over 5–6 million businesses in 28 cities faced sanctions for violations, including production halts, criminal detention, and condemnations. About 5% of large-scale recycling enterprises started focusing on domestic waste plastic collection, another 5% relocated to Southeast Asian countries or "waste origin" countries such as Europe, the United States, and Japan, while the remaining 80% to 90% were forced to shut down their operations. Additionally, in November 2016, the effectiveness of the household solid-waste (HSW) sorting strategy was examined in Nanjing City's Jianye District. The findings revealed that the HSW sorting policy successfully promoted waste categorization and increased recycling rates. Overall, around 52% of respondents supported the mandatory policy, and the sorting rate of HSW increased by approximately 49.7% (Chen et al., 2018). Investments in source-separated facilities, publicity, and enhanced supervision coverage significantly benefited the source-separated collection of HSW. For every 1 million Yuan invested in source-separated facilities and publicity, the sorting rate increased by 1.1%, and for every 1% increase in special supervision coverage, the sorting rate increased by 3.6% (Chen et al., 2018).

2.5 MALAYSIA

Malaysia is actively engaged in the global plastic industry, with around 1,300 plastic producers in operation today. In 2016, Malaysian exports in this sector amounted to RM 30 billion, resulting in the production of 2.26 million metric tonnes of resin. However, the accumulation of plastic waste has become a pressing environmental concern. Currently, Malaysia ranks eighth among the top 10 countries with the highest mismanaged plastic waste. Additionally, illegal plastic production in Malaysia reached approximately 0.94 million tonnes, of which 0.14 out of 0.37 million tonnes found their way into the Malaysian ocean (MESTECC, 2018).

a) No Plastic Bag Day Campaign

In 2011, the Ministry of Domestic Trade, Cooperatives, and Consumerism (MDTCC) implemented the No Plastic Bag Campaign Day as a means of safeguarding the environment and diminishing the reliance on plastic bags. The campaign was introduced at selected super- or hypermarkets, major shops, and shopping malls nationwide. On weekends, customers were charged MYR 0.20 for each plastic bag (Zen et al., 2013). The funds generated from the campaign were allocated to finance environmental projects aimed at addressing the issue of environmental debris. The plastic bag levy, or tax model, is an example of environmental taxation targeting pollution generated by producers, with the intention of offsetting the external costs imposed on the environment. The implementation of plastic levies primarily supports social changes among consumers, raises awareness of ecological impacts, and reduces littering and excessive plastic usage (Zen et al., 2013). Consumers were given the choice to avoid paying the plastic bag levy by bringing their own reusable bags. However, the campaign triggered various responses from the public, including consumers, environmentalists, policymakers, and plastic manufacturers

(Zen et al., 2013). The campaign, along with the plastic bag tax, was accompanied by educational and awareness initiatives aimed at influencing consumer behaviour and educating them about the importance of levies in reducing plastic usage. This aspect was examined to determine if the marketing message effectively reached consumers and improved their environmental knowledge, understanding, and attitude (Zen et al., 2013). Consumer support for the plastic bag ban campaign offers crucial insights for policymakers.

b) Roadmap Towards Zero Single-Use Plastics

The Malaysian Ministry of Energy, Science, Technology, Environment, and Climate Change launched the Roadmap towards Zero Single-Use Plastics (Chen et al., 2021). This policy introduced pollution charges for consumers and manufacturers. In 2019, the state governments of Selangor and all three federal territories in Malaysia implemented a ban on plastic-straw usage (Chen et al., 2021). A nationwide ban on straws was introduced in 2020, but enforcement has been lacking so far. However, reports from Kuala Lumpur, Putrajaya, and Selangor have indicated a negative response to the plastic straw ban. Similar to the No Plastic Bag Campaign, it may take time for people to adjust, as plastic straws are commonly seen as a necessary item in Malaysia. The government's incentives to reduce single-use plastics are relatively small compared to the costs associated with waste separation, recycling, and manufacturing plastics with recycling properties. Recycling expenses for plastics are higher than acquiring new plastic materials (Chen et al., 2021). This initiative follows a staged, evidence-based, and comprehensive approach outlined in the roadmap. The initiation spans from 2018 to 2030 and includes action plans. The vision of this initiative is to achieve "Towards zero single-use plastics for a cleaner and healthier environment in Malaysia by 2030." This concept encourages the involvement of stakeholders interested in addressing plastic pollution in Malaysia. The strategy aims to develop actions that can steer the current trajectory towards a more sustainable path, fostering a cleaner and more circular environment by 2030.

c) National Recycling Program

The implementation of the "National Recycling Program" in 2000 and the introduction of the "National Strategic Plan for Solid Waste Management (2000–2020)" in 2005 demonstrate Malaysia's strong commitment to waste reduction (Sreenivasan et al., 2012). As per Article 102 of the Act, the government placed the responsibility of product collection on manufacturers, distributors, importers, or retailers (Sreenivasan et al., 2012). These responsibilities align with the Waste Management Hierarchy, which aims to minimize the amount of waste entering landfills and disposal sites (Hashim et al., 2011). Among the hierarchy concepts, the 3Rs program – Reduce, Reuse, and Recycle – plays a crucial role. To foster a 3R culture in society, it is essential to conduct awareness programmes and train groups of people to implement the 3Rs initiative effectively.

The 3Rs approach not only enhances waste management systems but also reduces the environmental impact caused by human activities. It promotes improved economic operations, minimizes the environmental effects of waste disposal, reduces waste loss, and extends the lifespan of waste sites. Developed nations tend to have more success with the 3Rs compared to underdeveloped countries (Sreenivasan et al., 2012). In Malaysia, traders have endeavoured to align with the waste management hierarchy, starting with waste minimization, sorting, and recycling, followed by wastewater treatment such as incineration and composting, and ultimately, waste storage in designated disposal areas. This integrative approach requires the involvement of multiple stakeholders, including the government, industry, and waste management concessionaires.

In response to the pressing waste management issues, the Malaysian government launched the 6th Malaysia Plan for 1991–1995, which aimed to diversify waste management sources (Jereme et al., 2015). The subsequent 7th Malaysia Plan (1996–2000) also prioritized the quality of life and environmental concerns. However, with the population growth and limited availability of landfill sites due to objections from citizens and local leaders, the National Recycling Program was reintroduced in the 8th Malaysia Plan (2001–2005). The program aimed to normalize recycling practices among the population, reduce solid waste management costs, minimize landfill disposal, and decrease the use of raw materials, while also increasing stakeholder awareness and collaboration (Jereme et al., 2015). Recycling and solid waste disposal were privatized under the 9th and 10th Malaysia Plans, with the implementation of the Solid Waste Management Bill to improve services, alleviate financial burdens, and involve the private sector (Jereme et al., 2015). Waste management facilities are responsible for the collection, storage, transportation, treatment, and recycling of all non-hazardous waste in the country.

d) Eco-Labelling Certification Scheme

Consumer awareness and preference for environmentally friendly products are on the rise worldwide in industrialized nations (Rashid, 2009). Malaysia's government has also responded effectively to this issue. As a result, the Standards and Industrial Research Institute of Malaysia (SIRIM) introduced the national eco-labelling program, which validates products based on environmental criteria, including the use of environmentally compostable and non-hazardous plastic packaging materials (Suki, 2013). The Federal Agriculture Marketing Authority (FAMA) has the Malaysia Best logo for environmentally friendly agricultural products, and the Malaysian Energy Commission has the Malaysian Energy Commission emblem for energy-efficient electrical items (Suki, 2013). Eco-labels have the potential to serve as informative tools for consumers to make environmentally conscious choices, while also providing producers with a means to differentiate themselves in the marketplace and gain market share (Rashid, 2009). Eco-labels are generally categorized into two types: self-declaration claims and independent third-party claims (Rashid, 2009). Self-declaration claims are made by the producer, retailer, or marketer and may pertain to a single attribute or provide an overall assessment of the product. These claims may include phrases such as "environmentally friendly," "ozone friendly," "organic," "pesticide-free," "biodegradable," and "recyclable," often displayed on the product (Rashid, 2009). However, these claims are rarely independently verified.

On the other hand, independent third-party claims are based on meeting pre-defined criteria that have been independently validated by a competent authority. Typically, these criteria are based on a product's life-cycle perspective (Rashid, 2009). In Malaysia, a certified eco-labelling regime for green consumption has been established by the Malaysian Bioeconomy Development Corporation, previously known as BiotechCorp, and SIRIM Bhd. GreenTech's MyHijau Mark aims to consolidate the various eco-labels. However, GreenTech does not have statutory recognition like SIRIM Bhd. The lack of significant economic rewards for producers in creating environmentally friendly products has resulted in a lack of incentives. Moreover, obtaining multiple eco-labels is not cost-effective and time-consuming. Producers must make efforts and collaborate with multiple relevant organizations to obtain the labelling (Raj, 2019).

2.5.1 Progress in Malaysia

The "No Plastic Bag" campaign has had a widespread impact, eliciting various emotions, attitudes, and behaviours from the public, including representatives from the plastics industry, customers, market vendors, and store owners. A study on this campaign identified three forms of anti-consumer behaviour: complete opposition to consumption, partial opposition to consumption, and no anti-consumption behaviour (Saleh Omar et al., 2019). Research indicates that the public is generally supportive of the campaign and expresses positive reactions, leading to increased awareness and a reduction in plastic-bag usage for the betterment of the environment and future generations. However, there still exist individuals with partial opposition to consumption or no anti-consumption behaviour. For example, some consumers carry eco-friendly bags in their cars for emergency shopping, while others pay for plastic bags during purchases but feel guilty about the environmental impact. On the other hand, many consumers actively adopt ecologically friendly practices by bringing their own cloth bags while shopping in supermarkets and hypermarkets (Saleh Omar et al., 2019).

To support the "No Plastic Bag" campaign, several well-known retail malls have implemented initiatives. One such initiative is the AEON Green Fund, established in 2011, where customers requesting plastic bags on Saturdays are charged RM 0.20 per bag. A portion of the collected funds goes to the AEON Green Fund, supporting the company's environmentally and socially responsible initiatives. The fund has been used for reforestation, orangutan rehabilitation, sustainability education, beach clean-ups, landscape planting, tree planting for reforestation, reducing plastic bag usage, and celebrating Malaysia Day and Harvest Festival. AEON has also introduced the AEON reusable shopping bag, which is available at their outlets to promote the reduction of single-use plastic bags. AEON stores nationwide implemented a "No Plastic Bag Every Saturday" policy since October 2, 2010, aligning with the government's efforts to reduce plastic bag dependency. Tesco Malaysia introduced the "Unforgettable Bag" as part of their initiative to limit the use of single-use plastic bags. Customers are encouraged to return these bags to the store after each shopping trip. This initiative supports the government's Roadmap to Zero Single-Use Plastics 2018–2030, aiming to make Malaysia cleaner and more ecologically sustainable by

2030. Giant, one of the country's largest retail chains, also played a significant role in eliminating plastic bag usage nationwide. They conducted the Giant 1 Million Reusable Bags Giveaway campaign, distributing reusable bags during the Hari Raya Aidilfitri festival and National Day festivities. The initiative aimed to replace millions of plastic bags with a significantly reduced number of reusable bags, promoting sustainable habits among customers.

Other companies have also taken steps to reduce plastic bag usage. IKEA Malaysia eliminated single-use plastic bags by providing blue carrying bags and complimentary carton boxes to customers. 7-Eleven Malaysia initiated an initiative to transform post-consumer plastic bottle waste into eye-catching tote bags, encouraging Malaysians to adopt eco-friendly purchasing practices. Despite these efforts, a survey revealed that only 44% of Malaysians are aware of the country's "Roadmap Towards Zero Single-Use Plastic," indicating the need for more awareness and education. The effectiveness of the Malaysian government's National Recycling Program is also questioned, as the country failed to meet its 2020 objectives for waste diversion and recycling rates (Lacovidou and Ng, 2020).

2.6 SINGAPORE

Singapore, a developed country with the world's seventh highest GDP per capita, has experienced a significant rise in plastic consumption, as plastics have become affordable and widely available for daily use. Over the past three decades, shops and restaurants have been providing free plastic bags, containers, and disposable cutlery to customers, making single-use plastics an integral part of daily life. As a result, plastics have become the most common type of waste disposed of in Singapore, amounting to over 763,400 tons in 2018. Unfortunately, only 6% of the plastic waste produced was recycled (Tammy and Karlsson, 2020).

a) Singapore Packaging Agreement

The Singapore Packaging Agreement (SPA) was a collaborative initiative involving the government, industry, and non-governmental organizations (NGOs) aimed at reducing packaging waste, which accounts for about one-third of Singapore's domestic waste (NEA, 2012). The agreement was voluntary to allow the sector the freedom to develop cost-effective measures for waste reduction. During the review of the Singapore Green Plan (SGP) in 2012, the Clean Land Focus Group, responsible for waste management review, recommended that Singapore consider adopting the concept of Extended Producer Responsibility (EPR) to reduce waste, including packaging waste, which had proven effective in reducing plastic waste in other countries (Peck and Tay, 2010). A poll conducted during the assessment of the SGP 2012 revealed that 94% of respondents agreed that producers should take steps to minimize packaging consumption in their products (Peck and Tay, 2010). Subsequently, the National Environment Agency (NEA) conducted investigations into various packaging practices in countries such as Australia, Japan, and New Zealand. The NEA also engaged with industry representatives to understand their concerns. The industry believed that enacting laws would increase costs, which would eventually be passed on to consumers. Moreover, laws would limit the industry's flexibility to innovate product designs. Therefore, the parties agreed to initiate a voluntary program modelled after New Zealand's Packaging Accord (Peck and Tay, 2010). This strategy was based on product stewardship, holding all stakeholders involved in the product's lifecycle accountable for reducing its environmental impact. The program was expected to foster collaboration between the government, industry, and community, engaging the entire packaging supply-chain. It aimed to encourage industry to assume greater corporate responsibility, shifting the focus from compliance to continuous improvement.

In 2007, the NEA announced the first Singapore Packaging Agreement to address packaging waste. On World Environment Day, June 5th, 2007, numerous parties, including industry associations, individual enterprises, NGOs, the Garbage Management & Recycling Association of Singapore, public waste collectors, and the NEA, signed the agreement (Peck and Tay, 2010). Initiatives such as the Packaging Partnership Programme and Mandatory Packaging Reporting Regulations continued to support sustainable packaging waste management in Singapore after the SPA expired on June 30, 2020 (Peck and Tay, 2010). The primary goals of the SPA were to reduce waste from product packaging by optimizing production processes, improving packaging design, and increasing packaging waste reuse and recycling. The SPA also aimed to raise consumer awareness and provide education on waste reduction. Raising awareness is crucial as consumer behaviours, such as choosing items with less packaging and participating in recycling, directly impact the program's effectiveness. Initially, the agreement focused on food and beverage packaging, which constituted a significant portion of household packaging waste. When the agreement was first signed, it involved 32 signatories, including five industrial groups representing over 500 firms, 19 individual companies, two NGOs, the Trash Management and Recycling Association of Singapore, public waste collectors, and the NEA. In October 2009, the initiative expanded to include additional product packaging, such as detergents, household items, toiletries, and personal care products. By the end of 2009, the number of signatories had risen to 95, with additional signatories from various businesses, including hotel and shopping centre owners or managers (Peck and Tay, 2010).

The implementation of the SPA was overseen by a Governing Board. The Governing Board provided guidance to the signatories, assisted in resolving challenges, and ensured their active engagement in fulfilling the commitments of the SPA. Board members were strong advocates of the SPA and actively sought opportunities to engage with other industry players and partners to encourage their participation. For instance, the Governing Board organized monthly CEO luncheons, inviting both signatories and non-signatories to exchange best practices for reducing packaging waste and share the financial savings achieved through more environmentally sustainable methods. These activities also provided non-signatories with an opportunity to learn about the SPA and be encouraged to join the program. Since the signing of the agreement, the signatories have made significant progress in waste reduction.

b) National Recycling Plan

The national recycling plan was initiated in April 2001. As part of this program, all Housing and Development Board estates, private landed properties, and condominiums/private flats that participated in the public garbage collection scheme were required to provide recycling containers and recycling collection services (NEA, 2012). The program involved the collection of paper, plastic, glass, and metal recyclables in the same blue recycling container by Public Waste Collectors (PWCs) under the National Recycling Program. The plan primarily focused on households, schools, and workplaces. Waste minimization and recycling efforts in these locations were carried out based on the principles of the 3R concept, as outlined in the following table.

TABLE 2.5
National Recycling Plan (NEA, 2012)

Place	Plan
At home	*Reduce* a) Bring a reusable shopping bag or an alternative bag such as paper or cloth to decrease the use of plastic. b) Purchase items with less packaging that is made of plastic materials such as food items and grocery items. c) Avoid requesting shopping bags when shopping. *Reuse* a) Reuse medicine containers for travel toiletries. b) Before disposing of home garbage in waste bins or chutes, use unwanted plastic bags to confine it. c) Plastic bags are subsequently burnt with other rubbish in waste-to-energy incineration plants, which are designed to safely incinerate waste, recover energy from the incineration process, and fulfil air quality requirements through effective combustion and fuel gas emission control. d) Use reusable bags when doing grocery shopping.
At school	*Reduce* a) Avoid over-purchasing food and beverages for school activities and events. b) Instead of offering individual sachets of soap and refreshments, use refillable dispensers. c) Purchase durable devices and durable stationeries. d) Purchase refillable stationery. *Reuse* a) Reuse plastic for watering plants or as planting pots. b) Donate unneeded things that are still in good shape to charitable organizations. Recycle a) Plastics are advised to recycle under certain conditions. Steps to recycle at school: i. Collect recyclable goods in an undesired box/container labelled "Recycling Box." Remove any food or liquid residue; rinse if necessary. ii. Check that the recyclable container is empty for example, there is no food or liquid waste in the container. iii. Place recyclables such as old newspapers/magazines, ripped plastic files, and other materials in the Recycling Box. iv. Empty the recyclables from the box and place them in the school's recycling container.

Place	Plan
At work (office)	*Reduce* a) Avoid requesting for plastic carriers if food can be held by hand. b) Consider to bring own reusable lunchbox and cutleries for takeaways food items. c) Consider to use own mug to meetings to avoid using plastic or Styrofoam cups. *Recycle* a) Appoint an official to oversee the recycling program's implementation. b) Engage a recyclables collector. c) Place recycling containers in strategic areas with adequate labelling. d) Raise awareness and engage employees on a regular basis. e) Teach the cleaners how to keep recyclables and regular garbage separate during collection to minimize contamination. f) Monitor the plan on a frequent basis and keep employees informed by recording the recycling program's success. *Reuse* a) Effective management of natural resources, unfinished goods, finished goods, and waste streams. b) Companies can participate in the Packaging Partnership Programme to minimize consumer product packaging waste.
At work (industry)	*Reuse* a) Look for opportunities to use waste as a feed stock in another operation. b) Search for opportunities to keep debris from being contaminated so that it may be recycled as a raw material in the main process. *Recycle* a) In November 2003, JTC Industrial Estates initiated a recycling programme. Under this scheme, special corners in bin centres have been put up to collect wood debris for repurposing. Recycling bins are also strategically placed.

After being used, specialized recycling trucks are provided to collect the discarded garbage and transport it to Materials Recovery Facilities (MRF). Plastics are taken to a recycling plant, where they are sorted by type. Each type of plastic is crushed into smaller fragments and mixed together to form a homogeneous slurry. The combined material is heated and forced through a screen to create strands through a process known as extrusion. These plastic strands are then chopped into pellets and cooled, which can be used as raw material for manufacturing new products.

c) Extended Producer Responsibility (EPR) Laws

EPR is a strategic policy that holds producers accountable for the environmental impact of their products throughout their lifecycle. EPR ensures proper collection and disposal of products after use, promotes responsible manufacturing practices, and encourages waste reduction. Considering that plastics are the most common type of waste generated in Singapore and the recycling rate is low, it is crucial to consider upstream initiatives like EPR. Although many multinational corporations in Singapore have set their own targets for reducing their plastic footprint by 2025, these efforts may not lead to significant changes in the way Singaporeans consume and dispose of plastic, especially if they are one-time strategies such as switching to greener materials (Qiyun, 2020). Businesses often face limitations, whether physical or financial, that prevent them from making substantial changes to their supply chains. Therefore, EPR is essential in linking legal requirements to financial incentives, compelling companies to make necessary transformations and strengthening municipal-waste management. Singapore's Resource Sustainability Act (RSA), which focuses on packaging waste, is expected to be the first EPR in Southeast Asia by 2025 (Qiyun, 2020). This represents a significant step forward in Singapore's efforts to manage packaging waste, particularly after more than two decades of the Singapore Packaging Agreement (SPA), which aimed to raise awareness and reduce packaging waste but has been criticized for its ineffectiveness in promoting circular economy transitions.

In contrast, the RSA was passed in 2019 and includes regulatory measures targeting waste streams, particularly packaging with high generation and low recycling rates. Large manufacturers of packaging and packaged items are currently required to declare the amount and type of packaging they put on the market. By 2022, these producers will be required to outline their plans for reducing, reusing, and recycling these packaging materials. This information will serve as a starting point for the development of the EPR program (Qiyun, 2020).

2.6.1 Progress in Singapore

The Singapore Packaging Agreement (SPA) was a collaborative initiative involving the government, industry, and NGOs to reduce packaging waste. The SPA recognized signatories that achieved significant reductions in packaging waste between March 1, 2019 and February 29, 2020. Ten multinational corporations and major regional organizations, along with five small and medium-sized enterprises (SMEs), were acknowledged for their efforts in reducing, reusing, or recycling packaging waste and their use of recycled packaging materials. Over 100 signatories were honoured for their contributions to waste reduction. By June 30, 2020, the SPA has brought together more than 200 organizations in Singapore, resulting in the elimination of approximately 62,000 tonnes of packaging waste, valued at around $150 million. Notable businesses such as Tetra Pak, Boncafé, Resorts World Sentosa (RWS), F&N Foods, and Starbucks Coffee Singapore Pte Ltd actively cooperated with the National Environment Agency (NEA) to efficiently reduce packaging materials.

Firstly, Tetra Pak, a Swedish-Swiss food processing and packaging solutions company, was a signatory of the SPA. The company's largest packaging material facility in the world is based in Singapore, offering 132 different types of carton packaging to 32 markets. Tetra Pak implemented several environmental improvement projects under the SPA, including the reduction and reuse of polyethylene (PE) plastic waste used in laminate carton packaging. Previously, the excess PE trim from carton processing was compacted into bales for sale-to-waste dealers and shipped overseas for recycling. However, Tetra Pak purchased new machinery in 2009 to recover and repurpose the PE trim in the package-production process. This initiative resulted in a net reduction of 380 tonnes of PE waste per year with just one machine. Tetra Pak's efforts in minimizing packaging waste earned them the Distinction Award in 2009. Secondly, Boncafé International Pte Ltd, a local gourmet coffee company, witnessed a significant 33% increase in packaging material usage over the past decade as its business grew. Concerned about the environmental impact of packaging waste, Boncafé became a signatory of the SPA in 2006. The company tackled the issue by working with material suppliers to reduce the thickness of their packaging material without compromising its quality and appearance. Through design modifications, Boncafé successfully reduced the thickness of polyethylene (PE) material from 140 to 100 microns, resulting in a 14% reduction in plastic usage. With an estimated production of 900,000 packets per year, this change amounted to a reduction of 1,516 kg of packaging material annually.

Resorts World Sentosa (RWS) took progressive measures to reduce single-use plastic in their dining outlets, such as the Malaysian Food Street food court. They replaced disposable sauce dishes with biodegradable ones, switched to reusable and recyclable cups instead of single-use plastic cups, and substituted plastic takeout bowls with paper ones. Additionally, RWS transitioned from standard bins to Smart Bins that compress waste, reducing the need for plastic garbage bags. These efforts resulted in significant savings of packaging materials and plastic. F&N Foods redesigned their Ice Mountain Water PET bottles, reducing the packaging material by 2.5 g per bottle, leading to a reduction of 57 tonnes of plastic material per year. Similarly, their Ice Mountain Sparkling Water PET bottles underwent a new design, saving 4g of packaging material per bottle and resulting in a yearly reduction of 3.09 tonnes of material. Starbucks Coffee Singapore Pte Ltd incorporated recycled materials in their packaging, using plastic bags composed of 95% recycled or post-consumer fibre. They also actively engaged customers through their "Say YES to Waste Less" campaign, promoting the use of reusable cups and offering discounts for customers who brought their own tumblers and mugs. These efforts reached approximately 3 million individuals and contributed to the utilization of 103.9 tonnes of recycled plastic material in their carrying bags annually.

In terms of recycling efforts in Singapore, the National recycling plan initiated in 2001 mandated recycling containers and collection services in designated residential areas. Despite these efforts, the domestic recycling rate decreased over the years. To address this, the NEA launched the Recycle Right 2022 campaign, aimed at increasing public awareness of proper recycling practices. The campaign involved updating labels on recycling bins and chutes, distributing educational materials through schools, and introducing a mascot named Bloobin to raise awareness through augmented reality filters on Instagram. Overall, these various initiatives and collaborations between businesses, organizations, and the government have made substantial progress in minimizing packaging waste and promoting recycling in Singapore.

2.7 THAILAND

Thailand has witnessed a significant transformation in its plastic consumption culture over the past 50 years. Analysis of a sediment core retrieved from the Gulf of Thailand in the 1950s revealed the absence of microplastics, indicating that plastic usage began to emerge in the country during the 1960s. Subsequently, from the

1970s to the 1980s, Thailand experienced a sharp increase in plastic consumption, aligning with the global trend of widespread use of single-use plastic bags. However, a concerning aspect is that 80% of Thailand's marine plastic pollution is a result of inadequately managed land-based waste. According to the Pollution Control Department (PCD), a division of the Ministry of Natural Resources and Environment (MONRE), 26% of the waste generated in Thailand's 23 coastal districts, totalling 1.3 million tonnes, flowed into the ocean in 2016 (Marks et al., 2020). Furthermore, research conducted by Thailand's Department of Coastal and Marine Resources in 2017 revealed the composition of marine plastic debris: 12% polystyrene foam, 10% plastic wraps, 24% plastic bags, 20% plastic bottles, 18% glass bottles, 6% plastic caps, and 5% plastic straws (Marks et al., 2020).

a) Limitations on International Imports of Plastic Scraps

On May 25, 2021, the Pollution Control Department (PCD) released a notification titled "Notification of Pollution Control Department Regarding the Definitions of Plastic Wastes and Plastic Scraps B.E. 2564 (2021)" (Leungsakul, 2017). This notification established distinct definitions for plastic waste and plastic scrap, according to the PCD. Plastic waste encompasses any plastic object or its fragments that have been discarded after use, degraded to a non-functional state, or contaminated with other types of waste. On the other hand, plastic scrap refers to plastic pieces or useless residue. Thailand strictly imports "plastic scraps" as raw materials for the industrial sector, adhering to the customs tariff regulation on plastic importation, and does not permit the importation of "plastic wastes." In 2020, Thailand imported 150,807 tonnes of plastic scraps, and between January and April 2021, an additional 44,307 tonnes were imported (Leungsakul, 2017). The country is currently undergoing an adjustment period to avoid disrupting operators and gradually imposing restrictions on plastic scrap imports. The ultimate goal is to achieve a complete ban on the importation of plastic scraps within five years.

To facilitate efficient recycling processes of plastic waste within the country and promote its use as a high-quality raw material, which is in increasing demand in various industrial sectors, the government has fostered collaboration between public and private sectors to enhance public awareness of plastic waste separation. The Minister of Natural Resources and Environment, as the chairman of the Plastic and Electronic Waste Management Subcommittee under the National Environment Board, has expressed the Ministry's strong commitment to maximizing the utilization of plastic waste within the country (Leungsakul, 2017). Over the next five years, the government plans to completely prohibit the importation of foreign plastic scraps, as it has never had a policy of importing plastic garbage. Any unlawful smuggling of such materials will be addressed, with the Customs Department and affiliated organizations responsible for ensuring the prompt return of containers to their place of origin.

b) Development of Roadmap on Plastic Waste Management

During a cabinet meeting on April 17th, 2018, the Prime Minister of Thailand instructed the Ministry of Natural Resources and Environment to collaborate with all sectors and develop a set of action measures aimed at preventing and minimizing plastic waste throughout its entire lifecycle (Envilliance, 2018). Subsequently, the National Environment Board established a subcommittee dedicated to plastic waste management. Under the leadership of the subcommittee's secretary, a working group was formed to implement mechanisms for managing plastic waste. The working group drafted a comprehensive plastic-waste-management strategy spanning from 2018 to 2030, serving as a framework and guidance for addressing Thailand's plastic-waste problem (Envilliance, 2018). To create this roadmap, the working group sought the involvement of government agencies, the private sector, non-governmental organizations, international organizations, educational institutions, and other relevant parties. They conducted three meetings with the working group, three meetings with plastics industry entrepreneurs, and five meetings with the plastic-waste-management subcommittee to gather input and perspectives (Envilliance, 2018). The draft version of the roadmap was presented to and approved by the National Environment Board. Finally, during a cabinet meeting on January 4th, 2019, the strategy was officially acknowledged (Envilliance, 2018). Subsequently, based on feedback from stakeholders, the roadmap underwent revisions. On April 17th, 2019, the cabinet approved "the roadmap on plastic waste management, 2018–2030" (Envilliance, 2018). This plan has since served as a policy framework to tackle Thailand's plastic waste problem and aims to limit and eliminate plastic usage while promoting environmentally friendly alternatives. The roadmap is divided into three distinct phases.

The first phase commenced in 2019 with the prohibition of three plastic products: cap seals (plastic that covers bottle tops), OXO-degradable plastic, and microbeads. In the second phase, the use of four additional categories of single-use plastic will be banned in 2022, including Styrofoam food boxes, plastic straws, thin plastic bags with a thickness of less than 36 microns, and single-use plastic cups (Buakamsri, 2021). A widely circulated guide pamphlet has been disseminated through various media channels. The third phase of the roadmap, scheduled for completion by 2027, aims for 100% of plastic waste to be reusable. Furthermore, three action plan measures have been implemented, as indicated in the Table 2.6.

In late 2019, initiatives were launched through traditional and online media platforms to promote public engagement in reducing and eliminating plastic usage. The Ministry of Natural Resources and Environment foresaw that these campaigns would lead to a decrease in the amount of plastic waste by 0.78 million tonnes annually and generate savings of 3.9 billion baht in waste management costs each year. Furthermore, this strategy would contribute to a reduction in greenhouse gas emissions, amounting to 1.2 million tonnes of CO_2 equivalent, and help preserve 1,000 acres of landfill space (Jangprajak, 2021).

2.7.1 Progress in Thailand

According to a report by the Customs Department, the quantity of plastic waste entering the country in 2018 exceeded 500,000 tonnes, which is ten times higher than the pre-2015 figures. Due to the significant increase in plastic waste inflow, the responsible authorities for integrated e-waste and plastic waste management

TABLE 2.6
The Action Plan's Measures for Roadmap (Jangprajak, 2021)

Action plan	Measures
Plastic waste reduction at the source	• Minimize single-use plastics. • Eco-Design of packaging. • Use alternatives to replace single-use plastics. • Establish plastic product standard. • Green procurement. • Encouragement of eco-investment. Developing a plastic database. • Tax subsidies to encourage the use of biodegradable plastic packaging.
Reduce the consumption of single-use plastic	• Promote green consumption through education and outreach. • Stakeholders should work together to eliminate single-use plastics. • Establish laws, regulations, and procedures to prevent marine littering. • Develop a policy for the management of plastic trash through international collaboration.
Reduce the consumption of single-use plastic	• Promote green consumption through education and outreach. • Stakeholders should work together to eliminate single-use plastics. • Establish laws, regulations, and procedures to prevent marine littering. • Develop a policy for the management of plastic trash through international collaboration. Promote green consumption through education and outreach. • Stakeholders should work together to eliminate single-use plastics. • Establish laws, regulations, and procedures to prevent marine littering. • Develop a policy for the management of plastic trash through international collaboration.
Plastic waste management after consumption	• The municipal government issued laws and regulations for garbage separation based on the 3R concept. • Actively support a Circular Economy and waste to energy. • Capacity development for the informal sector and disposal buyers. • Create a law to avoid/resolve the issue of marine plastic trash. • Regulate the import of foreign plastic scraps.

had planned to implement a ban on plastic waste imports by September 2020 (Rujivanarom, 2021). However, despite the proposed ban, plastic waste continued to be imported into the country in subsequent years, surpassing the deadline. Statistics from the Customs Department reveal that, between 2017 and 2020, the total amount of shipped plastic waste exceeded 1 million tonnes. Furthermore, within the first five months of 2021 alone, over 58,000 tonnes of plastic waste were shipped to Thailand, indicating that the issue is far from being resolved (Rujivanarom, 2021).

The subcommittee has revised its policy to allow for an extension of the original deadline by gradually reducing the import of plastic waste by 20% each year until a complete ban is implemented in 2026. This decision is based on the recognition that the country's recycling industry still requires a significant amount of imported plastic waste, estimated at up to 685,190 tonnes per year (Rujivanarom, 2021). Environmentalists and municipal plastic waste collectors have launched a campaign to halt plastic waste imports by the end of 2021. The Ecological Alert and Recovery Thailand (EARTH) has expressed dissatisfaction with the current staged ban strategy because the government should immediately prohibit the importation of plastic waste to align with Thailand's goal of recycling 100% of plastic waste by 2027, as outlined in the plastic-waste-management plan (Bangkok Post Public Company, 2021). Furthermore, she refutes the notion that Thailand is compelled to import cleaned plastic waste due to a lack of sufficient and low-quality plastic waste within the country (Bangkok Post Public Company, 2021). Ms Saetang also points out that the plastic recycling industry is highly polluting, and relying on imported plastic waste is akin to allowing others to litter in one's own home.

While acknowledging that the quantity of household plastic waste generated in Thailand is substantial enough for the recycling industry, Ms Saetang notes that the waste-sorting and -collection systems are still inadequate for effectively separating recyclable plastic waste (Bangkok Post Public Company, 2021). The Pollution Control Department estimates that over 2 million tonnes of plastic waste are generated annually in Thailand, with only a quarter of that amount (approximately 500,000 tonnes) being properly collected and recycled. Additionally, the Ministry of Sector has reported an increased demand for recyclable plastic waste from the recycling industry, reaching 680,000 tonnes this year (Rujivanarom, 2021).

2.8 SOUTH KOREA

Korea has a high energy-consumption society, with 90% of its mineral resources and 97% of its energy being imported. In 2013, Korea imported raw materials worth an average of KRW 1 trillion per day, totalling KRW 371 trillion annually. According to Statista's research, South Korea produced approximately 1.77 million metric tonnes of engineering plastic resin in 2018, an increase from 1.70 million metric tonnes in 2017 (Sea Circular, 2019).

a) Extended Producer Responsibility (EPR)

Korea was one of the early adopters of the Extended Producer Responsibility (EPR) system in the waste-management industry. The EPR system was introduced in 2003 and required producers and importers to recycle a specified percentage of their products. Initially, the policy covered 15 products, but by 2008, it expanded to include 24 items, including packaging materials. The Korea Environment Corporation oversees the EPR system, ensuring compliance with reporting requirements and waste collection and recycling data. The federal government develops and enforces EPR rules, while local governments focus on effective waste collection, recycling, and reuse. Private recycling collectors are employed to collect and sell recyclable waste to recycling facilities. The EPR system includes labelling solutions to improve monitoring and provide information on recyclability and proper disposal. The Ministry of Environment proposed a partial amendment to expand the EPR system's scope in July 2021, and public input was invited until August 2021. The proposed amendment includes adding 17 additional products to the mandatory recycling list, with a phased implementation starting from 2022 (Kengo, 2021).

b) Framework Act on Resource Circulation (FARC)

The Framework Act on Resource Circulation (FARC) was implemented in 2018 to transform Korea into a resource-circulating society that maximizes resource efficiency and addresses energy and environmental issues. The FARC aims to shift from a production-oriented and wasteful economic structure to a sustainable and efficient resource-circulation model. It establishes a waste hierarchy and assigns responsibilities to key stakeholders, emphasizing waste reduction, reuse, recycling, energy recovery, and appropriate disposal. The FARC is particularly concerned about hazardous chemicals and provides provisions to assess their circular utilization and potential risks to human health and the environment. The law includes two main programmes: the Recyclable Resource Recognition Program (RRRP) (refer to Figure 2.3) and the Resource Circulation Performance Management Program (RCPMP). The RRRP recognizes waste products that meet specified criteria as "recyclable resources," exempting them from general waste regulations. Businesses involved in the collection, transportation, recycling, and distribution of recyclable resources face fewer regulatory constraints and can participate more effectively in the market for recycled products. The RRRP designation can be revoked if the waste products no longer meet the safety criteria. The FARC provides guidelines for waste restrictions on items that lose their recyclable resource classification (Kim and Jang, 2022).

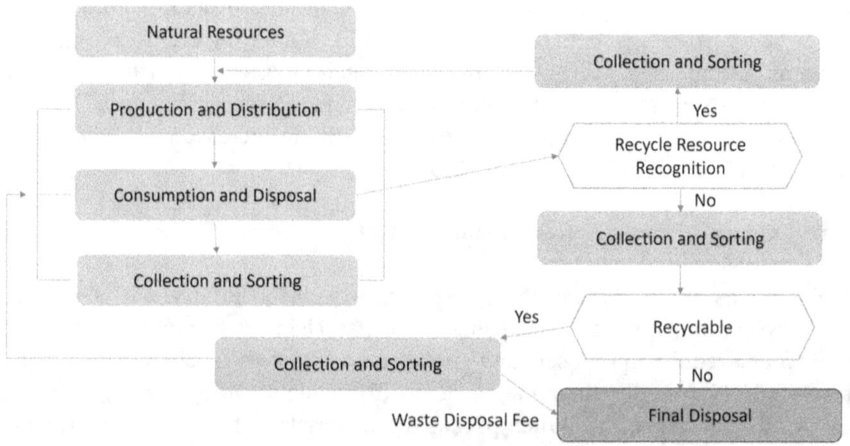

FIGURE 2.3 Summary of Recyclable resource Recognition Program (RRRP).

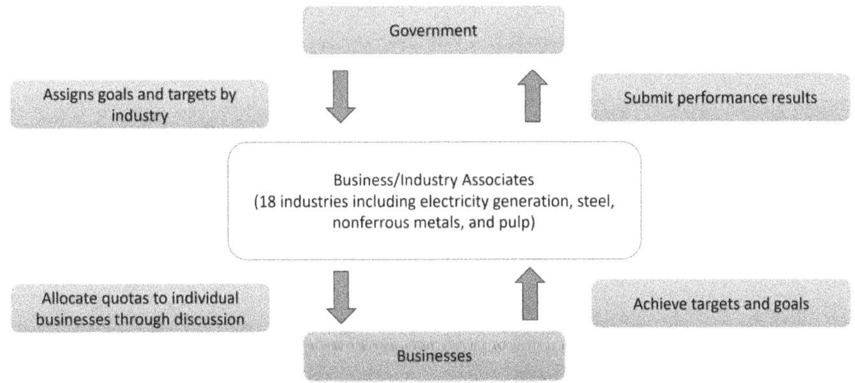

FIGURE 2.4 Summary of RCPMP (Kim and Jang, 2022).

As per the Framework Act on Resource Circulation (FARC), it is the responsibility of the Minister of Environment to regulate waste generation and promote resource circulation. This is achieved by setting national goals for final waste disposal, resource circulation, and energy recovery rates in the medium to long term and through phased implementation, as outlined in Article 14 (Kim and Jang, 2022). Municipal and provincial governments, along with businesses, are required to establish their own targets for final waste disposal and resource rotation rates. The effective implementation of the Resource Circulation Performance Management Program (RCPMP) is crucial for these provisions (refer to Figure 2.4). Under the RCPMP, the Ministry of Environment incentivizes businesses to improve their recycling performance by setting resource circulation goals for different sectors and providing financial and technological support to encourage their achievement. The RCPMP establishes mandatory recycling quotas for enterprises and organizations that generate significant amounts of waste, and their recycling performance is evaluated based on these quotas. The overview of the RCPMP is depicted in the following figure.

c) National Marine Litter Management Plan

In May 2021, the South Korean Ministry of Oceans and Fisheries (MOF) introduced the 1st Framework on Marine Debris Management for the period of 2021 to 2030. This framework outlined comprehensive policies and implementation strategies for effective marine debris management over the next decade (CEARAC, 2020). Until now, marine debris management in South Korea has primarily been governed by the Marine Environment Management Act, which focused mainly on collection and disposal. The "3rd National Marine Litter Management Plan 2019–2023" was established under this act in 2019 (CEARAC, 2020). The 3rd Management Plan aimed to assess the previous "2nd National Marine Litter Management Plan (2nd Management Plan) (2014–2018)" by analysing global and domestic trends in marine litter management and evaluating the extent of domestic marine litter (CEARAC, 2020). The evaluation of the second management plan contributed to the formulation of goals and strategies for the third management plan and guided its implementation.

The framework consists of five key implementation methods and 16 initiatives, with the objective of improving the entire life cycle of marine debris, including its generation, collection, and disposal, while fostering greater collaboration with relevant agencies. The strategy aims to reduce 60% of the total influx of marine plastic waste by 2030 and achieve zero influx by 2050 (CEARAC, 2020). The first strategy is focused on prevention and involves implementing a deposit system for buoys and fishing gear to address the root causes of debris production. It also includes expanding the distribution of eco-friendly buoys, preventing land debris from entering the oceans, and establishing preventive measures against overseas marine debris through international organizations and bilateral partnerships. Additionally, the second strategy emphasizes an improved collection and delivery system to address blind spots in debris collection. The Ministry plans to deploy seven cleaning vessels in island areas, establish an adequate number of drop-off stations for collected marine debris, and develop a responsive mechanism to handle large influxes of debris caused by natural disasters like typhoons and floods. The MOF also collaborates with relevant authorities to enhance the overall collection process and has implemented a prediction system that utilizes ICT, artificial satellites, and drones to anticipate the production and movement of maritime trash (CEARAC, 2020). Strategy 3 focuses on promoting the concept of recycling. It involves developing and enhancing the management infrastructure for marine trash through improved lifecycle installation and the capabilities of relevant authorities. The Ministry is designing an energy system specifically tailored to islands and fishing villages, conducting pilot projects for distributed models, and working towards the establishment of circular economy towns. Strategy 4 aims to strengthen the management foundation by establishing and operating an inter-ministerial committee at the government level to address marine pollutants. The governing authority for marine debris management is also being specialized with expanded expertise. Extensive studies are being conducted to address concerns about the impact of microplastic waste on marine ecosystems, and comprehensive countermeasures are being developed to consider all potential consequences and types of microplastics. Lastly, Strategy 5 focuses on raising public awareness about marine litter. The government, local authorities, and relevant agencies are encouraged to collaborate and engage in various public participation initiatives. Educational programmes and promotional materials targeting different social groups such as citizens, fishermen, and students have been designed and distributed to enhance public awareness (Ministry of Oceans and Fisheries, 2016).

2.8.1 Progress in South Korea

Through the implementation of environmental protection regulations, South Korea has effectively reduced the consumption of raw materials by promoting the reuse of discarded items that would otherwise be incinerated or sent to landfills. Since 2003, the Extended Producer Responsibility (EPR) system has played a crucial role in the recycling of various packaging materials, including plastic packaging. Recycling targets set by the EPR system have consistently been met, with recycling rates exceeding 100% in several years. In 2003, the recycling requirement was fulfilled completely, followed by 104% in 2004, 123% in 2006, 115% in 2007, and 113% in 2008, demonstrating consistent performance above the set targets. This trend indicates a steady increase in recycling rates, with consistently high levels of recycling being achieved. The quantity of packaging materials, including plastic films introduced in 2004, experienced a significant increase until 2005, rising from 642,000 tonnes per year to 853,000 tonnes per year. Subsequently, the quantity remained relatively stable, reaching 804,000 tonnes per year in 2006, 867,000 tonnes per year in 2007, and 865,000 tonnes per year in 2008 (Jang et al., 2020). This indicates that significant increases were observed until 2005, followed by a period of maintenance at similar levels thereafter.

Since the implementation of the Extended Producer Responsibility (EPR) system, the volume of recycled materials has increased by approximately 46%, rising from 928,000 tonnes in 2001 to 1,368,000 tonnes in 2008. This significant increase can be attributed to the expansion of EPR recycling products and the development of recovery and recycling infrastructure (Kim, 2012). The establishment of collection and transportation systems, along with advancements in recycling technology, has notably improved the recycling rates for packaging materials such as PET and Styrofoam. Previously, plastic film was primarily incinerated or sent to landfills, but since the introduction of the EPR system in 2004, 75% of total waste in this category has been recycled (Kim, 2012). The recycling of EPR goods has generated an estimated value added of around 1.7 trillion won, while also saving approximately 1.9 trillion won in processing fees for incineration and landfilling (Kim, 2012). Furthermore, the implementation of the EPR system has led to an increase in the production of eco-friendly products by companies. According to the Environmental Protection Agency, the number of products with environmental certifications has witnessed a nearly 18-fold increase between 2001 and 2008, rising from 326 products to 6,005 products. Additionally, the number of products certified as "Good Recycled" has grown by approximately 48%, from 166 goods to 245 goods (Kim, 2012).

The growth of the recycling infrastructure has been supported by the establishment of five regional Home Appliances Recycling Centres, Refuse Plastic Fuel Facilities, and mutual assistance association recycling facilities. The functioning of recycling project mutual help groups has also facilitated the initiation of recycling processes. Through the establishment of mutual assistance societies, small recycling enterprises have been able to manage excessive competition and improve the quality of recyclable materials. Furthermore, the development of recycling-assistance funds has resulted in an increase in the number of recycling enterprises, with funding rising from 29.3 billion won in 2001 to 43.9 billion won in 2008 (Kengo, 2021). Long-term, low-interest loans for the development of recycling facilities and technology, along with government-sponsored recycling-related research and development projects, have effectively stimulated the recycling sector. As a result, the scope of the recycling market has expanded from 47.2 billion won between 2000 and 2010 (Kengo, 2021).

2.9 INDONESIA

Indonesia is currently ranked as the second-largest contributor to plastic waste in the world's oceans, with an estimated 3.22 million tonnes of plastic garbage being dumped annually. This alarming statistic highlights the significant role Indonesia plays in the global plastic trash crisis. Moreover, it is not just the oceans that are affected by this plastic garbage problem; Indonesian rivers have also been heavily impacted. According to data published in Nature Communications, four of Indonesia's rivers, namely Brantas, Solo, Serayu, and Progo, are among the top 20 most contaminated rivers worldwide. The severity of this issue is further underscored by scientists' predictions that, by the year 2050, the weight of plastic in the oceans will surpass that of fish.

a) National Plastic Action Partnership (NPAP)

In March 2019, Minister Luhut B. Pandjaitan, the Coordinating Minister of Maritime Affairs and Investment in Indonesia, launched the National Plastic Action Partnership (NPAP) as part of the Global Plastic Action Partnership (GPAP). This initiative involves collaboration between businesses, civil society groups, and local stakeholders, with the aim of establishing a public-private partnership to address plastic pollution in Indonesia. With over 150 member organizations, the NPAP serves as Indonesia's primary platform for fostering collaboration and driving positive change towards a plastic-pollution-free country. The GPAP's objective is to accelerate Indonesia's transition to a circular economy by identifying viable investment ideas that can be recognized and replicated globally. As part of its ambitious national effort, Indonesia aims to reduce marine plastic litter by 70% and tackle solid-waste issues. The GPAP is hosted by the World Economic Forum and supported internationally by the governments of Canada and the United Kingdom, as well as prominent companies like The Coca-Cola Company, PepsiCo, Nestlé, and organizations including the World Resources Institute, the World Bank, Pew Charitable Trusts, and SYSTEMIQ (Yann, 2019).

Through the NPAP, the Indonesian government has set three key goals. Firstly, in collaboration with Indonesia's National Waste Management Policy and Strategy (Jakstranas), the National Action Plan on Marine Debris, and other initiatives, the NPAP aims to reduce marine debris leakage by 70% by 2025. Additionally, by implementing the NPAP Multistakeholder Action Plan, Indonesia aims to prevent 16 million tonnes of plastic waste from entering the ocean by 2040. Moreover, the suggested system reform initiatives under the NPAP are expected to contribute to Indonesia's economic development by creating over 150,000 direct jobs (GPAP, 2020).

b) Ecolabelling Scheme

The Indonesian Eco-label Logo and the Eco-label Accreditation and Certification Scheme were launched on June 5, 2004, which coincided with World Environment Day. The term "Ekolabel" was originally introduced by the Indonesian Ministry of Environment in a regulation that addressed environmental management, ecolabelling, clean manufacturing, and environmentally-oriented technology. Indonesia has established a national Ecolabel Scheme, under which products can receive a Type I Ecolabel known as "*Ramah Lingkungan*," meaning "environmentally friendly" (Figure 2.5) The Indonesian Eco-label, also referred to as Ekolabel Indonesia, aims to serve as an effective tool for environmental protection, human well-being, and the enhancement of product efficiency and competitiveness.

FIGURE 2.5 Type I Ecolabel "*Ramah Lingkungan*" (Razif and Persada, 2016).

FIGURE 2.6 Type II Ecolabel (Razif and Persada, 2016).

Indonesia's Environmental Protection and Management Act No 32/2009 seeks to enhance environmental quality by implementing measures to mitigate pollution and degradation and by shifting the focus from reactive approaches to preventive ones (Razif and Persada, 2016). As a result, the Indonesian Ecolabel Program is designed with several objectives in mind: promoting the integration of environmental impact reduction throughout the product lifecycle, fostering the demand and supply of environmentally friendly products, providing proactive guidance to industries for improving their products, and educating and assisting consumers in understanding and identifying environmentally friendly products.

The ecolabel program adopts technical references such as ISO 14020 General Principles of Environmental Labels and Declarations, ISO 14024 Guideline of Ecolabel Type I, and ISO 14021 Environmental Labels and Declarations – Self-Declared Environmental Claims (Type II Environmental Labelling) for its implementation (Figure 2.6). In Indonesia, this ecolabel can be found on retail items. The ecolabel certification criteria involve rigorous technical assessments of the environmental aspects of products throughout their life cycles. As of October 2013, eco-label standards have been established for 12 product groups, including plastic shopping bags (Razif and Persada, 2016).

The verification agencies authorized by the Ministry of the Environment are responsible for verifying the claims made by companies regarding their pro-environmental achievements. To qualify for displaying the ecolabel symbol on their products, manufacturing businesses must meet the prescribed ecolabel certification requirements. These requirements encompass a range of environmental management criteria, including the implementation of an environmental management system, the adoption of quality and standardized products, and the use of environmentally friendly packaging. During the certification process, all the necessary criteria are used as assessment factors for the ecolabel.

c) Waste Management Act (Law Number 18 of 2008)

The increasing diversity of waste, particularly hazardous packaging waste that poses challenges for natural decomposition, is a result of people's consumption patterns. On May 7, 2008, President Doctor Haji Susilo Bambang Yudhoyono approved Law Number 18 of 2008 on Waste Management in Jakarta. This law was subsequently published in the State Gazette of the Republic of Indonesia, Number 69, to raise awareness among the Indonesian population. The Supplement to the State Gazette of the Republic of Indonesia Number 4851 provides an explanation of the law. Law Number 18 of 2008 establishes the main policy framework for integrated and comprehensive waste management as well as defining the rights and obligations of the community and the duties and authority of the government and regional governments in delivering public services. The legal control of waste management, as stipulated by Law 18 of 2008, is guided by principles such as responsibility, sustainability, fairness, awareness, togetherness, safety, security, and economic value. Article 28H paragraph (1) of the Republic of Indonesia's 1945 Constitution recognizes everyone's right to a pleasant and healthy living environment. Consequently, the government is obligated to provide public waste management services, thereby becoming the authorized and accountable entity in the waste management sector, even though it may collaborate with businesses. Additionally, waste management activities can be carried out by solid-waste organizations and community groups operating within the solid waste sector (Jogloabang, 2019).

2.9.1 Progress in Indonesia

Indonesia is encountering challenges in meeting its waste management obligations at the local level. In 2010, the country generated approximately 85,000 tonnes of waste per day, a number projected to rise to 150,000 tonnes per day by 2025, indicating a 76% increase within a decade. Household waste accounts for around 40% of the total solid waste produced, primarily due to inadequate collection services and limited inter-municipal collaboration in certain regions. Consequently, local governments are emphasizing the need to establish an integrated waste management system that encompasses collection, recycling, and proper disposal (Akenji et al., 2020).

Insufficient capacity within local governments, coupled with a lack of confidence in the public sector, poses risks to the private sector and hampers essential investments in solid-waste management. The large number of informal collectors and the predominant role of the informal sector in recycling present significant challenges for local governments in implementing an effective system for separate waste collection. However, some cities, such as Banjarmasin City and Bogor City, have already implemented comprehensive bans on plastic bags, resulting in significant reductions in their usage. Banjarmasin City achieved an 80% reduction in plastic waste since the policy's introduction in 2016, while Bogor City saw a monthly reduction of 41 tonnes of disposable plastic bags from July 2018 (Ricky, 2019). Bogor City plans to further strengthen these regulations in the future.

Private sectors in Indonesia have also taken action to address the plastic issue, driven by its severity and advancements in technology for product design and waste recycling. Organizations like PRAISE, a non-profit dedicated to sorting and recycling packaging waste, promote recycling, public education, and awareness initiatives in the country (Borongan and Kashyap, 2018). Multinational corporations such as Unilever and Veolia have recognized the gravity of the problem and have entered into partnership agreements on sustainable packaging, initially focusing on Indonesia and India. These agreements involve the implementation of collection systems for used packaging, expanding recycling capacity, and establishing circular business strategies (Borongan and Kashyap, 2018). In addition, SC Johnson has initiated a pilot project in eight Indonesian towns, using blockchain technology in collaboration with Canadian firm Plastic Bank to collect and recycle plastic. This project enables people to exchange plastic waste for digital tokens at local recycling centres. Furthermore, companies like Henkel and Marks & Spencer have incorporated recycled social plastic into their products and packaging (Akenji et al., 2020).

2.10 TAIWAN

In Taiwan, the government has taken steps to reduce plastic waste and limit the use of disposable plastics, in accordance with global efforts. The Environmental Protection Administration (EPA), the central authority responsible for environmental matters, introduced a system to regulate and recycle certain types of plastic waste, such as plastic containers, from everyday garbage. This system was established based on the Waste Management Act (WMA) and has undergone various updates since 1997 (Tsai, 2022). In fact, Taiwan is one of the Asian countries which has comprehensive regulation on plastic-waste management, which was covered in the Waste Disposal Act as early as year 2006. Taiwan has announced regulation on the ban to use single-use plastic (Article 21) with a further amendment in 2023 (Taiwan Ministry of Environment, 2023a). This ban covers the government sector, schools, healthcare, retail supermarkets, distributors, fast food chains, and restaurants. No cup, bowls, plates, boxes, lids and plastic films are allowed. However, biomass made of packaging such as wood, sugarcane, rice husk, rice straws, etc., using natural fibres containing less than 10% of plastic are not limited. Meanwhile, single use plastics are still allowed for special circumstances such as water supply interruption, transmittable diseases, or the breakdown of the cleaning and sanitation system; the owners of businesses are required to notify the local authority for approval within 36 hours. Taiwan also introduced specific regulations on the single-use beverage cups which impose that at least NTD 5 need to be charged to the consumer for any demand of single-use beverage cups. Business owners need to put up signage about the charges on the premises; the wording of the signage needs to be at least 5 cm, while consumers have the alternative to borrow reusable cups for usage, and an upfront fee will be collected for borrowing purposes. The implementation performance, as mentioned earlier, needs to keep a record and report; the target of cup consumption needs to be reduced for 15, 18, and 25% for the years of 2023, 2024, and 2025, respectively (Taiwan Ministry of Environment, 2023b). In addition, Taiwan also has specifically banned single-use plastic straws and plastic carrier bags, starting in 2019. Plastic carrier bags are banned for government sectors, schools, supermarket, distributors, chain stores, fast food stores, healthcare appliances, home appliances, book stationery, bakeries, and beverages. However, plastic carrier bags are still allowed for the packing of seafood, meats, and vegetables, pharmacies for drugs packing, and packing for final products in factories (Taiwan Ministry of Environment, 2023c). Taiwan's government also initiated a regulation on Excessive Product Packaging Restrictions in 2005, according to which packaging needs to follow the allowable packaging ratio to minimize unnecessary plastic used for packing (Taiwan Ministry of Environment, 2023d). A latest regulation has been issued and will be implemented in Taiwan, regulating single-use tourism consumption and toiletries. This regulation will be enforced starting 2025, including a ban on any liquid personal-care products supplying less than 180 ml. The service provider cannot display uncontrollable usages of personal care items on the premises unless charges are imposed. This is to ensure the responsibility of tourists while promoting using self-prepared toiletries to minimize wastage due to negligence (Taiwan Ministry of Environment, 2023e).

The 4-in-1 Recycling Program, initiated by the EPA, involves different parties, including the public, local authorities, recycling companies, and a recycling fund. The program operates on the principle of extended producer responsibility, which means that manufacturers and importers of products provide funding for the program, which, in turn, supports the collection and recycling efforts carried out by licensed companies. As part of this program, citizens are required to separate their household waste into recyclable items, non-recyclable items, and organic waste (such as food scraps). The program also encourages recycling by allowing only non-recyclable items to be placed in special plastic bags. In Taipei city, citizens need to purchase these bags, and the price increases with the size of the bag.

According to data from the EPA, the overall recycling rate in Taiwan has significantly improved over the years, increasing from 5.8% in 2000 to 58.9% in 2020 (Tsai, 2022). However, plastic waste still accounts for 15–20% of the total non-recyclable garbage, indicating that plastic products, including single-use plastics, are still being excessively used in Taiwan. In response to this issue, the EPA has been implementing restrictions on the use of plastic bags since 2006 and has continued to introduce further limitations on plastic products, under the authority granted by the Waste Management Act (Tsai, 2022).

The legal framework for these restrictions on plastic products in Taiwan is established by the Waste Management Act and the Resource Recycling Act. These acts empower the EPA to prohibit or restrict the manufacturing, import, sale, and use of plastic items, packaging, and containers in order to address concerns related to environmental pollution. The EPA revised the Waste Management Act in October 2001 to strengthen the legal principles for restricting plastic product usage, in line with international environmental initiatives such as the Sustainable Development Goals (SDGs) and the need to combat marine pollution (Tsai, 2022).

Before officially implementing the regulations on plastic product restrictions, the EPA announced a policy in May 2002 to limit the use of plastic shopping bags and plastic disposable dishes. The policy was implemented in two stages. The first stage targeted government agencies, schools, hospitals, state-owned enterprises, and military organizations, which were required to restrict the use of plastic shopping bags and disposable dishes by specific dates in 2002. The second stage, starting from January 2003, extended the restrictions to businesses such as department stores, supermarkets, convenience stores, and fast-food chains. The EPA conducted a pilot program from 2003 to 2005 to evaluate the effectiveness of the restrictions and formally established the relevant regulations in 2006. To summarize, the EPA in Taiwan has been taking measures to reduce plastic waste and restrict the use of plastic products (refer to Table 2.7 and Table 2.8). These efforts are in line with international policies and aim to address environmental concerns. The EPA has introduced recycling programmes, encouraged citizen participation, and implemented regulations to promote the responsible use of plastic and minimize its negative impact on the environment (Tsai, 2022).

2.10.1 Progress in Taiwan

In the 1990s, Taiwan emerged as a major global hub for plastic-product manufacturing, driven by cost reduction and the need to mitigate the spread of diseases like hepatitis (Tsai, 2022). To address the environmental impact of plastic bags, Taiwan

TABLE 2.7
Chronology of Recycling Policies of Municipal Solid Waste in Taiwan

Implementation year	Main policies	Comments
1988	Extended producer responsibility (EPR) incorporated	Incorporated into the Waste Management Act (WMA).
1997	4-in-1 Recycling Program announced	A special feedback mechanism covers the public community, local authorities (cleaning teams), recycling enterprises, and recycling fund. The fees are collected from responsible manufacturers and importers.
2000	Trash collection fee per bag	Only implemented in Taipei city.
2001	Food waste (kitchen waste) recycling	Mainly reused as pig feed and organic compost.
2003	Bulk waste recycling	Bulk waste mainly includes discarded furniture, which may still be reused after minor repairs.
2005	Compulsory MSW sorting	MSW must be sorted into garbage (general waste), food waste, recyclable waste, and bulk waste.
2010	Sustainable material management (SMM) introduced	Using the cradle-to-cradle principle by lifecycle assessment and material flow analysis.
2017	Diversified MSW Treatment Program	Focusing on waste-to-energy promotion and food waste-to-biopower.

Source: Adapted from Tsai, 2022, with permission MDPI.

TABLE 2.8
Major Regulations on Plastic Product Use in Taiwan

Regulation	Announcement date	Latest revision announcement date	Effective date
Plastic Shopping Bag Restriction Targets, Implementation and Date of Implementation	June 9, 2006	August 15, 2017	January 1, 2018 (Penalty started January 1, 2019)
Plastic Disposable Tableware Restriction Targets, Implementation, and Date of Implementation	June 9, 2006	August 8, 2019	August 8, 2019
Restriction on the Use of Plastic Pallets and Packaging Boxes	March 28, 2007	December 23, 2011	January 1, 2012
Ban on Manufacturing, Import and Sale for Cosmetic and Personal Care Products Containing Plastic Microbeads	August 3, 2019	–	January 1, 2019 (Ban on Manufacturing, Import) July 1, 2019 (Ban on Sale)
Single-Use Plastic Straw Restriction Targets and Implementation	May 8, 2019	–	July 1, 2019 (Penalty started July 1, 2020)

Source: Adapted from Tsai, 2022, with permission MDPI.

implemented a plastic bag levy in 2002, resulting in a significant reduction of over 59% in plastic bag usage. Taiwan also developed a successful circular-economy model through its 4-in-1 Recycling Program, which started in 1997. Recent data from 2018 to 2020 revealed an increase in municipal solid-waste (MSW) generated by employees in various industries, with a rise from 7,871 thousand metric tonnes in 2017 to 9,741 thousand metric tonnes in 2018 (Tsai, 2022). The amount of recyclable waste, including various containers, such as metal, glass, Tetra Pak, paper, and plastics, also saw an upward trend. The reported amounts of plastic containers and other plastic products recycled by implementing agencies in Taiwan have consistently increased since 2016, unaffected by the COVID-19 pandemic. However, there has been a slight increase in the composition of plastics in garbage in recent years. This shift may be attributed to a decline in plastic-product recycling and the promotion of waste-to-power programmes under the "Diversified MSW Treatment Program." Despite regulatory measures in place for 15 years, the impact of restrictions on plastic product usage in Taiwan has not been significant. Figure 2.7 also revealed a notable increase in chlorine content in 2020, indicating the generation of kitchen waste containing salt, household waste with disinfectants (e.g., chlorine dioxide, sodium hypochlorite), or products made of PVC during the COVID-19 pandemic. Consequently, the incineration of MSW may lead to the release of higher amounts of toxic compounds such as dioxins (Tsai, 2022).

In summary, Taiwan's plastic product manufacturing industry has grown significantly, and efforts have been made to reduce plastic bag usage and promote recycling through the 4-in-1 Recycling Program. However, there has been a moderate increase in the proportion of plastics in garbage, possibly due to decreased recycling

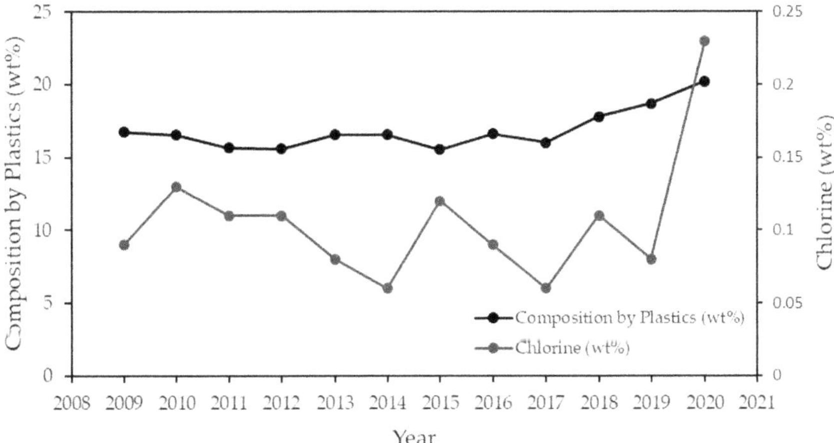

FIGURE 2.7 Variations in garbage compositions by plastics and elemental chlorine during the period of 2009–2020 [Reference Tsai, 2022]

and waste-to-power initiatives. The impact of plastic product usage restrictions in Taiwan has been limited, and the COVID-19 pandemic has contributed to an increase in chlorine content in waste, potentially leading to higher emissions of toxic compounds during incineration.

2.11 AUSTRALIA

In 2021, the Australian government published National Plastics Plan 2021 with a national commitment to tackling the plastic challenge. In fact, the Council of Australian Government has agreed to establish a timetable to ban the export of waste plastic, paper, glasses, and tyres. In addition, the National Waste Policy Action Plan (NWPAP) has been passed by Australia's environmental ministers. The mission of the Australian government to combat plastic pollution is shown in Table 2.9:

The National Plastics Plan 2021 is committed to minimizing plastic impacts with five main pillars of action, which are (1) Prevention, (2) Recycling, (3) Usage and Consumption, (4) Plastic Waste Disposal, (5) Research and Development. The summary of specific actions can be found in Box 2.1:

2.12 NEW ZEALAND

New Zealand has started to enforce the Waste Minimisation (Plastic and Related Products) Regulations 2022 on October 1, 2022, whereby it is illegal to provide, sell, or manufacture the following plastic products in New Zealand (Ministry for the Environment New Zealand, 2023a; Ministry for the Environment New Zealand, 2023b; Ministry for the Environment New Zealand, 2023c)

a) Single-use plastic drink stirrers for all plastic types. (Note: No exemptions are given. Suggestion is given to use reusable metal spoons, wooden fibre stirred without polyfluoroalkyl (PFAS) substances added.)

TABLE 2.9
Mission of the Australian Government to Combat Plastic Pollution

Year	Plans
2019	• Council of Australian Government established a timetable to ban export of waste plastic, paper, glass, and tyres • National Waste Policy Action Plan (NWPAP) agreed to by Australia's environment ministers
2020	• First National Plastics Summit • Passing of the Recycling and Waste Reduction Act 2020 • Microbeads phased out in rinse-off cosmetics, personal care, and cleaning products
2021	• First National Plastics Plan delivers on action 5.5 of the NWPAP • CSIRO's "A circular economy roadmap for plastics, tyres, glass, and paper in Australia" released (January 2021) • Regulation of unsorted, mixed plastic-waste exports (July 2021) • First review of National Environment Protection (Used Packaging Materials) Measure 2011 and the Australian Packaging Covenant to evaluate co-regulatory arrangements • National Plastics Design Summit
2022	• Regulation of unprocessed single polymer or resin waste plastic exports (July 2022) • Phase out non-compostable plastic packaging products containing additive fragmentable technology that do not meet relevant compostable standards (AS4736–2006, AS5810–2010, and EN13432) (July 2022) • Phase out expanded polystyrene (EPS) in loose fill and moulded consumer packaging (July 2022) and food and beverage containers (December 2022) • Phase out PVC packaging labels (December 2022) • Review progress of 2025 National Packaging Targets
2023	• At least 80% of supermarket products to display the Australasian Recycling Label (December 2023)
2025	National Packaging Targets for industry: • 100% of packaging is reusable, recyclable, or compostable • 70% of plastic packaging goes on to be recycled or composted • 50% average recycled content within packaging (20% for plastic packaging) • Problematic and unnecessary single-use plastic packaging phased out (target 5 of NWPAP)
2030	• Work with the textile and whitegoods sectors on an industry-led phase-in of microfibre filters on new residential and commercial washing machines by July 1, 2030

Source: Adapted from National Plastics of Australian Government, 2021.

b) Single-use plastic cotton buds for all plastic types. (Note: Exemptions are given when cotton buds are used for medical devices and not sold by retailers, veterinary clinics, commercial food laboratories for food sampling, laboratories for scientific investigation, or part of the testing kit such as COVID-19 rapid test kit.) (Suggested alternatives: cotton buds material includes non-synthetic fibre cotton buds with PFAS free or reusable cotton buds made of silicone, etc.)

BOX 2.1 AUSTRALIA'S NATIONAL PLASTICS PLAN: PILLAR OF ACTION.

Pillar of Action: Prevention- Source of Plastics

Stop Problematic and Unnecessary Plastics

Collaborate with industry to expedite the elimination of specific polymer types in various applications and consider implementing regulatory measures if industry-led phase-outs are not accomplished. The following actions will be taken:

- By July 2022, plastic packaging products utilizing additive fragmentable technology that fail to meet the relevant compostable standards (AS4736–2006, AS5810–2010, and EN13432) will be phased out.
- By July 2022, the use of expanded polystyrene (EPS) will be discontinued in loose packaging fill and moulded packaging for consumer goods. Additionally, EPS containers for consumer food and beverages will be phased out by December 2022.
- By December 2022, PVC packaging labels will be gradually eliminated.

Beaches Free from Plastics

Collaborate with the Boomerang Alliance to eradicate single-use plastics from Australia's beloved beaches and aid local businesses in transitioning to alternative products.

Encourage Industry Shift to Recyclable Plastics

The industry will be encouraged to shift towards more valuable plastics that are easily recyclable, such as PET, HDPE, LDPE, and PP. Furthermore, efforts will be made to promote the development of products that are designed to be more recyclable and facilitate the recycling process.

National Packaging Targets

The industry is expected to achieve four National Packaging Targets by 2025, with two specific goals focused on prevention. These targets are as follows:

- Ensure that 100% of packaging is reusable, recyclable, or compostable.
- Eliminate problematic and unnecessary single-use plastics from the market.

Pillar II—Recycling for Better Future

Waste Export Ban

Implement regulations to control the export of waste plastics by prohibiting the export of unsorted, mixed plastic starting from July 1, 2021. Additionally, the export of unprocessed single polymer or resin plastics will be banned from July 1, 2022.

Development of Recycling Capacity

The Recycling Modernization Fund is set to significantly boost Australia's recycling industry, generating a substantial $600 million investment in recycling initiatives. This will be further bolstered by the Modern Manufacturing Strategy, which allocates $1.5 billion and recognizes recycling as a crucial national manufacturing priority.

Industry to Engage Recycled Plastics in Production

Businesses are encouraged to make a commitment to enhance their utilization of recycled content by participating in the Member Pledge program facilitated by the Australian Packaging Covenant Organisation (APCO).

Government to Procure and Use More Recycled Plastics

The Australian government has enhanced the Commonwealth Procurement Rules to incorporate sustainability, including the utilization of recycled materials, as an integral part of the value for money evaluation for all its procurement activities.

Pillar III- Usages and Consumptions

Accessible Recycling Information

Collaborate with the industry to ensure that all APCO members earning annual revenue exceeding $500 million adopt the Australasian Recycling Label (ARL) by the end of 2023. This initiative aims to have 80% of supermarket products, including those with recycled content, display the ARL. The Australian government will also provide assistance to facilitate the adoption of the ARL by small to medium-sized enterprises (SMEs).

Against "Greenwashing"

In cases where companies engage in deceptive or misleading labelling and environmental claims, such as misrepresenting recyclability, the Australian government will refer these companies to the Australian Competition and Consumer Commission (ACCC) for investigation.

App- Recycle Mate

The Australian government is providing support for the nationwide implementation of the Recycle Mate App in 2021. This app assists consumers in determining the recyclability of products.

Consistent Container Deposit Schemes (CDS)

The Australian government will continue its collaboration with state and territory governments to enhance the alignment of various aspects of Container Deposit Schemes (CDSs).

Pillar IV Plastic Wastes Disposal

Fibre filters for washing machines that capture microplastics

Collaborate with the textile and whitegoods sectors to implement an industry-led gradual implementation of microfibre filters on newly manufactured residential and commercial washing machines. This phase-in process aims to be completed by July 1, 2030.

Environment Restoration Fund (ERF)

Over a span of four years, the Australian Government has allocated a funding of $100 million to safeguard the environment for the benefit of future generations. The projects initiated under the ERF (Ecosystems Restoration Fund) primarily concentrate on preserving threatened and migratory species as well as their habitats, encompassing Australia's coastlines, oceans, and waterways.

Partnership with Indonesia

Under the joint leadership of the CSIRO (Commonwealth Scientific and Industrial Research Organisation) and the Indonesian Ministry of Marine Affairs and Fisheries, an Indonesia-Australia Systemic Innovation Lab focused on marine plastic waste will be established.

Pacific Ocean Litter Project

Pacific Island countries will receive a funding of $16 million from the Australian government to combat the detrimental impact of single-use plastics on the ocean.

Shipping Waste

Efforts will be made to decrease shipping waste through the implementation of the Marine Litter from Ships and the Ship-Generated Garbage in the Pacific Action Plan set forth by the International Maritime Organization.

Pillar V: Research and Development

Waste Data Visualization Platform

Allocate a funding of $20.6 million to develop a publicly accessible Waste Data Visualization Platform.

Cooperative Research Centres Projects Grants (CRC-P)

Allocate a funding of $29.1 million to support research projects focused on showcasing innovative methods for recycling plastics and minimizing the amount of plastic waste sent to landfills.

Circular Economy Roadmap

The CSIRO's National Circular Economy Roadmap for Plastics, Glass, Paper, and Tyres, titled "Pathways for unlocking future growth opportunities for Australia," offers valuable insights to bolster the circular economy. This roadmap will serve as a crucial resource for governments, industry stakeholders, and researchers, informing their decision-making regarding investments, policy development, and research priorities in the future.

National Environmental Science Program (NESP)

Waste impact management will be designated as a cross-cutting priority within the National Environmental Science Program (NESP). This priority will provide support for policy development, program management, and regulatory processes in both marine and terrestrial environments.

c) Plastics with pro-degradant additives. (Suggested alternatives: recyclable plastic trays or containers using PET or PP or fibre-based without PFAS or compostable biodegradable plastic products.)
d) Certain PVC food trays and containers. (Suggested alternatives: recyclable plastic trays or containers using PET or PP or fibre-based without PFAS.)
e) Polystyrene takeaway food and beverage packaging. (Suggested alternatives: recyclable plastic trays or containers used PET or PP or fibre-based without PFAS.)
f) Expanded polystyrene food and beverage packaging. (Suggested alternatives: recyclable plastic trays or containers used PET or PP or fibre-based without PFAS.)

In order for the products to be claimed as a biodegradable and compostable type of plastic product, the products are required to fulfil requirements and get certified according to the following standards in Box 2.2.

BOX 2.2

Industrial compost certifications

- AS 4736—Australian seedling industrial composting
- EN 13432—Seedling industrial composting
- EN 13432—OK compost industrial composting
- EN 13432—Din industrial
- ASTM D 6400 or 6868—Biodegradable products institute/US composing council.

Home compost certifications

- AS 5810—Australian seedling home composting
- Variation of EN 13432—OK compost home composting
- AS 5810/NF T 51–800—Din home

Anaerobic degradation standards

- ASTMD5511—Standard test method for determining anaerobic biodegradation of plastic materials under high-solids anaerobic-digestion conditions

Starting on July 1, 2023, single-use plastic produced bags, plastic plates, bowls and cutlery, plastic straws, and plastic-produced labels are banned. However, an exception is still made for disabled people and medical uses of plastic straws, reusable plastic straws such as silicone straws, as well as single-use plastic drinking straws that are mechanically attached to beverage boxes (such as juice or milk boxes) or are considered an integral part of the packaging until January 1, 2026. Also, this ban does not include single-use plastic bags that serve as part of the packaging for pre-packaged produce. "Pre-packaged" refers to fruits or vegetables that are already bagged, whether sealed or unsealed, before being made available for sale. Therefore, the regulations do not prohibit the use of plastic bags that are integral to or form an essential part of the packaging. For instance, this exemption applies to items like pre-packaged mesclun lettuce, sealed bags of apples or potatoes, sugar snap peas in a sealed bag, leafy greens and herbs packaged in a plastic sleeve, or bagged lettuce. In addition, as mentioned previously about the ban of plastic produced labels, the labels used need to meet the criteria of either AS 5810–2010 Biodegradable plastics: Biodegradable plastic suitable for home compositing or NF T51–800 Plastics: Specifications for plastics suitable for home compositing. The regulations do not cover labels used on export-bound produce to prevent plastic contamination within New Zealand's environment. Additionally, produce labelled prior to July 1, 2023 and already in the supply chain is exempt. The requirement for home compostable-label adhesive is postponed until July 1, 2025, with label manufacturers globally expected to develop such adhesive by that time. Similarly, labels used on imported produce are not mandated to be home compostable until July 1, 2025, allowing for a transition period and avoiding the cost of removing labels from imported goods. Finally, all polyvinyl chloride and polystyrene food and beverage packaging will be phased out from mid-2025.

2.13 GULF COUNTRIES AND REGION

United Arab Emirates (UAE) Dubai has imposed of a tariff of 25 fils on single-use carrier bags at purchase, and this policy came into effect on July 1, 2022. All single-use plastics are banned at airports in Dubai, while Abu Dhabi's ban on the single-use plastics started June 1, 2022. In a recent press conference on June 23, 2022, the Ministry of Municipality announced a ban on single-use plastic bags in Qatar, effective from November 15, 2022. The Cabinet's approval led to this decision, which prohibits institutions, companies, and shopping centres from using single-use plastic bags for packaging, presentation, circulation, transportation, or the carrying of products and merchandise. Instead, they are required to use multi-use plastic bags, biodegradable bags, paper bags, "woven" cloth bags, or other biodegradable materials that meet the approved standard specifications. This step aligns with efforts to promote sustainable alternatives and reduce plastic waste in Qatar.

In addition to the various universal regulatory measures, leveraging coordination through pre-existing regional seas programmes can be an effective avenue to prevent the introduction of pollutants into the marine environment and mitigate their impacts. For the Arabian/Persian Gulf region, the Regional Organization for the Protection of the Marine Environment (ROPME) plays a crucial role as established by the Kuwait Regional Convention for Cooperation on the Protection of the Marine Environment from Pollution (the Kuwait Convention). The Convention, with contracting parties including Iran, Saudi Arabia, Iraq, Kuwait, Qatar, Oman, and the UAE, emphasizes the importance of preserving marine resources and coastal amenities during urban and rural development processes. It also highlights the need for integrated management approaches, research programmes, and monitoring to achieve environmental and developmental goals while addressing pollution from shipping, dumping, land-based sources, and other human activities (Stöfen-O'Brien et al., 2022).

To support the obligations set out in the Kuwait Convention, several protocols have been adopted, although most require additional signatures and ratification. While the Convention establishes relevant obligations, its focus on plastic marine debris has been limited. Regional cooperation efforts should encompass both sea-based and land-based activities, enabling targeted national action plans and addressing concerns in urban and rural environments (refer to Table 2.10). Another regional actor with the potential to address plastic marine debris is the Cooperation Council for the Arab States of the Gulf (GCC), comprising Bahrain, Kuwait, Oman, Qatar, Saudi Arabia, and the UAE. The GCC aims to enhance coordination, cooperation, and integration among member states, including joint environmental action, policy convergence, capacity-building, environmental awareness, and conservation of natural resources. However, it is important to note that the GCC does not encompass all littoral states in the region (Stöfen-O'Brien et al., 2022).

2.14 CONCLUSION

Plastic pollution is a major environmental problem, and many Asian countries have implemented policies to address it. These policies vary in approach, but they all aim to reduce the use of plastic and increase recycling rates. Some of the most common policies include plastic bag bans, extended producer responsibility (EPR) systems, and recycling programmes. Plastic bag bans have been successful in some countries, but they have also led to job losses in the plastic manufacturing industry. EPR systems require manufacturers to take responsibility for the recycling of their products, and they have been effective in some countries, but they can be challenging to implement. Recycling programmes have been successful in some countries, but they can be expensive and require a high level of public participation. In addition to these policies, some countries are also using other strategies to address plastic pollution, such as ecolabelling schemes and Break Free from Plastic movements. Ecolabelling schemes certify products that are made from recycled materials or that are designed to be easily recycled. Break Free from Plastic movements are calling on companies to reduce their use of single-use plastics. The effectiveness of these policies varies, but they all have the potential to make a significant difference in reducing plastic pollution. It is important to continue to evaluate these policies and to find new and innovative ways to address this growing problem.

TABLE 2.10
Overview of the Applicable Regulatory Framework Regarding Marine Debris in the Gulf Region

Empty cell	Approach	Framework	Potential application to the Arabian/ Persian Gulf region	Ratification
International Regulatory Framework	Sea-based sources of marine debris	**United Nations Convention on the Law of the Sea (UCLOS)**	Overall regulatory framework relating to any sea-based activity, not all countries have ratified UNCLOS	Bahrain (1985), Iraq (1985), Oman (1989), Kuwait (1986), Qatar (2002), Saudi Arabia (1996)
		The International Convention for the Prevention of Pollution from Ships 73/78 Annex V	Discharge prohibition of onboard generated waste, Persian Gulf and Gulf of Oman Special Area under Annex V, all countries have ratified this instrument, Riyadh Memorandum of Understanding	Bahrain (1991), Iran (2003), Iraq (2018), Oman (1988), Kuwait (2007) Qatar (2006), Saudi Arabia (2005), United Arab Emirates (2007),
		1972 London Convention and 1996 London Protocol	Depending on instrument, different dumping standards prevail, waste assessment guidelines exist, very fragmented status of ratification = patchwork regime	**LC:** Iran (1997), Oman (1984) and UAE (1975) **LP:** Iran (2016) and Saudi Arabia (2006)
	Land-based sources of marine debris	**United Nations Convention on the Law of the Sea - Global Programme of Action for the Protection of the Marine Environment from Land-based Activities (GPA)**	Overall regulatory framework related to land-based activities GPA not legally binding, provides an umbrella framework for land-based activities with relevance to marine debris	UNCLOS: See above.
Regional Regulatory Framework and/or Policy Framework	Integrated approaches	**Regional Organization for the Protection of the Marine Environment**	Mandate to address land-and sea-based sources of marine debris	Bahrain, Iraq, Iran, Kuwait, Oman, Qatar Saudi Arabia and UAE
		Cooperation Council for the Arab States of the Gulf (GCC)	General Principles of Environment Protection, Joint Environmental Actions envisaged, limited to GCC members	Bahrain, Oman, Kuwait, Qatar, Saudi Arabia, UAE

Source: Adapted from Stöfen-O'Brien et al., 2022. With permission of Elsevier.

REFERENCES

Akenji, L. et al. (2020). Circular economy and plastics. [online] *ASEAN Organisation*. Available at: https://environment.asean.org/wp-content/uploads/2020/02/Circular-Economy-gap-analysis-final.pdf [Accessed 27 Feb. 2022].

Aoki-Suzuki, C. (2016). International policy trends of resource efficiency circular economy. *Journal of Life Cycle Assessment, Japan*, 12(4), pp. 267–272.

Aravind, I. (2019). Just how bad is India's plastic problem?. [online] *The Economic Times*. Available at: https://economictimes.indiatimes.com/news/politics-and-nation/how-india-is-drowning-in-plastic/articleshow/69706090.cms?from=mdr [Accessed 29 Aug. 2021].

Astellas (2021). *News | Astellas Pharma Inc*. [online] www.astellas.com. Available at: www.astellas.com/en/news/17266 [Accessed 9 Feb. 2022].

Bangkok Post Public Company (2021). Plastic waste imports are "unwanted." *Bangkok Post*. [online] Available at: www.bangkokpost.com/thailand/general/2160703/plastic-waste-imports-are-unwanted [Accessed 22 Feb. 2022].

Bhatia, A. (2018). Plastic ban in Maharashtra: Five things you should know. *News*. [online] NDTV-Dettol Banega Swasth Swachh India. Available at: https://swachhindia.ndtv.com/plastic-ban-in-maharashtra-five-things-you-should-know-18502/ [Accessed 30 Jul. 2021].

Borongan, G. and Kashyap, P. (2018). *Managing Municipal Solid Waste and Packaging Waste*. [online] Berlin: Deutsche Gesellschaft Für Internationale Zusammenarbeit (GIZ) GmbH. Available at: www.giz.de/de/downloads/giz2018_Indonesia-Country-Profile_web.pdf [Accessed 26 Feb. 2022].

Brady, E. (2021). The effects of China's ban on imported scrap plastic on global recycling efforts. [online] *Earth.org—Past | Present | Future*. Available at: https://earth.org/china-ban-on-imported-scrap-plastic/ [Accessed 15 Feb. 2022].

Buakamsri, T. (2021). Thailand's plastic pollution crisis. [online] *Taragraphies*. Available at: https://taragraphies.org/2021/12/09/thailands-plastic-pollution-crisis/ [Accessed 13 Feb. 2022].

CEARAC (2020). *National Actions on Marine Microplastics in the NOWPAP Region*. [online] Available at: www.cearac-project.org/RAP_MALI/National_actions_on_MPs.pdf [Accessed 16 Dec. 2022].

Chen, F., Li, X., Ma, J., Yang, Y. and Liu, G.-J. (2018). An Exploration of the impacts of compulsory source-separated policy in improving household solid waste-sorting in pilot megacities, China: A case study of Nanjing. *Sustainability*, 10(5), p. 1327.

Chen, H. L., Nath, T. K., Chong, S., Foo, V., Gibbins, C. and Lechner, A. M. (2021). The plastic waste problem in Malaysia: Management, recycling and disposal of local and global plastic waste. *SN Applied Sciences*, 3(4).

Devashish, D. (2019). Catch 22: Hits of plastic ban. *Jamshedji Tata School of Disaster Studies*, [online] pp. 1–5. Available at: www.researchgate.net/publication/334308199_Catch_22_Hits_of_Plastic_Ban [Accessed 29 Jul. 2021].

Enviliance (2018). Thailand roadmap on plastic waste management—nviliance ASIA. [online] Available at: https://enviliance.com/regions/southeast-asia/th/th-waste/th-plastic-waste [Accessed 19 Feb. 2022].

Environmental Department (2018). *Notification*. [online] *MPCB GOV India*. Available at: www.mpcb.gov.in/sites/default/files/plastic-waste/rules/plasticwasteenglish119102020.pdf [Accessed 1 Aug. 2021].

Government of Kerala (2019). [online] Sanitation.kerala.gov.in. Available at: <http://sanitation.kerala.gov.in/wp-content/uploads/2019/12/output.pdf> [Accessed 13 July 2021].

GPAP (2020). *Indonesia*. [online] *Global Plastic Action Partnership*. Available at: https://globalplasticaction.org/countries/indonesia/ [Accessed 16 Nov. 2021].

Hashim, K. S. H., Mohammed, A. H. and Redza, H. Z. M. S. (2011). Persidangan Kebangsaan Masyarakat, Ruang dan Alam Sekitar (MATRA 2011) (16th – 17th November 2011) Conference Proceeding. Available at: https://core.ac.uk/reader/300388660. [Accessed 1 Jan. 2024].

Jang, Y. -C., Lee, G., Kwon, Y., Lim, J. and Jeon, J. (2020). Recycling and management practices of plastic packaging waste towards a circular economy in South Korea. *Resources, Conservation and Recycling*, [online] 158, p. 104798. https://doi.org/10.1016/j.resconrec.2020.104798

Jangprajak, W. (2021). National Action Plan on Plastic Waste Management in Thailand. Available at: www.iges.or.jp/sites/default/files/inline-files/S1-5_PPT_Thailand%20Plastic%20Action%20Plan.pdf. [Accessed 11 Jul. 2023].

Jereme, I. A., Sitar, C., Begum, R. A., Talib, B. A. and Allam, Md. M. (2015). Assessing problems and prospects of solid waste management in Malaysia. [online] 12(2). Available at: www.academia.edu/23595259/Assessing_Problems_and_Prospects_of_Solid_Waste_Management_in_Malaysia [Accessed 17 Jul. 2021].

Joe, T. (2021). China single-use plastic bag & Straw ban now in effect across major cities. [online] *Green Queen*. Available at: www.greenqueen.com.hk/china-single-use-plastic-bag-straw-ban-now-in-effect-across-major-cities/ [Accessed 6 Jul. 2021].

Jogloabang (2019). *UU 18 Tahun 2008 Tentang Pengelolaan Sampah*. [online]. Available at: https://www-jogloabang-com.translate.goog/pustaka/uu-18-2008-pengelolaan-sampah?_x_tr_sl=id&_x_tr_tl=en&_x_tr_hl=en&_x_tr_pto=op [Accessed 16 Nov. 2021].

Kaminaga, Y. (2018). The meaning of reducing plastic bags: A case of Nagoya. *The Journal of Yokkaichi University*, 30(2), pp. 59–60.

Kawaguchi, M. (2018). Tokyo homebuilder launches mass production of environmental friendly wooden straws—The Mainichi. [online] *The Mainichi*. Available at: https://mainichi.jp/english/articles/20181208/p2a/00m/0na/026000c [Accessed 17 Jul. 2021].

Kengo, Y. (2021). Korea announces legislative notice of amendment to add 17 items to EPR. [online] *Enviliance*. Available at: https://enviliance.com/regions/east-asia/kr/report_4036 [Accessed 25 Feb. 2022].

Kim, I. and Jang, Y. (2022). Material efficiency and greenhouse gas reduction effect of industrial waste by material circulation in Korea. *Journal of Cleaner Production*, 376, p. 134053. https://doi.org/10.1016/j.jclepro.2022.134053

Kim, J. H. (2012). Extended producer responsibility (EPR) and job creation in Korea. *Waste Management and the Environment VI*, [online] 163(1743–3541). Available at: www.witpress.com/Secure/elibrary/papers/WM12/WM12007FU1.pdf [Accessed 25 Feb. 2022].

Klein, C. (2021). Japan: Rate of plastic waste recycled. [online] *Statista*. Available at: www.statista.com/statistics/1169339/japan-rate-of-recycled-plastic-waste/ [Accessed 9 Feb. 2022].

Lacovidou, E. and Ng, K. S. (2020). Malaysia versus waste. *Brunel University London*. [online] www.brunel.ac.uk. Available at: www.brunel.ac.uk/news-and-events/news/articles/Malaysia-Versus-Waste [Accessed 18 Feb. 2022].

Leungsakul, S. (2017). *Situation of plastic scrap & E-waste import into Thailand*. [online] Available at: www.env.go.jp/en/recycle/asian_net/Annual_Workshops/2019_PDF/Session1/S1_11_Thailand_ANWS2019.pdf [Accessed 13 Feb. 2022].

Marks, D., Miller, M. A. and Vassanadumrongdee, S. (2020). The geopolitical economy of Thailand's marine plastic pollution crisis. *Asia Pacific Viewpoint*, [online] 61, pp. 1360–7456. Available at: www.researchgate.net/publication/338764240_The_geopolitical_economy_of_Thailand%27s_marine_plastic_pollution_crisis [Accessed 21 Mar. 2022].

Mesicek, V. (2021). Borealis' Bornewables portfolio of circular polypropylene solutions proven to substantially reduce carbon emissions—Borealis. [online] *Borealisgroup*. Available at: www.borealisgroup.com/news/borealis-bornewables-portfolio-of-circular-polypropylene-solutions-proven-to-substantially-reduce-carbon-emissions [Accessed 9 Feb. 2022].

MESTECC (2018). *Malaysia's Roadmap towards Zero Single-use Plastics 2018–2030.* [online] pp. 2–10. Available at: www.moe.gov.my/images/KPM/UKK/2019/06_Jun/Malaysia-Roadmap-Towards-Zero-Single-Use-Plastics-2018-2030.pdf [Accessed 5 Aug. 2021].

Ministry for the Environment New Zealand (2023a). *Guidance on Plastic Products Banned from October 2022.* Available at: https://environment.govt.nz/publications/plastic-products-banned-from-october-2022/#descriptions-of-banned-products-exemptions-and-alternatives [Accessed 11 Jul. 2023].

Ministry for the Environment New Zealand (2023b). *Guidance on Single-use Plastic Products Banned or Phased out from July 2023.* Available at: https://environment.govt.nz/publications/plastic-products-banned-from-july-2023/#single-use-plastic-produce-bags [Accessed 11 Jul. 2023].

Ministry for the Environment New Zealand (2023c). *Guidance on Plastic Products Banned from Mid 2025.* Available at: https://environment.govt.nz/publications/plastic-products-banned-from-mid-2025/ [Accessed 11 Jul. 2023].

Ministry of Oceans and Fisheries (2016). *Key Components of the 1st Framework on Marine Debris Management(2021~2030)—Ministry of Oceans and Fisheries.* [online] www.mof.go.kr. Available at: www.mof.go.kr/en/page.do?menuIdx=1480 [Accessed 16 Mar. 2022].

Ministry of the Environment (2014a). *History and Current State of Waste Management in Japan.* [online] Available at: www.env.go.jp/en/recycle/smcs/attach/hcswm.pdf.

Ministry of the Environment (2014b). *The State of Play on Extended Producer Responsibility (EPR): Opportunities and Challenges.* [online] Available at: www.oecd.org/environment/waste/Global%20Forum%20Tokyo%20Issues%20Paper%2030-5-2014.pdf.

Ministry of the Environment (2017). *The Act on Promotion of Procurement of Eco-friendly Goods and Services by the State and Other Entities.* [online] Available at: www.env.go.jp/policy/hozen/green/attach/gpp%20pamphlet_eng.pdf [Accessed 16 Nov. 2021].

Ministry of the Environment (2019). *Establishing a Sound Material-cycle Society and Measures to Address Marine Litter.* Tokyo: Japan Environment Quarterly.

National Plastics of Australian Government (2021). Available at: https://www.agriculture.gov.au/sites/default/files/documents/national-plastics-plan-2021.pdf [Accessed 2 Jan. 2024].

NEA (2012). Singapore packaging agreement. [online] *Nea.gov.sg.* Available at: www.nea.gov.sg/programmes-grants/schemes/singapore-packaging-agreement [Accessed 11 Jan. 2022].

Neo, P. (2018). Thailand to stop all foreign plastic waste imports by 2021 following Malaysia, Vietnam. [online] *foodnavigator-asia.com.* Available at: www.foodnavigator-asia.com/Article/2018/11/19/Thailand-to-stop-all-foreign-plastic-waste-imports-by-2021-following-Malaysia-Vietnam [Accessed 9 Mar. 2022].

Ogunmakinde (2019). A review of circular economy development models in China, Germany and Japan. *Recycling*, 4(3), p. 27.

Peck, T. G. and Tay, M. G. (2010). Chapter 2 Singapore packaging agreement and the 3R packaging awards. [online] *ERIA*, pp. 22–39. Available at: www.eria.org/uploads/media/Research-Project-Report/RPR_FY2009_10_Chapter_2.pdf [Accessed 8 Jan. 2022].

Preetha, M. S. (2020). State bans plastic for primary packaging too. *The Hindu.* [online] 9 June. Available at: www.thehindu.com/news/national/tamil-nadu/tn-bans-plastic-for-primary-packaging-too/article31790446.ece [Accessed 18 Jul. 2021].

Qiyun, W. (2020). Beyond Plastic Recycling: A look at Extended Producer Responsibility in Singapore. [online] *Singapore Policy Journal.* Available at: https://spj.hkspublications.org/2020/09/08/beyond-plastic-recycling-a-look-at-extended-producer-responsibility-in-singapore/ [Accessed 13 Feb. 2021].

Raj, S. (2019). Having one eco-label will simplify certification. *New Straits Times.* [online] NST Online. Available at: www.nst.com.my/opinion/letters/2019/08/511887/having-one-eco-label-will-simplify-certification [Accessed 23 Aug. 2021].

Rashid, N. R. N. A. (2009). Awareness of eco-label in Malaysia's green marketing initiative. *International Journal of Business and Management*, 4(8).

Razif, M. and Persada, S. F. (2016). Environmental impact assessment (EIA) framework for ekolabel certification initiative in indonesia: Case study of a rattan-plywood based furniture industry. *International Journal of ChemTech Research*, [online] 9(0974–4290). Available at: www.researchgate.net/publication/303789920_Environmental_impact_assessment_EIA_framework_for_ekolabel_certification_initiative_in_indonesia_Case_study_of_a_rattan-plywood_based_furniture_industry [Accessed 16 Nov. 2021].

Ricky, M. N. (2019). Bogor city plastic bag ban sued, refuses to back down. [online] *Tempo*. Available at: https://en.tempo.co/read/1200541/bogor-city-plastic-bag-ban-sued- [Accessed 10 Mar. 2022].

Rujivanarom, P. (2021). Thailand's plastic waste conundrum | Heinrich Böll Foundation | Southeast Asia Regional Office. [online] *Heinrich-Böll-Stiftung*. Available at: https://th.boell.org/en/2021/10/26/thailands-plastic-waste-conundrum [Accessed 20 Feb. 2022].

Saleh Omar, B., Quoquab, F. and Mohammad, J. (2019). "No plastic bag" campaign of Malaysia. *Green Behavior and Corporate Social Responsibility in Asia*, pp. 113–119.

Saori, Y. (2020). Mannequin-maker takes on plastic waste. *NHK WORLD-JAPAN News*. [online] NHK WORLD. Available at: https://www3.nhk.or.jp/nhkworld/en/news/backstories/864/ [Accessed 17 July 2021].

Sawaji, O. (2019). "Plastics smart" campaign. *Highlighting Japan*, June 2019 [online] Gov-online.go.jp. Available at: www.gov-online.go.jp/eng/publicity/book/hlj/html/201906/201906_09_en.html [Accessed 12 July 2021].

Sea Circular (2019). Country Profile – South Korea. Available at: https://www.sea-circular.org/wp-content/uploads/2020/05/SEA-circular-Country-Profile_SOUTH-KOREA.pdf#:~:text=South%20Korea%20has%20one%20of,with%2097.7%20kg%20per%20capita [Accessed 1 Jan. 2024].

Shrivastav, R. (2020). Kerala has taken a bold step with its ban on single-use plastics. [online] *Downtoearth.org.in*. Available at: www.downtoearth.org.in/news/waste/kerala-has-taken-a-bold-step-with-its-ban-on-single-use-plastics-68760 [Accessed 22 July 2021].

Sreenivasan, J., Govindan, M., Chinnasami, M. and Kadiresu, I. (2012). Solid waste management in Malaysi—move towards sustainability. *Intechopen.com*. [online] Available at: www.intechopen.com/books/waste-management-an-integrated-vision/solid-waste-management-in-malaysia-a-move-towards-sustainability [Accessed 17 Aug. 2021].

Stöfen-O'Brien, A., Naji, A., Brooks A. L., Jambeck, J. R. and Khan, F. R. (2022). Marine plastic debris in the Arabian/Persian Gulf: Challenges, opportunities and recommendations from a transdisciplinary perspective. *Marine Policy*, 136, p. 104909.

Suki, N. M. (2013). Green awareness effects on consumers' purchasing decision: Some insights from malaysia. *Green Awareness Effects*, [online] 9(2), pp. 50–63. Available at: http://ijaps.usm.my/wp-content/uploads/2013/07/Art3.pdf [Accessed 20 Aug. 2021].

Suzuki, T. (2021). All retailers in Japan required to charge fee for plastic bags from July 1—The Mainichi. [online] *The Mainichi*. Available at: https://mainichi.jp/english/articles/20200630/p2a/00m/0na/007000c [Accessed 18 Jul. 2021].

Taiwan Ministry of Environment (2023a). Available at: https://oaout.moenv.gov.tw/law/LawContent.aspx?id=GL006481&kw=%e5%a1%91%e8%86%a0 [Assessed 23 Oct. 2023].

Taiwan Ministry of Environment (2023b). Available at: https://oaout.moenv.gov.tw/law/LawContent.aspx?id=GL007806&kw=%e5%a1%91%e8%86%a0 [Assessed 23 Oct. 2023].

Taiwan Ministry of Environment (2023c). Available at: https://oaout.moenv.gov.tw/law/LawContent.aspx?id=GL006482&kw=%e5%a1%91%e8%86%a0 [Assessed 23 Oct. 2023].

Taiwan Ministry of Environment (2023d). Available at: https://oaout.moenv.gov.tw/law/EngLawContent.aspx?lan=E&id=5 [Assessed 23 Oct. 2023].

Taiwan Ministry of Environment (2023e). Available at: https://oaout.moenv.gov.tw/law/LawContent.aspx?id=GL007865&kw=%e5%a1%91%e8%86%a0 [Assessed 23 Oct. 2023].

Tammy, H. and Karlsson, H. (2020). *IPSOS Briefing Singapore a Singapore Perspective on Plastics and the Opportunity for Brands to Drive Change*. [online] Available at: www.ipsos.com/sites/default/files/ct/publication/documents/2020-03/singapore-perspective-on-plastic-pollution-ipsos.pdf [Accessed 11 Jan. 2022].

The Economic Times (2007). Restriction boosts Kerala's plastic industry. *The Economic Times*. [online] 23 September. Available at: https://economictimes.indiatimes.com/industry/indl-goods/svs/packaging/restriction-boosts-keralas-plastic-industry/articleshow/2395868.cms?from=mdr [Accessed 10 Feb. 2022].

The Hindu (2019). A year into plastic ban, Tamil Nadu still struggles to kick the habit. *The Hindu*. [online] 29 December. Available at: www.thehindu.com/news/national/tamil-nadu/plastics-down-but-not-out/article30422345.ece [Accessed 12 Feb. 2022].

The Nippon Foundation (2020). Plastic bottle collection machines being installed at 7-eleven stores in Yokohama. [online] *The Nippon Foundation*. Available at: www.nippon-foundation.or.jp/en/news/articles/2020/20201030-51019.html [Accessed 10 Feb. 2022].

The Times of India (2019). Plastic ban will show results in 3 months: Tamil Nadu. *Chennai News—Times of India*. [online] Available at: https://timesofindia.indiatimes.com/city/chennai/plastic-ban-will-show-results-in-3-months-tamil-nadu/articleshow/86905336.cms [Accessed 9 Mar. 2022].

Tsai, W.-T. (2022). Environmental policy for the restriction on the use of plastic products in Taiwan: Regulatory measures, implementation status and COVID-19's impacts on plastic products recycling. *Environments*, 9, p. 7.

UNEP (2017). Comparative analysis of green public procurement and ecolabelling programmes in China, Japan, Thailand and the Republic of Korea: Lessons learned and common success factors. [online] *Green Growth Knowledge*. Available at: www.greengrowthknowledge.org/sites/default/files/downloads/resource/UNEP_green_public_procurement_ecolabelling_China_Japan_Korea_Thailand_report.pdf [Accessed 16 Feb. 2022].

Wang, H. and Jiang, C. (2020). Local nuances of authoritarian environmentalism: A legislative study on household solid waste sorting in China. *Sustainability*, 12(6), p. 2522.

Wen, Z., Xie, Y., Chen, M. and Dinga, C. D. (2021). China's plastic import ban increases prospects of environmental impact mitigation of plastic waste trade flow worldwide. *Nature Communications*, [online] 12(1), p. 425. Available at: www.nature.com/articles/s41467-020-20741-9.

Wong, S. (2021). China: Global plastics production share 2014–2019. [online] *Statista*. Available at: www.statista.com/statistics/1247211/plastics-production-share-of-china/ [Accessed 5 Aug. 2021].

Xinhua (2017). China announces import ban on 24 types of solid waste – China – Chinadaily.com.cn. [online] www.chinadaily.com.cn. Available at: https://www.chinadaily.com.cn/china/2017-07/21/content_30194081.htm [Accessed 10 Aug. 2021].

Yamakawa, D. (2014). *Extended Producer Responsibility*. Kyoto: Kyoto Prefectural University.

Yann, Z. (2019). Indonesian government and partners announce next steps to tackle plastic pollution. [online] *World Economic Forum*. Available at: www.weforum.org/press/2019/03/indonesian-government-and-partners-announce-next-steps-to-tackle-plastic-pollution/ [Accessed 16 Nov. 2021].

Zen, I. S., Ahamad, R. and Omar, W. (2013). No plastic bag campaign day in Malaysia and the policy implication. *Environment, Development and Sustainability*, 15(5), pp. 1259–1269.

Zhang and Laney (2017). *China: National Plan on Banning "Foreign Garbage" and Reducing Solid Waste Imports*. [online] Washington, DC: Library of Congress, 20540 USA. Available at: www.loc.gov/item/global-legal-monitor/2017-08-08/china-national-plan-on-banning-foreign-garbage-and-reducing-solid-waste-imports/ [Accessed 15 Aug. 2021].

3 Policies of Plastic Use in European Countries

3.1 INTRODUCTION

According to a communication from the Commission to the European Parliament, The Council, The European Economic and Social Committee, and The Committee of The Regions (2018), the plastic industry overall employed about 1.5 million people and added almost EUR 340 billion to the economy in the year 2015 in the European Union. However, with such a massive supply and demand for plastics in Europe, it resulted in around 25.8 million tonnes of plastic waste being generated annually. Similar to common scenarios around the world, the recycling rate of plastics is relatively low when compared with other materials such as paper or metals. Plastic wastes are either left to degrade in landfills, openly burnt, or even dumped into the sea, causing serious marine pollution. Indeed, the overall demand of recycled plastics only accounts for 6% of the total plastic demand in Europe. Numerous of investments in plastic recycling capacity are unable to proceed due to the industry's low profitability as well as volatility in the market. The European Commission, starting in the year 2018, has formed the European Strategy for Plastics in a Circular Economy, with the working timeline in ANNEX 1 and ANNEX II, listing the detailed of implementation of measures on plastic wastes as shown in Table 3.1 and Table 3.2.

In addition, all of the 27 EU Member States have committed to the blueprint called European Green Deal to reduce emissions by at least 55% by 2030, compared to 1990 levels, in order to turn the EU into the first climate-neutral continent by 2050. In addition to this European Green Deal, the EU has implemented the Plastic Tax, also known as Levy. However, such a tax is a different common understanding about paying based on commercial profit. The Plastic Tax is a contribution from the Member States to the EU based on the amount of non-recycled plastic packaging waste produced by each Member State. Beginning January 1, 2021, the contribution is calculated by the weight of non-recycled plastic packaging waste with a uniform rate of EUR 0.80 per kilogram. The contributions are calculated based on Eurostat data, as contributed by Member States' early years for the existing reporting obligations (specifically, the Packaging and Packaging Waste Directive and its Implementing Decision [Decision (EU) 2019/665]). Through the standard practice for any sources of revenue in the EU budget, Member States will pay their contributions on a monthly basis based on forecasts. This contribution is adjusted after July 2023, when the final data will be available. The detail of the Plastic Tax is covered in the next section, together with other policies of the countries.

DOI: 10.1201/9781003387862-3

TABLE 3.1
List of Future EU Measures to Implement the Strategy

Measures	Timeline
Improving the economics and quality of plastics recycling	
Actions to improve product design:	Q1 2018 onwards
• preparatory work for future revision of the Packaging and Packaging Waste Directive: Commission to initiate work on new harmonized rules to ensure that, by 2030, all plastics packaging placed on the EU market can be reused or recycled in a cost-effective manner.	Q1 2018 onwards ongoing
• follow-up to COM (2018) 32: "*Communication on the implementation of the circular economy package: options to address the interface between chemical, product and waste legislation*": improve the traceability of chemicals and address the issue of legacy substances in recycled streams	
• new eco-design measures: consider requirements to support the recyclability of plastics	
Actions to boost recycled content:	Q1–Q3 2018
• launching an EU-wide pledging campaign targeting industry and public authorities	Q1 2018 onwards ongoing 2018
• assessment of regulatory or economic incentives for the uptake of recycled content, in particular, in the context of the:	2018 onwards
• revision of the Packaging and Packaging Waste Directive (see foregoing)	
• evaluation/review of the Construction Products Regulation	
• evaluation/review of End-of-life Vehicles Directive	
• as regards food-contact materials: swift finalization of pending authorization procedures for plastics recycling processes, better characterization of contaminants and introduction of monitoring system	
• development of quality standards for sorted plastics waste and recycled plastics in cooperation with the European Standardization Committee	
• Ecolabel and Green Public Procurement: further incentivize the use of recycled plastics, including by developing adequate verification means	
Actions to improve separate collection of plastic waste:	2019
• issue new guidelines on separate collection and sorting of waste	ongoing
• ensure better implementation of existing obligations on separate collection, including through ongoing review of waste legislation	
Curbing plastic waste and littering	
Actions to reduce single-use plastics:	ongoing
• analytical work, including the launch of a public consultation, to determine the scope of a legislative initiative on single-use plastics	
Actions to tackle sea-based sources of marine litter:	Q1 2018
• adoption of a legislative proposal on port reception facilities for the delivery of waste from ships	2018 onwards
• development of measures to reduce loss or abandonment at sea of fishing gear (*e.g., including recycling targets, EPR schemes, recycling funds, or deposit schemes*)	
• development of measures to limit plastic loss from aquaculture (*e.g., possible Best Available Techniques Reference Document*)	

TABLE 3.1
Continued

Measures	Timeline
Actions to monitor and curb marine litter more effectively:	2018 onwards
• improved monitoring and mapping of marine litter, including microplastics, on the basis of EU harmonized methods	
• support to Member States on the implementation of their programmes of measures on marine litter under the Marine Strategy Framework Directive, including the link with their waste/litter management plans under the Waste Framework Directive	
Actions on compostable and biodegradable plastics:	Q1 2018 onwards
• start work to develop harmonized rules on defining and labelling compostable and biodegradable plastics	Q1 2018 onwards ongoing
• conduct a lifecycle assessment to identify conditions where their use is beneficial and criteria for such application	
• start the process to restrict the use of oxo-plastics via REACH	
Actions to curb microplastics pollution:	ongoing ongoing
• start the process to restrict the intentional addition of microplastics to products via REACH	Q1 2018 onwards ongoing
• examination of policy options for reducing unintentional release of microplastics from tyres, textiles, and paint (e.g., *including minimum requirements for tyre design (tyre abrasion and durability, if appropriate) and/or information requirement (including labelling, if appropriate), methods to assess microplastic losses from textiles and tyres, combined with information*	
• development of measures to reduce plastic pellet spillage (*e.g., certification scheme along the plastic supply chain and/or Best Available Techniques reference document under the Industrial Emissions Directive*)	
• evaluation of the Urban Waste Water Treatment Directive: assessing effectiveness as regards microplastics capture and removal	
Driving investment and innovation towards circular solutions	
Actions to promote investment and innovation in the value chain:	2019
• commission guidance on the eco-modulation of EPR fees	in mid-2018
• recommendations by the recently launched "*Circular Economy Finance Support Platform*"	By mid-2019 ongoing
• examine the feasibility of a private-led investment fund to finance investments in innovative solutions and new technologies aimed at reducing the environmental impacts of primary plastic production	2018 onwards Q2 2018
• direct financial support for infrastructure and innovation through the European Fund for Strategic Investment and other EU funding instruments (e.g., structural funds and smart specialization strategies, Horizon, 2020)	
• pursue work on life-cycle impacts of alternative feedstocks for plastics production	
• development of a Strategic Research Innovation Agenda on plastics to guide future funding decisions	

(Continued)

TABLE 3.1
Continued

Measures	Timeline
Harnessing global action	
Actions focusing on key regions: • project to reduce plastic waste and marine litter in East and Southeast Asia to support sustainable consumption and production, the promotion of the waste hierarchy and extended producer responsibility, and improve recovery of fishing gear • examining options for specific action to reduce plastic pollution in the Mediterranean, in support of the implementation of the Barcelona Convention • cooperation on plastic-waste prevention in major world river basins	2018 onwards
Actions in support of multilateral initiatives on plastic: • renewed engagement on plastics and marine litter in fora such as the UN, G7, G20, the MARPOL convention, and regional sea conventions, including the development of practical tools and specific action on fishing and aquaculture • support to action under the Basel Convention, particularly for the implementation of the toolkit on environmentally sound waste management	2018 onwards
Actions relating to bilateral cooperation with non-EU countries: • promote a circular plastics economy in non-EU countries through policy dialogues on trade, industry, and environment as well as economic diplomacy • use bilateral, regional, and thematic funding in EU development, neighbourhood, and enlargement policies to support the plastics strategy by preventing and appropriately managing waste and supporting the circular economy; through programmes and instruments including "Switch to Green" and the External Investment Plan	2018 onwards
Actions relating to international trade: • support the development of international industry standards on sorted plastic waste and recycled plastics • ensure that exported plastic waste is dealt with appropriately in line with the EU Waste Shipment Regulation • support the development of a certification scheme for recycling plants in the EU and in third countries	2018 onwards

Source: Fully adapted from ANNEX I Communication from the Commission to the European Parliament, the Council, the European Economic and Social Committee and the Committee of the Regions A European Strategy for Plastics in a Circular Economy (European Commission, 2018).

3.2 AUSTRIA

Austria is one of the successful recyclers in the European Union. This success can be attributed to the implementation of the extended producer responsibility (EPR) system, which has been active for more than 25 years. The EPR aims to encourage manufacturers to design and produce products by considering environmentally

TABLE 3.2
List of Measures Recommended to National Authorities and Industry

Key measures to improve the economics and quality of plastics recycling
National and regional authorities are encouraged to:

- favour reusable and recycled plastics in public procurement
- make better use of taxation and other economic instruments to:
- reward the uptake of recycled plastics and favour reuse and recycling over landfilling and incineration
- step up separate collection of plastics waste and improve the way in which this is done
- put in place well-designed EPR schemes and/or deposit systems, in consultation with the relevant sectors
- make voluntary commitments in support of the strategy's objectives; in particular, as regards the uptake of recycled plastics

Industry is encouraged to:

- take concrete steps to improve dialogue and cooperation across the value chain; in particular, on material and product design aspects
- make voluntary commitments in support of the strategy's objectives; in particular, as regards the uptake of recycled plastic

Key measures to curb plastic waste and littering
National and regional authorities are encouraged to:

- raise awareness of littering and consider fines, where they do not exist already; promote beach clean-up activities
- step up waste collection, particularly near the coasts, and improve coordination between the authorities responsible for waste management, water, and the marine environment
- step up efforts to eradicate illegal and non-compliant landfills
- develop national monitoring of marine litter on the basis of harmonized EU methods
- engage in regional seas conventions; in particular, to develop regional plans against marine litter
- consider introducing EPR; in particular, to provide incentives for collecting discarded fishing gear and recycling agricultural plastics
- consider introducing deposit refund schemes; in particular, for beverage containers

Industry is encouraged to:

- promote existing alternatives to single-use plastic items (e.g., in catering and take-aways), where these are more environmentally beneficial
- pursue and implement cross-industry agreements to reduce the release of microplastics in the environment
- put in place measures to avoid spillage of plastic pellets

Key measures to drive investments and innovation towards circular solutions
National, regional and local authorities are encouraged to:

- make better use of economic instruments, especially to raise the cost of landfilling and incineration and promote plastic waste recycling and prevention
- make greater use of public procurement and funding to support plastic waste prevention and recycling of plastics

Industry is encouraged to:

- increase infrastructure and R&D investment in areas of direct relevance to achieving the strategy's objectives
- contribute to work on setting up a private investment fund to offset the environmental externalities of plastic production

(Continued)

TABLE 3.2
Continued

Key measures to harness global action

National and regional authorities, including in non-EU countries, are encouraged to:

- engage in international fora to develop a global response to the increase in marine litter
- take domestic action to reduce the leakage of plastics in the environment, prevent plastic waste, and increase recycling

Industry is encouraged to:

- play an active part in supporting an integrated, cross-border, circular plastics economy, including through the development of a global protocol for plastics

Source: Fully adapted from ANNEX II communication from the Commission to the European Parliament, the Council, the European Economic and Social Committee, and the Committee of the Regions A European Strategy for Plastics in a Circular Economy (European Commission, 2018).

friendly aspects, where producers are obliged to manage the waste being generated at end of life. The EPR system was introduced in the year 1993 in Austria, which was proven a success in terms of improving the collection rates of wastes. From 2023, the EPR provisions are extended to apply to several single-use plastics. The requirements come with higher recycling and collection targets for the said products, and the manufacturers are obliged to pay for awareness campaigns and the cost of clean-up of littering. The rates charged mainly depend on the degree to which packaging is recovered. In November 2020, the law consolidated federal law: Complete legislation for the Waste Management Act 2002, version of 03/12/2023 (*Bundesrecht konsolidiert: Gesamte Rechtsvorschrift für Abfallwirtschaftsgesetz* 2002, *Fassung vom* 12.03.2023) stated that the proportion of reusable packaging sold in retail needs to achieve 25% in 2025 to at least 30% by 2030; beer 15%, water 15%, juice 10%, non-alcoholic soft drinks 10%, and milk 10%. In order to ease the retailers to meet the requirement of the legislation, there are two options for implementing the law.

Option 1: The manufacturer who has products of 10, for instance, should have 10–15% marketed in returnable bottles or two products, with an exception made for bottles or cans with a small volume of 0.5 litres below immediate consumption. However, the products need to be part of the deposit return system.

Option 2: The manufacturer needs to ensure 25% of the total volume sold per beverage category or every fourth litres sold are made using reusable bottles or cans. In all categories, at least one product is sold in reusable bottles.

For the deposit return scheme as stated previously, starting in year 2025, the national deposit return system for disposable (single-use) beverage packaging made of plastic or metal, the deposit return scheme is:

a) 0.25 EUR deposit for all plastic bottles and all cans ranging from 0.1 litre to 3 litres.

b) A central organizing entity is formed with the stakeholders from industry and supermarkets.
c) Such a return scheme can be accessible publicly, either in supermarkets, shops, or common places like train stations.

Meanwhile, Austria has started a new levy system on producers and importers of plastic packaging at the average fee of 0.80 EUR per kg of plastic packaging based on the nature of recyclability of the products or contents of recycled material. In order to promote consumers choosing environmentally friendly products, the sellers of the beverage packaging are obligated to clearly state on the label whether it is single-use "ONE WAY" wordings or reusable beverage containers "REUSABLE" wording. This approach is also applicable to online retailers, yet small stores of less than 400 m^2 of space are exempted. Another related regulation of Austria is consolidated federal law: Complete legislation for the Packaging Ordinance 2014, version of 03/20/2023 (*Bundesrecht konsolidiert: Gesamte Rechtsvorschrift für Verpackungsverordnung* 2014, *Fassung vom* 20.03.2023). The objectives of the regulation are to ensure that the reuse of packaging is the main approach to avoid packaging waste; otherwise, recycling and recovery are the next approaches so that the amount of waste can be reduced. While promoting the reusing and recycling of packaging, food hygiene and consumer safety and health are still the foremost concerns to be taken care of. The risk of plastic's exposure to the surrounding environment needs to be prevented and mitigated when there is a transition to a circular-economy model with sustainable business activity and products are manufactured. In this regulation, Austria has set the targets of household recycling rates (Table 3.3) as follows.

Another important policy that has been implemented to encourage other means of waste treatment by the Austrian government is the landfill tax, which was introduced in 1989. The tax is paid by the landfill operators on a tonnages basis and depending on the type of landfill. In other words, landfills equipped with better and more environmentally friendly technology are subjected to lower tax rates compared to those without any anti-pollution provisions. The Austrian landfill tax is known as "*Altlastensanierungsbeitrag*," or "contaminated site contribution." At the starting points of the landfill tax, hazardous wastes were charged at ATS 200 (EUR 14.53) per tonne for hazardous wastes and ATS 40 (EUR 2.91) per tonne for all

TABLE 3.3
Austria's Household Recycling Rates and Targets

	from 2022	from 2023	from 2025	from 2030
Paper, cardboard, cardboard, and corrugated cardboard	80%	80%	80%	85%
Glass	80%	80%	80%	85%
ferrous metals	50%	60%	65%	75%
aluminium			65%	75%
plastics	60%	75%	80%	85%
beverage carton	50%	60%	80%	80%

wastes. Later, starting in 2004, the tax was amended to a different rate, depending on the landfill technology, with a surcharge being applied for low-standard technology. From 1997 to 2008, the municipal solid-waste landfilled at a lower standard was EUR 87 per tonne from 2006 to 2008, plus a surcharge of EUR 29 per tonne, where there was no impermeable liner or no vertical enclosure and a further EUR 29 per tonne where there was no landfill gas capture and treatment system (Ettlinger and Bapasola, 2016). In addition, beginning 2006, such a waste treatment tax has further included an incineration tax of EUR 7 per tonne. In 2008, the tax further was revised to include landfill for construction or inert waste and soil excavation: EUR 9.20 per tonne; residual waste landfills: EUR 20.60 per tonne; and mass or hazardous-waste landfills, including output from MBT: EUR 29.80 per tonne (Ettlinger and Ayesha, 2016). Untreated MSW that is stored or exported for disposal in a lower-standard landfill is taxed at EUR 87 per tonne, and the incineration tax is EUR 8 per tonne. Although the landfill tax was not aimed directly to reduce the landfilling of plastic packaging wastes, since separated packaging waste is rarely landfilled, the levy has a supportive effect on packaging wastes that eventually ends up in mixed households and commercial wastes. In simple terms, the policy aims to divert packing wastes from ending up in landfills directly while channelling towards recovery.

3.3 BELGIUM

Historically, since 1993, Belgium has introduced a Packaging Levy in conjunction with other ecotaxes marked as a notable policy move. This initiative aimed to transform consumer behaviour by emphasizing reusability through deposit-refund systems and recycling by manipulating the relative pricing of goods. At that time, for all types of beverage containers, the tax was uniformly set at 15 francs (EUR 0.37), independent of their potential for reusability. Meanwhile, some exemptions were selectively given in relation to the product recycling or re-use rates. In the effort to achieve these recycling goals, the Belgian industry established Fost Plus, which is a non-profit organization in Belgium that operates as an extended producer responsibility (EPR) system for packaging waste in the country. Fost Plus was found to manage and promote the recycling and recovery of household packaging waste in Belgium. Fost Plus was established in 1994 as a response to the European Union's Packaging and Packaging Waste Directive by initiating a collaboration spectrum of stakeholders, including producers, retailers, local authorities, and waste management companies. This is important because Fost Plus plays roles to ensure the collection, sorting, and recycling of packaging waste. It implements recycling programmes, raises public awareness about recycling practices, and supports the development of recycling infrastructure in Belgium. In order to achieve financial sustainability, producers of packaged goods marketed in Belgian market are required to contribute funds to Fost Plus based on the amount and type of packaging on the market. These contributions are used to run the organization's activities related to packaging waste collection, sorting, and recycling. Fost Plus plays a crucial role in helping Belgium achieve its recycling and environmental sustainability goals by ensuring that packaging waste is managed responsibly and diverted from landfills.

Thereafter, in 2004, the Packaging Levy underwent revision, resulting in the exemption of all beverage containers from a value-added tax (VAT), and subsequently, introducing substantially elevated tax rates, particularly for non-reusable containers. Such a move was to diminish the cost of reusable packaging while upholding higher price-levels for other container types. However, the prioritization of reusable containers over recyclable counterparts encountered legal challenges in 2005 and 2007. Subsequently, Belgium implemented the "Environmental Levy" in 2007, representing the second tax affecting packaging, yet public scepticism surrounding its underlying motives, and hence, the original comprehensive levy plan was scaled down to encompass economic disincentives aimed at four specific product groups: single-use carrier bags (EUR 3 per kilogram), single-use plastic (EUR 2.70 per kilogram), aluminium foil (EUR 4.50 per kilogram), and disposable plastic cutlery (EUR 3.60 per kilogram). Importantly, biodegradable bags were exempted from this taxation. On January 1, 2015, the Environmental Levy was discontinued, attributed to its successful realization of set objectives. The inception of the Environmental Levy aimed to tackle the consumption of single-use disposable products. For instance, the levy notably raised the cost of aluminium foil. It is noteworthy that, while this marked a substantial transition, the imposition of taxation on plastic bags, a component of this levy, evolved as a natural progression following communication campaigns and voluntary agreements aimed at mitigating plastic bag consumption. These efforts had already resulted progressive reductions in plastic bag usage prior to the tax's imposition, surpassing the targets outlined in the voluntary agreement. However, the tax's impact was more pronounced within larger supermarket chains compared to smaller retail establishments, where the practice of providing bags without charge continued.

Besides that, the recycling industry in Belgium faces complexities, particularly within the polyolefin recycling chain, due to limited applications for recycled polyolefin. Beyond regulatory measures, the general public remains sceptical of recycled plastics, often necessitating their export from Belgium. Notably, from 2014 to 2016, an average net mass of 200,000 tonnes of plastic waste was exported annually from Antwerp Port to China, accounting for approximately 59% of total plastic waste exports (Simon, 2018). According to a European Commission survey measuring Europeans' attitudes towards waste management, over 70% of respondents in Belgium sorted their household waste properly. To reduce household waste, 57% of Belgians believed manufacturers held the responsibility. In regard to plastic waste, 97% of respondents sorted plastic bottles and related items, while 63% believed higher taxes on unsorted household waste would encourage better separation. A majority preferred a pay-as-you-throw model based on unsorted waste quantity, with 54% of Belgians agreeing. Ultimately, 68% preferred purchasing new products over remanufactured ones.

Recently, Belgium's Single-Use Plastics (SUP) Directive was made to strengthen mitigating and diminishing the environmental impact of specific plastic products, while concurrently fostering the transition towards a circular economy. As of January 24, 2022, a Royal Decree has enforced the prohibition of utilizing catering products and non-food related items. Violations could incur penalties ranging from €50 to €100,000, in addition to the possibility of imprisonment for periods spanning from eight days to two years of prison time (Card, 2016).

a) Starting from January 17, 2023, Belgian legislation has effectively banned the use of shopping bags and plastic cups, both categorized as single-use plastic items.
b) Commencing July 3, 2024, plastic caps and lids must be affixed to plastic bottles and beverage containers.
c) By the year 2025, beverage bottles composed of PET material must comprise a minimum of 25% recycled plastics. This requirement is slated to increase to a minimum of 30% by 2030 for all plastic beverage bottles up to three litres in size.
d) Further notable measures on the horizon encompass the implementation of extended producer responsibility, along with the establishment of specific targets for the separate collection of single-use plastic bottles.

3.4 UNITED KINGDOM

Following the Parliamentary approval starting in October 2023, England implemented a ban on single-use plastic plates, trays, bowls, and cutlery as well as plastic balloon sticks and specific types of polystyrene cups and food and drink containers. The ban involves items made from plastic materials that are bio-based, biodegradable, or compostable as well. The announcement was delivered by the UK Environment Secretary, Therese Coffey, with the official title of Environmental Protection (Plastic Plates etc. and Polystyrene Containers etc.) (England) Regulations 2023. However, certain exemptions still applied; for instance, plates, trays, and bowls employed as packaging for takeout or pre-packaged food items (like salads and items from takeaways) remain unaffected by this ban, whereas all these items fall under the new, extended producer responsibility legislation, which will take effect in 2024. Meanwhile, the UK government previously banned single-use plastic straws, stirrers, and cotton buds in England since 2020 (Keller and Heckman, 2023).

Additionally, Wales and Scotland also announced comparable bans on single-use plastics. In Wales, the Environmental Protection (Single-use Plastic Products) (Wales) Bill received approval on December 6, 2022. In Scotland, the Environmental Protection (Cotton Buds) (Scotland) Regulations 2019 prohibits the supply and manufacture of plastic-stemmed cotton buds. Such regulation, also introduced in Scotland, called Environmental Protection (Single-use Plastic Products) (Scotland) Regulations 2021, came into effect on June 1, 2022. These new regulations prohibit the manufacturing and/or supplying certain specified single-use plastic products. Meanwhile, there are also several related regulations in the United Kingdom about waste management.

a) Producer Responsibility Obligations (Packaging Waste) Regulations 2007 and the Packaging (Essential Requirements) Regulations 2015, which implement the Packaging and Packaging Waste Directive.
b) Waste (England and Wales) Regulations 2011, which implement the Waste Framework Directive and address waste obligations under the Environmental Protection Act 1990.

c) Environmental permitting requirements, outlined in the Environmental Permitting (England and Wales) Regulations 2016, as well as the Waste Management Licensing (Scotland) Regulations 2011 in Scotland, governing waste management practices and offenses.
d) Powers within Part 3 of the Environment Act 2021, which enable the establishment of Extended Producer Responsibility (EPR) schemes and Deposit Return Schemes (DRS), harmonize recycling systems nationwide, enhance controls on plastic waste exports, and introduce charges for single-use plastics.
e) Packaging Waste (Data Reporting) (England) Regulations 2022.

Since the UK already exited from the European Union (EU), in order to ensure the continuity of UK environment and waste legislation, various statutory instruments have been introduced by amending EU-derived domestic legislation. The Northern Ireland Protocol, amended by the Windsor Framework, ensures the application of certain EU laws to Northern Ireland, while devolved administrations may adopt their own approaches. In line with the fact that the UK is no longer an EU Member State, the transposition of the Single-Use Plastics Directive into domestic law is not obligatory. Nevertheless, the Northern Ireland Protocol necessitated certain provisions of the directive to be transposed into the state by January 1, 2022. These included measures to reduce plastic cup and food container consumption, restrictions on certain single-use plastic products, product-specific requirements (primarily for plastic bottles), and new labelling mandates for specific plastics. For England, as of May 21, 2021, businesses must charge customers a minimum of 10p for single-use plastic bags. Comparable charges are 10p in Scotland, 5p in Wales, and 25p in Northern Ireland. Additionally, in England, the distribution and sale of single-use plastic drinking straws, plastic-stemmed cotton buds, and plastic drink-stirrers were banned from October 1, 2020. Further bans on single-use plastic plates, trays, bowls, cutlery, balloon sticks, and expanded and extruded polystyrene containers are set for October 2023, in alignment with bans in Scotland and the EU. Scottish regulations extend to oxo-degradable plastic products. On the other side, the Welsh government plans to restrict the sale of certain single-use plastics, including stirrers, drink straws, and plastic-stemmed cotton buds, from summer or autumn 2023. The government is also considering enforcement tools such as monetary penalties and compliance notices. The Environment Act 2021 came into force on November 9, 2021, aiming to revolutionize resource and waste management approaches. Its central goal is to reshape the production and consumption patterns of government, businesses, and individuals. Notably, Part 3 of the Act establishes the groundwork for resource and waste management, providing the necessary legal structure for the realization of the Resources and Waste Strategy's objectives. This segment empowers the implementation of Extended Producer Responsibility (EPR) and Deposit Return Schemes (DRS), sanctions the imposition of charges on specific single-use plastics, delineates protocols for segregating recyclable materials during collection, and facilitates the adoption of electronic waste tracking.

Scotland's deposit return initiative for beverage packaging, initially planned for July 2022, has been rescheduled to commence on March 1, 2024. Consumers will be charged a 20p deposit on specified beverage containers. However, the launch of a similar deposit return system in England, Wales, and Northern Ireland has been postponed and might start by October 1, 2025. Anticipated features include mandatory labelling standards, and the schemes are likely to resemble the Scottish model, except for glass containers. Also, a plastic packaging tax was introduced in April 2022. This tax escalated to £210.82 per tonne on April 1, 2023. The UK has universally banned microbeads in rinse-off personal care items.

3.5 BULGARIA

In Bulgaria, the Waste Management Act (WMA) has existed since July 2012 as the country's comprehensive legal framework, mainly to play a role as a fundamental guide for regulating measures and control mechanisms aimed at safeguarding the environment and human health (CMS, 2023a). In general, the Act serves to prevent and reduce waste generation, mitigation of the adverse impacts associated with waste generation and management, and the overall reduction of the environmental impact of resource utilization with better efficiency. The WMA also focuses on provisions concerning products that produce hazardous and/or ordinary waste during their production or after their final use. This Act sets forth criteria for extended producer responsibility, particularly concerning these products, with the intention of encouraging practices such as waste reutilization, prevention, recycling, and other methods of waste recovery. In the meantime, Bulgaria also enforces additional regulations alongside the WMA; specifically, addressing plastic bags and product packaging, as follows:

a) The Ordinance on Packaging and Packaging Waste ("OPPW").
b) Ordinance No. 2 of 23 January 2008 on Plastic Materials and Articles Intended to Come into Contact with Food.
c) Ordinance No. 2 of 23 July 2014 on Waste Classification.
d) Ordinance for Determining the Order of Payment and the Amount of the Product Fee.
e) Ordinance on reduction of the impact of certain plastic products on the environment ("ORICPPE").

Importantly, these regulations collectively contribute to Bulgaria's comprehensive waste management strategy, emphasizing responsible resource utilization, waste reduction, and proper handling of plastic bags and packaging materials. Within Bulgaria, the Ordinance Governing Payment Order and Product Fees mandates charges on plastic shopping bags and products that can become major wastes in the local market. For specific cases, such as the originators of such plastic bags on the market being unidentifiable, identical obligations are extended to distributors, including those who sell to end consumers. According to Article 59, paragraph 1 of the Waste Management Act (WMA), individuals who distribute specific products within Bulgaria and do not fulfil their obligations or meet targets for separate collection, reuse, or recycling of resultant waste are required to pay a product tax (CMS, 2023a). Product fees associated with plastic shopping bags (except for very thin ones without a grip) are solely levied when consumers purchase goods at the point of sale. As for plastic packaging, the product fee stands at BGN 2.33 per kilogram and covers several categories, including packaged items and packaging materials used for goods at the sales location and even oxidizable plastic shopping bags. On the other hand, there are selective plastic shopping bags that are exempt from product fees, including those that fulfil specific prescribed conditions, very thin bags without a grip, and plastic bags conforming to the EN 13432 standard, bearing Bulgarian inscriptions that fulfil particular labelling requisites.

The ordinances related to mitigating the impact of specific plastic products on the environment transposes the stipulations of EU Directive 2019/904 into Bulgarian law. This ordinance spells out the measures to be implemented by producers, importers, and business operators to curb the ecological impact of certain plastic goods (CMS, 2023a). These measures include:

a) Prohibitions on specific, single-use plastic products, including cotton swabs, disposable cutlery, plates, straws, plastic carrier bags with a thickness below 25 microns, and others, including oxo-degradable plastic products.
b) Introduction of marking requirements for producers, importers, and distributors of personal hygiene and tobacco products. These markings are intended to inform users about waste management and disposal options, plastic content, and potential environmental repercussions.
c) Mandatory registration for those placing certain plastic products on the market, such as food and beverage containers up to three litres, lightweight plastic carrier bags with a thickness below 50 microns, and more.
d) Implementation of extended producer responsibility measures in specific cases, involving raising consumer awareness and payment of a product fee for certain items.
e) Enactment of measures to reduce the use of single-use plastic products in the distribution of food and beverages. This includes information signage detailing product prices and environmental impact, with consumers incurring additional fees specified by ORICPPE.
f) From July 3, 2024, plastic beverage containers with capacities up to three litres and plastic lids or covers are only permissible on the market if these lids and covers remain attached during product use. In Bulgaria, the "Ordinance determining the procedure for payment and the amount of product fees" outlines charges applicable when introducing products that turn into mass waste after their use to the Republic of Bulgaria's market. A product fee is levied for specific categories; notably, for packaged goods or materials used for packaging goods at the point-of-sale location. However, certain instances exempt the need for a product fee:

    a) Plastic shopping bags, provided they meet the criteria of the bag's thickness is at least 25 microns (μm). The bag has minimum dimensions of 390 mm by 490 mm when unfolded. The bags bear inscriptions in Bulgarian, either on each bag's packaging or directly on the bags. These inscriptions must include name, registered office, and address of the entity placing the bags on the market with the designation: "reusable bag."

b) Very thin plastic shopping bags without a grip.
c) Plastic shopping bags adhering to the EN 13432 standard, marked with Bulgarian inscriptions that contain name, registered office, and address of the entity placing the bags on the market. Labelling: "the bag is biodegradable," date of manufacture, and expiration.

3.6 CROATIA

Generally, the most notable legislation in Croatia is the Waste Management Act, the Bylaw on Waste Management, the Bylaw on Packaging and Packaging Waste, and the Regulation on Management of Packaging Waste (CMS, 2023b). The Waste Management Act plays more roles to shape strategies to curb the adverse impacts of waste on human health and the environment, primarily through waste volume reduction. The importance of this Act is that it concerns waste management with a wide angle by avoiding processes that can risk human health and the environment while leveraging the valuable attributes of waste and aligning with the objectives of the European Green Deal. The Bylaw on Waste Management governs a spectrum of areas, including the content of diverse reports (related to by-products and waste reuse centres), the execution of waste management procedures, and the roster of procedures necessitating a waste management permit.

The Bylaw on Packaging and Packaging Waste establishes criteria for the collection, storage, and processing of packaging waste, alongside provisions for averting packaging waste, encouraging reuse, recycling, and other recovery processes, and reducing the ultimate disposal of packaging waste. Meanwhile, the Regulation on Management of Packaging Waste intricately delineates the approach for attaining specified targets in packaging waste management, elaborating on the packaging waste management fee and the return fee (CMS, 2023b). The Waste Management Act has emphasized that specific items, as follows, are banned:

a) Single-use plastics, including cotton bud sticks, straws, cutlery, plates, beverage stirrers, support sticks for balloons, food containers crafted from expanded polystyrene, and beverage containers and cups made from expanded polystyrene.
b) Products composed of oxo-degradable plastic.
c) Lightweight plastic carrier bags, excluding very lightweight variants.
d) Beverage containers (liquid) with capacities up to 3 litres, encompassing plastic caps or lids, and multi-layer (composite) beverage packaging, along with their plastic caps and lids, except if these caps or lids detach during the intended usage of the products. It is noteworthy that metal caps or lids possessing a plastic seal are not classified as plastic. Implementation of point (d) will commence on July 3, 2024.

Furthermore, the Act mandates that single-use plastics, such as pads, tampons, tampon applicators, wet tissues, and tobacco products accompanied by filters, must feature a conspicuous, easily readable, and non-erasable label on their packaging. In the case of drinking cups, this label should be placed directly on the product. This marking serves to inform users about the plastic content in the product, the detrimental environmental consequences of littering, and the appropriate waste management procedures.

An operational refund fee system is also in effect, specifically for managing single-use packaging made from PET, Al/Iron, and glass with a volume equal to or exceeding 0.20 litres, intended for beverages. The refund fee operates as an incentive to encourage waste holders to segregate beverage packaging waste from other waste and deliver it to the vendor or the entity overseeing recycling facilities to receive a reimbursement. This refund fee is a sum that producers placing single-use beverage packaging on the market pay to the Environmental Protection and Energy Efficiency Fund. Indeed, producers include this fee amount in the product price charged to buyers. The end user or consumer is entitled to receive the fee back from the seller or recycling facility operator upon returning the beverage packaging waste. The Fund reimburses the seller or recycling facility operator the refunded fee amount, subtracted from the producer's payment to the Fund. It's important to note that paying the refund fee does not fall under waste trading, and no tax is levied on the refund fee. Presently, the refund fee stands at HRK 0.5 (approximately EUR 0.07) per beverage packaging unit.

Pertinent regulations encompass the Commission Regulation (EU) No. 10/2011 from January 14, 2011 concerning plastic materials and articles meant for food contact, along with the recent Commission Regulation No. 2022/1616 of September 15, 2022, addressing recycled plastic materials and articles intended for food contact and repealing Regulation (EC) No. 282/2008. Croatian legislation further specifies details in line with Regulation No. 10/2011 to involve competent bodies responsible for implementing the regulation and penalties for producing and introducing objects and materials designed for direct food contact that do not adhere to the mentioned regulation. While the Act on materials and objects in contact with food addresses these details, it has remained unamended since 2018, and thus, does not address particulars compliant with Regulation No. 2022/1616. However, it does encompass provisions aligned with the now-repealed Regulation No. 282/2008 concerning recycled plastic materials and articles for food contact. The government intends to revise this act in the current year to bring it in line with the updated regulations.

3.7 FRANCE

The regulation of waste prevention and management in France falls under the jurisdiction of the French Environmental Code (FEC), which has been amended by the Circular Economy Law (Law No. 2020–105 of 10 February 2020) and its associated decrees (CMS, 2023c). These legal changes have brought forth fresh requirements concerning plastic waste, leading to a shifting in both production and consumption practices. The main objective behind the Circular Economy Law is to promote a "circular" economic model, emphasizing the eco-friendly design of products, conscientious consumption, prolonging product lifespan, product reuse, and waste recycling. This approach has resulted in the imposition of numerous new obligations on companies, inevitably accompanied by a range of potential penalties, primarily in the form of administrative fines. The core of waste prevention and management in France is governed by the French Environmental Code, as amended by the Circular Economy Law (Law No. 2020–105 of February 10, 2020), and its associated regulations. These legal changes have introduced novel requirements related to plastic waste, resulting in significant implications for both production processes and consumption behaviours.

Over the past years, the French legislative body has been actively combatting plastic pollution, particularly through targeted bans on various plastic items. Notable examples include the discontinuation of providing single-use plastic checkout bags at no cost (effective from January 1, 2016) and the prohibition of selling disposable plastic cups and plates (implemented from January 1, 2020). To advance these efforts, the Circular Economy Law introduced new measures that both bolstered and expanded upon existing initiatives. These measures include the following key actions:

a) A phased ban on additional single-use plastic products, including straws, disposable lids for cups, plastic cutlery, expanded polystyrene containers for on-site or on-the-go consumption (like kebab boxes) starting from January 1, 2021. Additionally, non-biodegradable plastic tea or herbal tea bags were prohibited from January 1, 2022.
b) The prohibition of importing and producing single-use plastic bags intended for sale or complimentary distribution.
c) Imposed restrictions on plastic usage.

Furthermore, commencing January 1, 2023, restaurants with seating for up to 20 patrons are obligated to utilize reusable cups, plates, containers, and cutlery for serving food and beverages meant to be consumed on the premises. Anticipated regulations set to take effect from January 1, 2025 include the banning of plastic containers used for cooking, heating, and serving food within specific healthcare establishments. France has upheld an approach of Extended Producer Responsibility (EPR) for an extended period. This approach follows a "pay-or-play principle," whereby entities involved in producing, importing, or distributing items encompassed by an EPR programme are mandated to either facilitate waste management for their products or financially contribute to such efforts by paying an eco-contribution to an eco-organization. This system has largely pertained to the management of waste stemming from household packaging, used batteries, accumulators, paper, and electrical and electronic equipment (EEE). Guided by French legislation, EPR adheres to the following foundational principles:

a) Regulation or even prohibition of the manufacture, sale, or provision of waste-generating products to simplify waste management.
b) Imposition of minimum recycled material incorporation levels for certain product and material categories to meet recycling targets.
c) Obligation for producers, importers, or exporters to demonstrate that the waste originating from their products can be managed under specific conditions. By no later than January 1, 2030, entities introducing over 10,000 units of a product to the market annually and declaring a turnover exceeding EUR 10 million must prove that their waste can effectively enter a recycling program.

The concept of a producer encompasses "any individual or legal entity engaged in the development, manufacturing, handling, processing, sale, or import of products that generate waste, as well as the components and materials used in their creation." Starting from January 1, 2022, these producers are required to register using a unique identifier. Modifications are introduced to the eco-contribution system, employing a structure of rewards and penalties based on environmental performance benchmarks. These criteria incorporate factors such as the integration of recycled materials, product longevity, ease of repair, potential for reuse or continued use, and the absence of ecotoxicity.

Commencing from January 1, 2022, the principle of Extended Producer Responsibility (EPR) is progressively broadened to encompass emerging product categories, resulting in the establishment of novel "new EPR schemes." These encompass diverse domains like textile products, toys, items associated with sports and leisure, tools for DIY and gardening, automobiles, chewing gums, and more. In the forthcoming years, the scope of the EPR principle will also extend to novel waste categories, including:

a) Non-biodegradable synthetic chewing gums (effective from January 1, 2024).
b) Professional packaging (non-household packaging) and plastic-containing fishing items (starting from January 1, 2025).

It's noteworthy that a recent legislative proposal, endorsed by the Senate on April 13, 2023, plans to merge the waste streams of paper and household packaging. These provisions are intended to apply retroactively from January 1, 2023. Producers within the ambit of EPR initiatives are now obligated to conceive and implement a five-year plan for prevention and eco-design, which may be carried out through collaboration with eco-organizations. The successful implementation of an EPR scheme necessitates several specific components, including:

a) Mandated waste-management responsibilities.
b) Adjustment of eco-contributions based on the environmental performance standards of products.
c) Guidelines for the establishment of individual systems by producers.
d) Provision of funds to facilitate the repair, further use, and reuse of products.
e) Stipulations for the return of used products by distributors.

Since January 1, 2018, the use of microbeads, which are solid plastic particles found in cosmetic products for exfoliation or cleansing, has been prohibited.

The prohibition on microplastics is set to expand to encompass various domains:

a) In-vitro medical devices will face the ban from January 1, 2024.
b) The ban will extend to all rinse-off cosmetics starting from January 1, 2026.
c) For cleaning products and items subject to the European Chemicals Agency's proposed restrictions, the ban will take effect from January 1, 2027.

However, there exist exceptions. For instance, microplastics can still be used in the production of medicinal products for human or veterinary use, but specific instructions must accompany these products. Industrial sites engaged in the production, handling, and transportation of plastic granules must be outfitted with equipment and protocols designed to prevent the loss and leakage of these granules. Furthermore, these sites will be subject to periodic evaluations by certified external entities. A significant development involves new washing machines sold in France. As of January 1, 2025, these machines are mandated to incorporate a mechanism that captures the plastic microfibres released from clothing during the washing process. Additionally, detailed in Decree No. 2022–748 concerning consumer information about the environmental attributes and features of products that generate waste, specific guidelines are outlined for textile products like clothing, linens, and shoes. Information regarding the presence of plastic microfibres is communicated as the proportion of synthetic fibres in the product's mass. This disclosure is requisite if the synthetic fibre content surpasses 50%. The statement "releases plastic microfibers into the environment during washing" accompanies this information. The rollout of this information begins on January 1, 2023, 2024, or 2025, contingent upon the annual turnover and quantity of units of products introduced to the French market.

3.8 GERMANY

Currently, there are no specific regulations in place for handling other types of plastic waste. Instead, the disposal processes are governed by the general waste laws outlined in the German Circular Economy Act (*Kreislaufwirtschaftsgesetz*) (CMS, 2023d). Particularly, these laws emphasize practices like separate waste collection and adherence to the waste hierarchy. Guidelines regarding packaging waste are detailed within the German Packaging Act (*Verpackungsgesetz*), which outlines the following key responsibilities:

a) General Packaging Requirements: These involve criteria concerning packaging dimensions, weight, reusability, recyclability, and the utilization of secondary raw materials.
b) Restrictions on Hazardous Substances: Regulations set limits for substances like lead, cadmium, mercury, and chromium VI in both packaging and its components.
c) Packaging Labelling Standards: While labelling isn't obligatory, certain packaging materials are required to carry specific labels.
d) Producer/First Distributor Responsibilities: Those handling system-relevant packaging filled with goods must register with the packaging register LUCID and partake in a take-back system while reporting relevant data. LUCID is "*Leichtes und umweltfreundliches Container-Informationssystem Deutschland*," which means to "Light and Environmentally Friendly Container Information System Germany." LUCID is the packaging register established in Germany as part of the German Packaging Act (*Verpackungsgesetz*). It serves as a centralized database that registers producers and distributors of packaging and ensures compliance with the extended producer

responsibility requirements outlined in the Packaging Act. Companies that place packaging on the German market are required to register with LUCID and provide information about the type and quantity of packaging they use. This information is used to calculate the contributions that these companies need to make to support the recycling and disposal of packaging waste. LUCID also helps monitor and track the recycling and disposal activities related to packaging materials.

e) Responsibilities for Non-System-Relevant Packaging: Producers and distributors of non-system-relevant packaging filled with goods are generally obligated to take back and appropriately dispose of specific packaging types.
f) Deposit and Return Regulations: Regulations stipulate deposit and return requirements for disposable beverage packaging. Last distributors must indicate whether packaging is disposable or reusable.
g) Ecological Criteria for Dual Systems: The Packaging Act mandates dual systems to factor in ecological considerations when calculating license fees, providing incentives for eco-friendly packaging.
h) Non-Compliance Penalties: Violations of the Packaging Act's obligations can result in distribution bans and administrative fines, reaching up to EUR 200,000.

In November 2020, the Federal Parliament (Bundestag) passed an amendment to the Packaging Act, aimed at enacting the provisions of Directive (EU) 2018/852 that amend Directive 94/62/EC on packaging and packaging waste (CMS, 2023d). The objective of the amendment is to diminish the use of lightweight plastic carrier bags and it came into effect on February 9, 2021. However, the legal prohibition on introducing lightweight single-use plastic bags to the market will only take effect from January 1, 2022. This ban encompasses disposable plastic bags with a thickness of less than 50 micrometres, with certain lightweight plastic bags having a thickness of under 15 micrometres being exempted. These latter bags are predominantly used for carrying loose fruits and vegetables. Prohibiting these bags as well could lead to an increased reliance on more complex packaging, which would undermine the purpose of the law. Violation of this planned ban will be treated as an administrative offense, resulting in fines of up to EUR 100,000. The legal prohibition is designed to replace the previous voluntary agreement with retailers to not distribute plastic bags for free. Since 2016, German retailers have charged customers for nearly all plastic bags, leading to a reduction of approximately two-thirds in their consumption. This successful voluntary agreement is now set to continue through the imposition of a ban on lightweight plastic carrier bags. Consequently, retailers that had not previously engaged in the voluntary agreement will also be subject to the prohibition.

Furthermore, the German Ordinance on Single-Use Plastics, effective from July 3, 2021, implements the provisions of Directive (EU) 2019/904. This ordinance bans certain single-use plastic products that have eco-friendly alternatives available. Items like plastic cotton swabs, disposable cutlery, plates, straws, stirrers, cotton buds, balloon wands, and certain containers (like foamed expanded polystyrene) are among the products prohibited. Additionally, the new German Ordinance on the characteristics and labelling of specific single-use plastic products implements requirements from Directive (EU) 2019/904 (CMS, 2023d). It mandates that single-use plastic products listed in the annex, with plastic closures and lids, must remain attached to containers during use. Starting from July 3, 2021, the guidelines for labelling will be enforced. As of July 3, 2024, the introduction of single-use plastic beverage containers to the market will be restricted to those equipped with attached plastic lids and closures during their usage period.

In addition, German Circular Economy Act (*Kreislaufwirtschaftsgesetz*) are broad directives in relevant to producer responsibility. In the context of packaging, these responsibilities are outlined specifically in the German Packaging Act, whilst the producers have the option to undertake their obligations voluntarily, including voluntary reclamation of products after they've been used. In October 2020, the German Circular Economy Act was adapted to align with the stipulations of Directive (EU) 2019/904 and Directive 2008/98/EC (amended by Directive 2018/851/EU). This adaptation granted the authority to create regulations that might enforce cost-sharing commitments on producers of certain products for the purpose of litter cleanup. It also established a duty of care for producers to ensure their products remain usable and don't become waste. The obligation for cost-sharing is limited to single-use plastic products outlined in Part E of Directive (EU) 2019/904. The duty of care complements product responsibility and surpasses the provisions of European regulations. Its primary objective is to prevent the disposal of returned goods. Specific regulations concerning cost-sharing obligations and the duty of care are yet to be defined in further ordinances awaiting approval from the Federal Government. The precise scope, i.e., the goods and companies to which these obligations apply, is still pending determination. To adhere to Directive 2008/98/EC on waste, the German Circular Economy Act also incorporates measures to prevent food waste (CMS, 2023d).

Recycling quotas set at the EU level extend to various aspects, including plastic packaging, in Germany. Furthermore, in response to amendments in the Packaging Act, the marketing of lightweight plastic carrier bags with wall thicknesses ranging from 15 to 50 micrometres, intended for point-of-sale filling, will be banned in Germany starting from January 1, 2022. This move aims to achieve a substantial reduction in the use of lightweight plastic carrier bags. Germany's strategy includes banning disposable plastic articles and expanding producer responsibility to cover cleanup expenses. Additionally, the promotion of increased recycle products utilization supports sustainable plastic management. Furthermore, efforts to fortify the reusable system and enhance limitations on plastic residues in organic waste (e.g., compost) are being pursued. In the latest development where the Packaging Act is further pending amended in order to align with the Single-use Plastics Directive (EU) 2019/908 in national law. This amendment encompasses the following areas among others:

a) A mandate for restaurants, bistros, and cafés selling on-the-go food or To-Go beverages to offer reusable packaging alternatives by 2023. These alternatives must not exceed the cost of disposable packaging. Reusable cups for all drink sizes must be available. Businesses with five employees or fewer and shop areas under 80 square metres are exempt, yet they still need to offer customers the option of using their reusable containers.

b) Imposing a mandatory deposit on all disposable plastic beverage bottles and beverage cans from 2022. Only milk and milk products will undergo a transition period until 2024.
c) Online marketplace operators and fulfilment service providers are required to ensure that packaged goods producers on their platforms are listed in the packaging register (LUCID) and participate in at least one dual system. These obligations come into effect on July 1, 2022.
d) By 2025, PET beverage bottles must comprise at least 25% recycled plastic. This quota will rise to a minimum of 30% by 2030 and will encompass all single-use plastic bottles, except for those with glass or metal bodies and plastic caps, lids, labels, stickers, or wrappers. Producers can decide to meet this quota per bottle or spread it out over a year in relation to their entire bottle production.

In the European Union, recycled plastics used for food contact must undergo a safety assessment and authorization process. This process is overseen by the European Food Safety Authority (EFSA) and the European Commission. The authorization process ensures that recycled plastics used for food contact are safe for human health. Specifically, Regulation (EC) No. 282/2008 outlines the requirements for recycling processes that can be used to produce recycled plastics for food contact. This regulation also specifies that materials and articles made from recycled plastics for food contact must be obtained from processes that have been authorized by the EFSA and the European Commission. In Germany, applications for authorization of recycling processes for recycled plastics for food contact must be submitted to the Federal Office of Consumer Protection and Food Safety (*Bundesamt für Verbraucherschutz und Lebensmittelsicherheit*) (CMS, 2023d).

3.9 HUNGARY

In Hungary, Government Decree 349/2021. (VI. 22.) is the enactment that enforces the provisions of Directive (EU) 2019/904 issued by the European Parliament and the Council on June 5, 2019. The directive focuses on mitigating the environmental impact of specific plastic products. This decree pertains to plastic-containing fishing gear and various single-use plastic items listed within it (CMS, 2023e). It outlines specific criteria for these products, mandates labelling and information dissemination, and places responsibilities on manufacturers, including registration and record-keeping obligations. The decree will be implemented in several stages, and the ban on the free distribution of single-use plastic products went into effect on January 1, 2023.

An amendment to Act CLXXXV of 2012 on Waste, effective from March 3, 2021, has incorporated the concept of an extended producer responsibility scheme in line with Directive 2008/98/EC. This amendment establishes the underlying framework and fundamental principles of the extended producer responsibility scheme. In essence, the scheme compels producers to take financial or financial and organizational accountability for waste management (CMS, 2023e).

Government Decree 80/2023. (III. 14.) provides comprehensive guidelines for the functioning of the extended producer responsibility scheme. This new scheme, commencing in July 2023, applies to various sectors such as packaging, specific single-use plastic items, electrical and electronic equipment, batteries and accumulators, motor vehicles, tyres, office and promotional stationery, cooking oil and fat, certain textiles, and wooden furniture. In essence, the regulations governing packaging and plastics waste are predominantly governed by the following legal texts (CMS, 2023e):

a) Act CLXXXV of 2012 on Waste: This legislation establishes overarching principles and regulations for waste management, including the concepts of waste prevention, extended producer responsibility, and the "polluter pays" principle.
b) Act LIII of 1995 on the General Rules of Environmental Protection: This act empowers the government to create implementing regulations that restrict or prohibit the production, market placement, use, and trade of single-use items. The principles of environmental protection outlined here include precaution, prevention, and restoration.
c) 442/2012 (XII. 29) Government Decree on packaging and packaging waste management activities: This government decree provides detailed guidelines concerning packaging and packaging waste linked to products introduced into the Hungarian market, including the collection of packaging and packaging waste.

Subsequent to the enforcement of Act XCI of 2020, which banned the distribution of single-use plastics starting from July 1, 2021, the implementation of this act embodies (CMS, 2023e):

a) Directive (EU) 2015/720 of the European Parliament and the Council of April 29, 2015, amending Directive 94/62/EC concerning the reduction of lightweight plastic carrier bag consumption.
b) Directive (EU) 2019/904 on lessening the environmental impact of specific plastic products.

There is currently no deposit return scheme (DRS) for packaging or plastic in Hungary. However, a new law will be implemented on January 1, 2024, which will form DRS. This new law, Act II of 2021, modifies the Act LIII of 1995 on the General Rules of Environmental Protection and the Act CLXXXV of 2012 on Waste (CMS, 2023e). The new DRS will be called the "return fee scheme" in Hungary. It will require distributors to charge a return fee for certain products, such as plastic bottles. This fee will be refunded to consumers when they return the empty bottles to designated collection points. The government of Hungary is introducing a return fee scheme as part of its Climate and Environmental Action Plan. The goal of the scheme is to reduce environmental pressures and pollution by encouraging people to return empty bottles and other containers. Under the scheme, distributors of certain products, such as plastic bottles, will be required to charge a return fee. This fee will be refunded to consumers when they return the empty containers to designated collection points. The government hopes that the return fee scheme will increase recycling rates and reduce litter. The specific products that will be covered by the scheme and the amount of the return fee will be determined by a government decree (CMS, 2023e).

The Act II of 2021 is a Hungarian law that aims to implement the government's Climate and Environmental Action Plan. One of the key elements of this plan is to ensure that plastic bottles can be returned. The Act II of 2021 amends the Act CLXXXV of 2012 on Waste, which introduces provisions relating to the binding return fee scheme. This scheme will require distributors to charge a return fee for certain products, such as plastic bottles. The return fee will be refunded to consumers when they return the empty bottles to designated collection points. The State will set up and operate the binding return fee scheme. The scheme will be used to establish a nationwide, single, integrated waste management system for the products covered by the scheme (CMS, 2023e).

The Environmental Product Levy is a different law that applies to a list of products, including packaging products and other plastic products. The amount of the levy is based on the customs tariff number and the weight of the products. The Environmental Product Levy is a tax imposed on certain products, including plastics and packaging. The amount of the levy varies depending on the type of product and its weight (CMS, 2023e).

a) For plastics in general, the levy is HUF 57/kg (approximately EUR 0.16).
b) For plastic packaging bags, the levy is HUF 1900/kg (approximately EUR 5.30).
c) For packaging made from other materials, the levy varies between HUF 19/kg (approximately EUR 0.05) and HUF 57/kg (approximately EUR 0.16).
d) The levy amount applicable to biodegradable plastic bags is HUF 500/kg (approximately EUR 1.4).

The Environmental Product Levy is intended to discourage the use of certain products that are harmful to the environment. It also helps to fund the collection and recycling of these products.

Hungary has approved the new own resources decision of the European Union, which is a system of how the EU raises its own funds. This decision was made by the Hungarian parliament through Act XLV of 2021. The new own resources decision includes a plastic packaging waste own resource, which is calculated based on the weight of plastic packaging waste generated in each Member State that is not recycled. A uniform call rate of EUR 0.80/kg is applicable. Government Decree 301/2021 (VI. 1.) prohibits the placing on the market of products made from oxo-degradable plastic. Oxo-degradable plastic is a type of plastic that contains special additives that make it break down into smaller pieces over time. However, these smaller pieces of plastic can still pollute the environment. Oxo-degradable plastic is often marketed as being environmentally friendly, but it can actually be more harmful than traditional plastic. When oxo-degradable plastic breaks down, it releases microplastics that can enter the food chain and harm wildlife (CMS, 2023e).

3.10 LUXEMBOURG

In Luxembourg, the overarching legal framework in Luxembourg that governs packaging and plastics is established through the following legislations (CMS, 2023f):

a) The “2017 Law,” which was amended, is the law of March 21, 2017 concerning packaging and packaging waste.
b) The “2022 Law,” enacted on June 9, 2022, addresses the reduction of the environmental impact of specific plastic products.

The primary focus of the 2017 Law is to prevent disposable waste and encourage recycling as a means of reducing overall waste. This law was recently updated by the 2022 Law, which aligns with the European Parliament and Council Directive 2019/904 of June 5, 2019. The 2022 Law targets products designed for short-term or one-time usage, aiming to combat marine litter and its environmental repercussions. The implementation of the 2022 Law will be phased in progressively, culminating by December 31, 2024. The 2017 Law encompasses various objectives (CMS, 2023f), including:

a) Reducing the annual consumption of lightweight plastic bags to a maximum of 40 bags per person by December 31, 2025 (excluding very lightweight plastic bags).
b) Prohibiting the free distribution of plastic bags at merchandise points from December 31, 2018 (very lightweight plastic bags excepted).
c) Banning the free provision of certain single-use products like beverage cups, food containers, and plastic bags at points of sale starting from January 1, 2025.

The 2022 Law introduces further measures (CMS, 2023f) as follows:

a) Aiming to decrease single-use plastic product consumption by 20% compared to 2022 levels by 2026, followed by annual 10% reductions thereafter.
b) Prohibiting the sale of specific products where more cost-effective alternatives exist as well as products made of oxo-degradable plastic.
c) Mandating the display of unpackaged fresh fruit and vegetables, unless in batches of 1.5 kilograms or more.
d) Requiring beverage containers’ caps and lids to remain attached during their use.
e) Setting requirements for beverage bottles’ recycled plastic content (25% by 2025, 30% by 2030).
f) Enforcing labelling on single-use products like tampons and wet wipes; educating consumers about waste management and plastic presence.
g) Applying extended producer responsibility to single-use plastic products and fishing gear containing plastic.
h) Mandating producers to cover costs for awareness campaigns, litter cleaning, waste collection, transportation, and treatment.

i) Establishing a deposit return scheme for packaging of liquids and comestible goods without direct sales, covering items like glass bottles and gas bottles.
j) Subjecting beverage packaging for human consumption on the Luxembourg market to a national deposit scheme.
k) Noting that Luxembourg lacks specific taxes on packaging or plastics. However, a new EU levy on non-recycled plastic waste was introduced in 2021, with Luxembourg government funding it from its own budget. A potential plastic tax to recover this EU levy might be explored in the future.
l) Highlighting that plastic bags can no longer be provided free of charge, and a 17% VAT rate (temporary 16% for 2023) applies to specific plastic bags given to consumers.

3.11 THE NETHERLANDS

The primary objective of the Dutch legislative framework is to decrease litter and safeguard both the public interest and the environment. Numerous laws and directives govern the management of packaging and plastic waste (CMS, 2023g), including:

1. General guidelines are outlined in the Environmental Management Act ("*Wet milieubeheer*").
2. The subsequent Decrees elaborate on additional responsibilities and obligations:

   a) The Commodities Act Decree regarding packaging and consumer products ("*Warenwetbesluit verpakkingen en gebruiksartikelen*").
   b) The Commodities Act regulation concerning packaging and consumer products ("*Warenwetregeling verpakkingen en gebruiksartikelen*").
   c) The Decree pertaining to single-use plastic products ("*Besluit kunststofproducten voor eenmalig gebruik*").
   d) The Regulation for single-use plastic products ("*Regeling kunststofproducten voor eenmalig gebruik*").
   e) The Decree about extended producer responsibility ("*Besluit regeling voor uitgebreide producentenverantwoordelijkheid*").
   f) The 2014 Decree on packaging ("*Besluit beheer verpakkingen* 2014").
   g) The Regulation on packaging ("*Regeling beheer verpakkingen*").
   h) The Decree addressing plastic drinking bottles ("*Besluit maatregelen kunststof drankflessen*").
   i) The Decree related to metal beverage packaging ("*Besluit maatregelen metalen drankverpakkingen*").

Starting from July 3, 2021, the marketing of certain types of single-use plastic products is prohibited in the Netherlands. This includes items like cotton swabs, cutlery, plates, straws, beverage stirrers, and the attachments for balloon supports. An exception is made for medical device cotton swabs and industrial/professional balloon sticks (CMS, 2023g). Moreover, products made from oxo-degradable plastics will also face a prohibition. Under the extended producer responsibility (EPR) measures in the Netherlands, producers or importers of single-use plastic products such as food containers, bags, drink packages, drinking cups, lightweight plastic bags, wet wipes, consumer balloons, and filtered tobacco products are responsible for the costs of cleaning, transporting, and processing litter. They must also fund measures to raise awareness about the environmental impact of these products. The implementation date of these measures varies depending on the product (CMS, 2023g). For disposable cups and containers containing plastic, customers will need to pay a fee, starting July 1, 2023. Plastic-containing disposable items will be prohibited for on-site consumption starting January 1, 2024. Plastic caps must be attached to plastic bottles and beverage containers starting July 3, 2023. By 2025, PET bottles must be made with at least 25% recycled plastic, increasing to at least 30% by 2030 (CMS, 2023g). Producers or importers marketing beverages in plastic bottles with a capacity of three litres or less must ensure that 90% of the total weight of bottles, caps, and lids is separately collected annually, starting July 1, 2022. Similarly, producers or importers selling beverages in metal containers with a capacity of three litres or less must ensure that 90% of the total weight of marketed metal packaging is separately collected annually, starting January 1, 2024.

The Dutch government has the authority to introduce additional regulations to reduce plastic pollution. One of these regulations is a new decree that modifies the 2014 Packaging Decree to establish minimum deposit levels for beverage packaging. Producers and importers of water or soda in plastic bottles with a volume of three litres or less must implement a deposit system (CMS, 2023g). This means that consumers will pay an additional fee when they buy these drinks, and they will get a refund when they return the empty bottles to designated locations, such as stores. The costs of this system are covered by the producers and importers. This deposit system was also be extended to metal beverage packaging as of December 31, 2022.

Starting April 1, 2023, a deposit of €0.15 will be charged for beverages sold in metal containers with a capacity of three litres or less. This is part of the Dutch government's efforts to reduce plastic pollution. The deposit will be refunded when consumers return the empty containers to designated collection points. Metal drink packages must bear a logo indicating the applicable deposit. This logo is designed to help consumers identify eligible containers and to encourage them to return them for recycling (CMS, 2023g). The Commodities Act Decree and regulation on packaging and consumer products outline general rules for packaging materials. These rules emphasize the quality of packaging materials in relation to public health. There are no specific regulations on recycled materials in food packaging, but the European Food Safety Authority recommends that at least 95% of recycled materials in new food packaging come from food packaging to prevent contamination with hazardous substances (CMS, 2023g).

3.12 POLAND

In Poland, the regulatory framework concerning packaging and plastics is established through a combination of both EU and national legislation. Under EU law, these regulations encompass Directive 94/62/EC of December 20, 1994 concerning packaging and packaging waste as well as Directive (EU) 2019/904 of June 5, 2019, which focuses on mitigating the environmental impact of specific plastic products. At the national level in Poland, the rules governing packaging and plastics waste are outlined in the following legislative acts (CMS, 2023h):

a) The Act of December 14, 2012 on waste (published in the Journal of Laws of 2022, item 699, as subsequently amended) establishes the fundamental regulatory structure for waste management in Poland. This act implements Directive 2008/98/EC, also known as the Waste Framework Directive.
b) The Act of June 13, 2013 on the management of packaging and packaging waste (published in the Journal of Laws of 2023, item 160) complements the aforementioned legislation with specific provisions relating to packaging and packaging waste. It aligns with Directive 94/62/EC, commonly referred to as the Packaging and Packaging Waste Directive.
c) The Act of May 11, 2001 on the obligations of entrepreneurs regarding the management of specific waste and product fees (published in the Journal of Laws of 2020, item 1903) currently addresses requirements for managing particular types of waste like oils and tyres. This act will also incorporate provisions concerning single-use plastics.

On April 14, 2023, an amendment to the Act on the obligations of entrepreneurs for the management of certain types of waste and on product fees was signed into law by the Polish President on April 27, 2023 (CMS, 2023h). This amendment will come into effect within 14 days of its official announcement. However, certain aspects of the regulations will take effect at later dates, specifically in 2024 and 2025. The Act introduces systematic measures aimed at diminishing the volume of waste derived from single-use plastic products, especially within marine ecosystems. It mandates annual escalation of mandatory recovery and recycling rates for packaging manufacturers. Additionally, it enforces restrictions on disposable packaging and plastic, aligning with Directive 2019/904. The Act incorporates penalties for non-compliance and outlines administrative penalties for entities violating its provisions (CMS, 2023h)

At present, Poland has initiated a limited extended producer responsibility system and is working on harmonizing it with EU standards. As previously mentioned, efforts towards the new, extended producer responsibility system are in preliminary stages, projecting numerous alterations. As the existing regulations, the producer responsibility framework involves these key principles (CMS, 2023h):

a) Producers must register within a dedicated database (Products, Packaging, and Waste Management Database or BDO), excluding those distributing packaged products and those exporting or recycling packaging waste.
b) Producers must ensure packaging quality, designing them for reusability and subsequent recycling or alternative forms of recovery.
c) Proper labelling of packaging is optional.
d) Producers must attain specified recovery levels for packaging waste, including recycling. Failure leads to product fees.
e) Implementation of an information and registration obligation system.

Significant changes are anticipated in the producer responsibility regulations:

a) Introducing a ban on single-use plastic products and oxo-degradable plastic products, as defined in separate legislation.
b) Additional requirements for food service entities, necessitating records of plastic packaging and single-use plastic products purchased and issued.
c) Imposing new fees for single-use plastic product placement to cover waste-management costs.
d) Implementing enhanced supervision and control over recycling processes.
e) Mandating visible, legible, and indelible markings on single-use plastic products' packaging or the products themselves.

In the current Polish context, a product fee is levied for failing to meet the required recycling and recovery benchmarks. Alternatively, this obligation can be fulfilled by appointing an authorized representative (such as a recovery scheme) in Poland, responsible for satisfying the pertinent criteria (CMS, 2023h). The fee is determined based on applicable rates corresponding to the type of packaging, ranging approximately from EUR 0.1 to 0.6 per kilogram, and the packaging's weight placed on the market within a given year. The Act dated April 14, 2023, amending the Act on the responsibilities of entrepreneurs for waste management and product fees, introduces a new fee applicable to single-use plastic packaging placed on the market. Entrepreneurs introducing single-use products listed in Section II of Annex 9 to the Act (wet wipes and balloons) are obliged to pay an annual fee, covering waste cleanup, transportation, and treatment costs associated with products of the same nature as those they've marketed (CMS, 2023h).

The aforementioned fees are computed by multiplying the fee rate established in delegated acts by the weight or number of disposable plastic products an entrepreneur placed on the market during a given calendar year. This pertains to single-use plastic products listed in Annex 9 to the Act, including items like food containers, packets, wrappers, beverage containers with a capacity up to three litres, cups for beverages (including their covers and lids), lightweight plastic carrier bags, wet wipes, balloons, and tobacco products. For single-use plastic products outlined in Sections II and III of Annex 9 to the Act (wet wipes, balloons, tobacco products), these are considered products assembled in the pre-packaging in which they're presented to end-users (CMS, 2023h). Maximum charge rates are as follows (CMS, 2023h):

a) PLN 0.20 per 1 kg for each type of marketed plastic disposable product listed in Section I of Annex No. 9 to the Act (food containers, packets, wrappers, beverage containers with a capacity of up to three litres, cups for beverages including their covers and lids, lightweight plastic carrier bags).
b) PLN 0.03 per unit for each type of plastic single-use product placed on the market, listed in Sections II and III of Annex 9 to the Act (wet wipes, balloons, tobacco products).

Specific rates for the above fees will be established in a separate implementing regulation subsequent to the adoption of the aforementioned Act. Entrepreneurs marketing single-use plastic products outlined in (CMS, 2023h):

a) Sections I and III of Annex 9 to the Act (food containers, packets, wrappers, beverage containers with a capacity of up to three litres, cups for beverages including their covers and lids, lightweight plastic carrier bags, tobacco products) or
b) Section II of Annex 9 to the Act (wet wipes, balloons) are required to remit the fee to a dedicated bank account maintained by the province marshal no later than March 15 of the year following the applicable calendar year.

If an entrepreneur places plastic disposable products on the market (listed in Annex No. 9 to the Act) and does not pay the fee or pays an amount lower than required, the province marshal will determine the fee arrears through a decision. The fee rate used will be the one from the respective calendar year. If the decision is not executed, the voivodship marshal will establish an additional fee equivalent to 50% of the unpaid amount. The fees must be paid within 14 days from the day the decision determining their amount becomes final. Entrepreneurs introducing single-use plastic disposable products listed in Annex 9 to the Act must maintain a register of products placed on the market during a calendar year. This register must record the respective weights or quantities of products, either in paper or electronic form, for a period of five years (CMS, 2023h).

3.13 PORTUGAL

The primary legal framework comprises Decree-Law No. 152-D/2017, dated December 11, 2017, which underwent recent revisions through Decree-Law No. 86/2020, dated October 14, 2020. This decree-law encompasses key aspects pertaining to various activities, encompassing the transport, elimination, and recovery of packaging and plastics. Another pertinent document is Ministerial Order No. 202/2019, dated July 3, 2019, which revolves around a pilot project for a deposit return system (CMS, 2023i). Up to date, Directive (EU) No. 2019/904, issued by the European Parliament and Council on June 5, 2019, related to reducing the environmental impact of specific plastic products, has not been incorporated into Portuguese national legislation (CMS, 2023i). According to a publicly available draft Decree-Law aimed at transposing this Directive, this action is slated for completion on July 1, 2021. In line with the Directive, Member States are obliged to ensure compliance with its provisions by July 3, 2021.

Consistent with the foregoing, Decree-Law No. 152-D/2017 of December 11, 2017 endorsed certain measures for implementation. Particularly, Law No. 76/2019, dated September 2, 2019, which previously mandated the cessation of single-use plastics in food, beverage, and retail activities, necessitated suppliers to adjust within a year (by September 3, 2020) (CMS, 2023i). This deadline was recently extended to July 1, 2021 due to the COVID-19 pandemic through two Decree-Laws (No. 10-A/2020 and 22-A/2021), dated March 13, 2021. Moreover, Law No. 77/2019, dated September 2, 2019 established the prohibition of ultralight, single-use plastics and food trays starting from June 1, 2023, with commercial establishments being obligated to provide alternatives (CMS, 2023i). Law 76/2019, dated September 2, 2019, dictates the discontinuation and unavailability of single-use plastic tableware in all catering and/or drinking sector establishments, other premises, non-sedentary activities, and retail.

Additionally, Resolution No. 141/2018, dated October 26, 2018 by the Council of Ministers, aimed at fostering the reduction of plastic products within the Public Administration. Resolution No. 190-A/2017, dated December 11, 2017 (recently amended in July 2019) endorsed the Action Plan for a Circular Economy. The revised Decree-Law No. 152-D/2017, dated December 11, 2017, outlines a comprehensive management system wherein packers and importers must adhere to either: i) an individual system where they personally undertake waste management responsibility or ii) an integrated system where this responsibility is transferred to a duly licensed management entity responsible for waste treatment and recovery (CMS, 2023i).

In accordance with the revised Decree-Law No. 152-D/2017, dated December 11, 2017, two existing Deposit Return Scheme (DRS) models are in place (CMS, 2023i):

a) The DRS for reusable packaging involves consumers paying a deposit upon acquiring such packaging, which is only refundable upon returning the same packaging. The application of a deposit for other products' packaging is optional.
b) The DRS pilot project pertains to non-reusable plastic beverage packaging. This project, scheduled to continue until June 30, 2021, offers a financial incentive upon delivering plastic packages to containers within commercial malls. The specific terms of this pilot project are detailed in Ministerial Order No. 202/2019, dated July 3, 2019.

Taxes have been introduced within the Budget Law to encourage reduced usage of plastic bags, encompassing: i) EUR 0.08 plus VAT on lightweight plastic bags (implemented in 2015 to curb consumption) and ii) EUR 0.30 per package (created for 2021), mandated to be indicated on the invoice, for single-use packaging made of plastic, aluminium, or multi-material with plastic or aluminium, intended for ready-to-eat meals, take-away, or home delivery CMS, 2023i). Portugal has set ambitious targets, including achieving 65% reuse and recycling of waste by 2025 and 70% by the end of 2030. The revised Decree-Law No. 152-D/2017, dated December 11, 2017, outlines specific objectives:

a) A pilot project for returning non-reusable plastic beverage packaging, in operation until June 30, 2021.
b) Starting from January 1, 2022, a mandatory deposit system for plastic, glass, ferrous metal, and aluminium beverage containers with non-reusable deposits.

Furthermore, the Portuguese Pact for Plastics unites 55 entities, all committed to ensuring that 100% of plastic packaging is either reusable, recyclable, or compostable by 2025.

While reusable tableware is governed by the revised Decree-Law 76/2019, dated September 2, 2019, regulations for the use of recycled materials in food packaging are not currently in place. Such regulations will be established upon the transposition of Directive (EU) No. 2019/904, dated June 5, 2019, includes these measures (CMS, 2023i):

a) Prohibition of introducing single-use plastic products like expanded polystyrene food containers onto the market.
b) Starting from January 1, 2023, establishments utilizing single-use food and beverage packaging for ready-to-eat, take-away, or home-delivered meals must provide reusable alternatives, requiring a deposit that is refundable upon packaging return.

3.14 ROMANIA

Romania's national waste management policy aligns with the goals of European waste prevention policy, focusing on reducing resource consumption and adhering to the waste hierarchy's practical application. The principle of preventive action is a fundamental aspect of Government Emergency Ordinance No. 195/2005 on environmental protection, which has been subsequently revised (CMS, 2023j). Additionally, Directive 2008/98/EC on waste, transposed into Romanian law by Law No. 211/2011 on waste management, establishes the waste hierarchy, prioritizing prevention, preparation for reuse, recycling, other waste management methods, recovery (including energy recovery), and disposal. Packaging and plastics waste are primarily governed by the following legal documents (CMS, 2023j):

a) Law No. 249/2015 concerning packaging and packaging waste management, with subsequent amendments.
b) Law No. 211/2011 on waste management regulations, with subsequent amendments.
c) Law No. 50/2015 for the endorsement of Government Ordinance No. 20/2010, which sets measures for consistent application of EU product marketing rules.
d) Government Emergency Ordinance No. 196/2005 on the Environmental Fund, subject to further revisions.
e) Government Emergency Ordinance No. 195/2005 on Environmental Protection, with subsequent changes.
f) Government Decision No. 856/2002 on waste management records and approval of the list of waste, including hazardous waste, with subsequent amendments.
g) Starting from July 1, 2018, Romania has prohibited the distribution of thin and very thin plastic bags with handles, giving preference to biodegradable or reusable alternatives.
h) Commencing January 1, 2019, the sale of thin and very thin plastic bags with handles has been prohibited as an additional measure to discourage single-use plastics.
i) Various municipalities announced since 2019 that single-use plastics would be restricted at public events, while local retailers, such as IKEA Romania and Kaufland, have already phased out single-use plastics. By the end of 2020, the hospitality sector, exemplified by the ACCOR group, also implemented comparable measures.

j) Market restrictions and product marking will take effect from July 3, 2021, with bottle design requirements coming into play from July 3, 2024. Extended producer responsibility measures are slated to start from December 31, 2024. The Romanian Government General Secretariat (SGG) anticipates that the transposition of Directive 2019/904 will be incorporated into Romanian law by July 2021. However, a public consultation draft bill is yet to be released.

The legislative modifications introduced in 2018 have significantly transformed the business landscape in the field of packaging waste recycling in Romania. Alongside the implementation of measures outlined in Government Emergency Ordinance No. 74/2018, companies engaged in recycling, producers placing packaged goods on the market, and local authorities have assumed new responsibilities related to waste recycling. Producers, facilitated by organizations that enforce extended producer liability obligations, are currently required to directly meet recycling objectives through local authorities and bear the costs of collecting and sorting packaging waste within their scope (CMS, 2023j).

The National Waste Management Plan, published in January 2018, outlines the objective of implementing mandatory extended producer liability schemes for all packaging by 2024. Starting from 2018, manufacturers, facilitated by Organizations Implementing Extended Producer Liability Obligations (OIREP), are currently mandated to meet recycling targets directly through local authorities. They are also required to cover the expenses related to collecting and sorting packaging waste within their jurisdiction. This obligation also takes into consideration any potential revenue generated from the positive value of the waste. Concurrently, local authorities are obligated to establish contracts or collaboration protocols with OIREP to ensure traceability of packaging waste, thereby aiding in the fulfilment of annual recycling targets set for producers (CMS, 2023j).

In March 2020, the Ministry of Environment released a study, developed by the Romanian Economical Studies Academy, addressing the implementation of a deposit-refund system. The study proposes the retailer's role as an intermediary between producers and consumers and as a collector of deposits from consumers. In December 2020, the Ministry introduced a draft bill to establish a deposit-return system (DRS) for non-reusable primary packaging, such as glass, plastic, or metal, with volumes between 0.1 to 3 litres. This system is intended for products like beer, alcoholic beverages, juices, and more. Expected to commence on April 1, 2022, the DRS would apply to domestically produced items as well as imports under non-discriminatory conditions. Producers introducing such packaging must register with the DRS, maintain records, submit relevant information, collect fees from distributors and retailers, and properly label the packaging. The minimum annual return targets are 65% for plastic from April 1 to December 31, 2022, increasing to 80% in the subsequent year and reaching 90% thereafter. Retailers have additional responsibilities under these regulations (CMS, 2023j).

An eco-tax, amounting to RON 0.15 (approximately EUR 0.03), applies to all types of plastic bags, except ecologically friendly alternatives like biodegradable and compostable options made from starch or cellulose. At the beginning of 2021, the Minister of Environment announced that the DRS draft bill would also introduce a packaging tax to enhance recycling. According to this bill, consumers would pay RON 0.5 (approximately EUR 0.1) for each plastic, aluminium, or glass package with a capacity of up to three litres, excluding dairy packaging. The funds would be collected through the DRS system. The European Council has proposed a tax of EUR 0.8 per kilogram of non-recycled plastic packaging waste, set to begin in 2021 as part of the post-COVID-19 economic recovery package totalling EUR 750 billion for the period 2021 to 2027. However, Romanian authorities have not yet revealed any legislative initiatives or implementation strategies for this EUR 0.8 per kilogram tax. Additionally, the alignment of this new tax with the existing RON 2 per kilogram contribution for the disparity between recovery goals and actual recycling quantities remains unclear. As for the plastic tax on non-recycled plastic packaging waste, it remains ambiguous whether this tax would be sourced from the state budget or imposed as a new levy on Romanian companies (CMS, 2023j).

3.15 SLOVAKIA

In Slovakia, the primary legislative framework governing packaging and plastics is the Waste Act (Act No. 79/2015 Coll.), which serves as the central source of regulations concerning these matters in the Slovak Republic (CMS, 2023k). The obligations and responsibilities outlined in the Waste Act are further detailed and supplemented by other legal instruments, including:

a) The Single-use Beverage Packaging Act (Act No. 302/2019 Coll.), which pertains specifically to single-use beverage packaging.
b) The Eco-design Act (Act No. 529/2010 Coll.), which addresses environmental design and the utilization of products.
c) Regulations issued by the Ministry of Environment of the Slovak Republic:
d) Regulation No. 373/2015 Coll. on Extended Producer Responsibility for Specified Products.
e) Regulation No. 371/2015 on Implementing Certain Provisions of the Waste Act.
f) Regulation No. 365/2015 Coll. establishing the Catalogue of Waste.
g) Regulation No. 366/2015 Coll. on Registration Duty and Reporting Obligation.
h) Regulation No. 347/2019 Coll. on Implementing Certain Provisions of the Single-use Beverage Packaging Act.

The Waste Management Plan of the Slovak Republic, although not a legislative source itself, guides the decisions and actions of state authorities in accordance with waste management goals (CMS, 2023k). Currently, a draft of the Waste Management Plan for 2021–2025 is undergoing the legislative process, while the Waste Management Plan for 2016–2020 remains in effect. Slovakia's legislative framework concerning packaging and plastics aligns with European trends, prioritizing environmental protection by minimizing waste generation, including plastics from packaging. Producers of packaging are obligated to charge customers for plastic bags provided with purchased goods, equal to the procurement price of the bag.

Effective from July 3, 2021, Slovakia has implemented a ban on the placement of single-use plastic products and packaging made from oxo-degradable plastics on its market. The specific items prohibited as single-use plastics are listed in Annex 7a of the Waste Act and include items like plastic straws, cutlery, cotton buds, and straws. Products falling under this prohibition, if placed on the Slovak market before its enforcement, can be distributed until December 31, 2021. This regulation aligns with the provisions of Directive (EU) 2019/904, which aims to reduce the environmental impact of certain plastic products (CMS, 2023k).

Additionally, according to the Single-use Beverage Packaging Act, single-use beverage packaging made of plastic or metal that is ineligible for the deposit-return scheme should not be introduced to the Slovak market after December 31, 2022. The legislation pertaining to packaging and plastics imposes several obligations on entities that manufacture packaging or packaged products and place them on the Slovak market. Some of the most noteworthy obligations include the extended responsibility scheme for producers. Under this scheme, producers are responsible for fulfilling duties aimed at waste prevention, collection, and recovery. Producers can meet these obligations individually, subject to statutory conditions, or through contracts with Producers Responsibility Organizations (CMS, 2023k). If producers based outside Slovakia are involved, they must fulfil their obligations through a designated Slovak-based responsible representative, who is subject to the same obligations as domestic producers. The overarching principle is that producers bear responsibility, even for the post-consumption phase of their products, and they must cover costs associated with their products becoming waste through the mechanism of Producers Responsibility Organizations outlined in the Waste Act (CMS, 2023k).

When it comes to producers of single-use beverage packaging, they bear certain responsibilities, both under the extended producer responsibility, as mandated by the Waste Act, and additionally, under the Single-use Beverage Packaging Act. Furthermore, producers are compelled to adhere to the specifications laid out in the Eco-design Act pertaining to material composition and packaging design, including specific plastic products. This means that producers must ensure that their products are designed and manufactured in accordance with environmental standards and requirements. The Single-use Beverage Packaging Act introduces a deposit return scheme (DRS) for plastic bottles and metal cans used for single-use beverage packaging. This scheme comes into effect on January 1, 2022 and brings forth various obligations for packaging producers and distributors engaged in the sale of beverages. Here are some noteworthy obligations (CMS, 2023k):

Producer Responsibilities: Producers are required to clearly mark their packaging to indicate its participation in the deposit system and maintain separate accounting records for product sales and deposit amounts. Additionally, they must reimburse costs related to the DRS, as stipulated in their contract with the Administrator. Importantly, producers must also adhere to the requirements established in the Eco-design Act for the material composition and design of their packaging and specific plastic products, aligning their production with environmental standards.

Administrator of DRS: The DRS is overseen by a non-profit organization known as the Administrator, which is responsible for coordinating and financing the scheme. The Administrator carries out tasks such as operating and funding the DRS, establishing minimal deposit amounts, arranging for the transport, reuse, and recovery of beverage packaging, and organizing education and information campaigns.

Beverage Packaging Distributors: Distributors, particularly those with sales areas exceeding 300 square meters (excluding distributors selling beverages as complementary goods), have obligations within the DRS. They are required to collect single-use beverage packaging either at their place of operation or within 150 metres. Other duties include indicating the deposit amount in the product's sales price, clearly marking the packaging to signify its participation in the DRS, and maintaining separate accounting records for both product sales and deposit amounts.

The implementation of the DRS aims to facilitate the purchase and subsequent return of beverage packaging by end-consumers. Relationships between the administrator, producers, and distributors are governed by contracts regulated by the Single-use Beverage Packaging Act and the Act itself. These regulations ensure that the DRS operates smoothly and in line with its intended objectives. Note that the DRS does not strip the beverages packaging producers of obligations and undertaking under the extended responsibility scheme. Taxes specifically targeting packaging and plastics do not exist. However, within the framework of Act No. 582/2004, which covers Local Taxes and Local Fees for Municipal Waste and Minor Construction Waste, municipalities have the authority to impose fees related to waste generated within their territorial jurisdiction (referred to as "*miestny poplatok za komunálne odpady a drobné stavebné odpady*" in Slovak) (CMS, 2023k).

Producers who fall under the extended responsibility scheme are obligated to make contributions to the Producers Responsibility Organisation. The Waste Act has introduced a concept known as "eco-modulation," which is designed to allow the Producers Responsibility Organization to request lower contributions from producers who utilize more recyclable and reusable materials in their production processes. Unfortunately, the Ministry of Environment of the Slovak Republic has not provided guidelines for calculating these contributions based on eco-modulation, rendering the concept impractical and unimplemented.

3.16 SLOVENIA

The primary legislative framework in Slovenia includes several recently enacted laws and decrees, such as the Environmental Protection Act, Act on the Prohibition of Certain Single-Use Plastic Products in the Slovenian Market, Decree on the Prohibition of Certain Single-Use Plastic Products and Labelling of Specific Plastic Products, Decree on the Reduction of the Environmental Impact of Certain Plastic Products, Decree on Waste, Decree on Waste Landfill, Decree on Packaging and Packaging Waste, Decree on Packaging and Packaging Waste Handling (partially in effect), Decree on Environmental Tax for Packaging Waste Pollution, and Decree on Environmental Tax for Pollution from Waste Landfilling. All these legislative acts are aligned with relevant EU regulations (CMS, 2023l).

It's important to note that the Environmental Protection Act includes provisions related to Extended Producer Responsibility (EPR), originally scheduled to take effect from January 1, 2023. These provisions required the establishment of a non-profit entity responsible for EPR obligations, to be set up by producers themselves. However, these provisions were temporarily suspended by a unanimous decision of the Slovenian Constitutional Court. Slovenia has also developed a waste-management programme and a waste-prevention programme, in accordance with various decrees, such as the Decree on Waste, Decree on Waste Landfill, and Decree on Packaging and Packaging Waste (CMS, 2023l). Slovenian legislation imposes specific requirements on packaging to ensure environmental protection. These requirements apply at every stage of the supply chain. Generally, there are limitations on the number of substances used in packaging production, with a focus on minimizing their negative environmental impact and reducing packaging waste generation.

Moreover, Slovenia implemented Directive (EU) 2019/904 through the Act on the Prohibition of Certain Single-Use Plastic Products in 2021. This directive regulates the reduction of the environmental impact of specific plastic products. The act prohibits the placement of single-use plastic products and products made from oxo-degradable plastic on the market, aligning with Part B of the Directive's Annex. The Market Inspectorate monitors compliance and can impose administrative fines of up to EUR 15,000 on violators. Additionally, Slovenian law mandates that single-use plastic products listed in Part D of the Directive's Annex must display clear and permanent markings on their packaging or the product itself. These markings inform consumers about appropriate waste-management options and the environmental impact of improper disposal. Non-compliance with this requirement can result in administrative fines. Slovenian legislation encompasses a range of requirements aimed at ensuring environmental protection, particularly in relation to packaging. These requirements and obligations vary for entities at different stages of the supply chain. Here are the most relevant aspects (CMS, 2023l):

a) Limiting Substances and Waste: Producers are generally required to minimize the use of substances in packaging production that have adverse environmental effects and reduce the amount of packaging waste generated.
b) Recovery and Recycling Obligations: Businesses that trade packaged products must ensure the recovery and recycling of packaging waste of the same type as the packaging they introduce to the market. This waste management can be conducted by the entity itself or through an authorized waste management company, with the costs borne by the entity.
c) Extended Producer Responsibility (EPR): Slovenia mandates entities to fulfil their EPR obligations for various waste products, including packaging and single-use plastic products.
d) Record-Keeping: Packaging producers are required to maintain records of the quantity of packaging, categorized by the materials used. These records must be registered with the Slovene Environmental Agency ("ARSO"), and reports for the previous year must be submitted to ARSO by March 31.
e) Environmental Duty: An environmental duty is imposed on the introduction of packaging onto the Slovenian market. Entities must register with financial authorities and pay this duty quarterly. The duty is determined based on environmental impact.

The Slovene Committee for Infrastructure, the Environment, and Spatial Planning supports expanding producer responsibilities throughout a product's entire lifespan in a cost-effective manner. However, the implementation of provisions requiring a single non-profit entity to carry out joint EPR obligations has been suspended by the Slovenian Constitutional Court. Until the court reaches a final decision, the previous regime, allowing multiple entities within a product type, remains in place. The new Environmental Protection Act allows for the establishment of Deposit-Return Systems (DRS), with the government authorized to adopt a decree defining the collection of waste. While no decree has been adopted yet, there are initiatives in this direction, suggesting the potential introduction of such a scheme in the future.

Additionally, Slovenian law imposes an environmental tax on packaging waste pollution and pollution from landfilling of waste. This tax applies to legal entities and sole traders involved in packaging and packaged goods. The tax is levied when packaged goods are first placed on the Slovenian market, with the calculation based on the environmental burden per kilogram of packaging, the number of units, and the packaging quantity. The tax rates vary based on the type of packaging material. The Slovene Value Added Tax Act excludes the costs of returnable packaging from the tax base, provided they are documented in the supplier's records (CMS, 2023l).

3.17 SPAIN

According to reports produces by the local industry, Spain's generation of plastic products for the year 2018 was estimated to be at 5.2 million tonnes, where plastic wastes from packaging alone comprised of 40% of the total waste. To date, 56.7% wastes that are generated in the country end up in landfills, while only 18.3% are sent to be recycled, 13.5% are used to generate energy, and the rest is used up for composting. These numbers, however, are still way below the current hierarchy of priorities set by the European Union. Currently, Spain is way below the average standard for recycling of the Member States in the European Union, where the average is 52%. The regions in the country that generate the largest share of wastes per resident annually are the Balearic Island (800.6 kg), Canary Island (594.1 kg), and Cantabria (532.6 kg), with the lowest being the Community of Madrid (363.8 kg). The wastes produced by most regions in the country kept rising over the years, and when compared to the data produced for the year 2015 and 2013, the region that had the highest increase in waste generation was the region of Extremadura, which managed to produce 409.4 kg annually per resident for the year 2013; the figure increased by 11.8% in just two years. As the production and demand for plastics keep rising globally, it is estimated that plastic production will double over the next 20 years, which will naturally lead to larger generation of plastic waste (43.3% of waste in Spain is recycled or reused, 8.7 points below the average of the European Union, 2018). Spanish national legislation concerning plastic packaging and waste comprises the following key components (CMS, 2023m):

a) Act 7/2022 of April 8 on waste and contaminated soils for a circular economy ("Act 7/2022"): This Act is designed to regulate waste management by introducing measures aimed at preventing waste generation and minimizing the adverse impacts on human health and the environment associated with waste generation and management. It also seeks to enhance resource efficiency. Additionally, this law includes a comprehensive set of obligations related to extended producer responsibility.

b) Royal Decree 1055/2022, of December 27, on packaging and packaging waste ("RD 1055/2022"): This Royal Decree is focused on preventing and reducing the environmental impact of packaging and packaging waste throughout their lifecycle. It introduces new obligations related to labelling, registration, single-use plastics, and the recyclability of packaging.
c) Royal Decree 293/2018, May 18, on reducing the consumption of plastic bags ("RD 293/2018"): The purpose of this Royal Decree is to implement measures to decrease the consumption of plastic carrier bags. It aims to prevent and reduce the adverse environmental impact caused by the waste generated by these bags, particularly in aquatic ecosystems, as well as in economic activities like fishing and tourism. The regulation also seeks to prevent the loss of material and economic resources due to littering and the dispersion of plastic bags in the environment.

With the publication of RD 1055/2022, the primary goal is to progressively phase out the marketing of single-use plastic packaging. This regulation promotes the use of reusable bags and encourages the optimized use of single-use bags to reduce unnecessary consumption. Several measures have been adopted with the publication of this law (CMS, 2023m), including:

a) Requirement for event organizers to implement alternatives to selling drinks in single-use containers and cups, ensuring access to non-packaged drinking water at festive, cultural, and sporting events, starting from July 1, 2023.
b) Compliance with separate collection targets for single-use plastic bottles as stipulated in the law.
c) Adherence to labelling obligations outlined in Commission Implementing Regulation (EU) 2020/2151 of December 17, 2020.
d) Producers of products are obligated, as part of their extended producer responsibility, to finance and organize the collection and treatment of single-use packaging waste, either wholly or partially.
e) Establishment of a deposit, return, and refund system for single-use packaging.
f) Requirement for food retailers with a sales area of 400 m^2 or more to allocate at least 20% of their sales area to offer products without primary packaging, including bulk or reusable packaging, effective from January 1, 2023.
g) Starting from July 3, 2024, only single-use plastic products listed in Part C of Annex IV to Act 7/2022, which includes single-use plastic products subject to eco-design requirements, may be placed on the market. This includes products whose lids and caps remain attached during intended use, with exceptions for metal caps and closures with plastic seals.
h) Producers of single-use plastics will bear various costs and expenses related to waste management as part of the extended producer responsibility scheme.

Act 7/2022 outlines the scope of responsibility for producers and defines the obligations they may be subjected to in the design and production phases of their products as well as in the management of waste resulting from their use. It also provides guidance on how to fulfil these responsibilities, whether individually or through collective systems. Moreover, this Act stipulates that producers of products that eventually become waste, following the "polluter pays" principle, are actively engaged in waste prevention and management organization (CMS, 2023m). They are encouraged to manage waste in accordance with the principles underlying the legislation. Royal Decree 1055/2022 (RD 1055/2022) further expands on these regulations, specifying the obligations that can be imposed on product producers through Royal Decree. This includes the establishment of a deposit, return, and refund system and defines the waste management aspects to be financed by producers. The decree also outlines control mechanisms for both individual and collective systems established for this purpose. Producers are given the choice between participating in an individual or collective extended producer responsibility system. In Spain, several extended producer responsibility systems exist for different types of packaging (CMS, 2023m), including:

a) Phytosanitary products and fertilizers packaging: Managed by AEVAE.
b) Lightweight packaging and paperboard: Administered by ECOEMBES.
c) Glass packaging: Overseen by ECOVIDRIO.
d) Medicinal product packaging and expired medicines: Handled by SIGRE.

RD 1055/2022 establishes that consumers of packaged products are required to return packaging waste covered by the deposit, return, and refund system to retailers and distributors. The return should meet the specified conditions of conservation and cleanliness as defined by the extended producer responsibility systems. This regulation aims to establish effective waste management systems and promote responsible consumer behaviour (CMS, 2023m):

1. For producers introducing reusable packaging to the market, the following obligations are imposed:
    - They must charge their customers, including the end consumer, a deposit fee for each unit of packaging sold (amounting to EUR 10 cents or more per unit).
    - They are required to accept the return of used packaging that matches the type, format, or brand they originally placed on the market, refunding the same amount initially charged.
    - Producers and distributors are only obligated to accept the return of packaging for products they themselves have placed on the market.
    - Consumers wishing to return used packaging must do so to the distributors or the product producers, adhering to the preservation and cleanliness conditions specified by the producers.
    - When reusable packaging reaches the end of its useful life, product producers must arrange for its separate disposal by material to an authorized waste manager and bear the associated financial costs.

2. A dedicated deposit, refund, and retrieval system is established specifically for plastic bottles up to 3 litres, in case the recycling and recovery objectives outlined in Article 10 of RD 1055/2022 are not met. Please refer to Article 47 of RD 1055/022 for details. This system must be operational within two years of non-compliance.
3. For other single-use packaging not covered by Article 47, product producers who introduce such packaging to the market have the option to voluntarily establish a deposit, return, and refund system (without the obligation of setting a minimum amount of EUR 10 cents).

The aforementioned obligations can be fulfilled through individual or collective extended producer responsibility schemes, such as Ecoembes. Two Royal Decrees related to recycled materials in food packaging should be taken into account (CMS, 2023m):

a) Royal Decree 846/2011, dated June 17, 2011, establishes the conditions that recycled polymer-based raw materials must meet for use in materials and articles intended to come into contact with food. This Royal Decree permits the marketing and use of raw materials based on recycled polyethylene terephthalate (PET) produced in Spain for plastic materials and articles designed for contact with packaged drinking water. However, certain conditions must be met, including:

    - The final containers must contain a minimum of 50% virgin PET.
    - Manufacturers of packaging using recycled PET must notify the competent health authorities in the Autonomous Community where the facility is located to facilitate official inspections.

b) The marketing and use of raw materials based on recycled polymeric material for food contact materials and articles, which are legally authorized in other EU Member States, is authorized for the same purpose, with identical restrictions and limitations as those in place in those countries, in accordance with the principle of mutual recognition. This authorization is governed by Royal Decree 847/2011, dated June 17, 2011, which establishes the positive list of substances allowed for the manufacture of plastic materials intended for food contact.

3.18 SWITZERLAND

Especially when compared to regulations within the European Union, plastic waste and packaging are currently subject to limited and indirect regulation in Switzerland. The Federal Council possesses the legal authority to enact regulations that can: (i) prohibit the introduction of products intended for single or short-term use if their benefits do not justify the pollution they cause or (ii) compel manufacturers to find ways to avoid waste production when there is no known environmentally compatible disposal process (article 30a, letters a and b, of the Federal Act on Environmental Protection (EPA); SR 814.01) (CMS, 2023n). Nevertheless, the Federal Council has not exercised this authority thus far.

The Federal Council has the power, according to article 30a of the EPA, to ban products from the market if their environmental pollution outweighs their utility. This also encompasses single-use and short-term packaging. Presently, no mandatory measures for single-use plastics are in place. Switzerland presently relies on voluntary measures implemented by the industry and retailers based on producer responsibility and collaboration (CMS, 2023n). Notably, the "Collection 2025" initiative is noteworthy, wherein private organizations spanning the entire value chain are collaborating on establishing a circular economy for plastic packaging and beverage cartons. They are currently in the process of creating a nationwide collection and recycling system for plastic packaging and beverage cartons, complete with an associated organizational and financial framework. As of mid-2023, 70 organizations have already signed the "closing cycles for plastic packaging and beverage packaging" pact (CMS, 2023n).

Switzerland does not currently operate a deposit system. However, Switzerland does have a functional PET recycling system that operates entirely on a voluntary basis. Thanks to this recycling system, over 80% of PET bottles used in Switzerland are recycled. According to Article 8, paragraphs 1 and 2, of the Ordinance on Beverage Containers, if the minimum recycling rate of 75% for PET beverage bottles is not met, the Federal Department of the Environment, Transport, Energy and Communications (DETEC) has the authority to introduce a deposit system based on the Ordinance on Beverage Containers (BCO) (CMS, 2023n). Other types of plastics are collected and partially recycled in specific regions of Switzerland on a voluntary basis, but there is currently no nationwide collection or recycling system for them. There are no specific taxes imposed on packaging or plastics in Switzerland. Moreover, in March 2021, the Council of States rejected a legislative proposal from the Commission for the Environment, Spatial Planning, and Energy (ESPEC) that sought to introduce a tax by 2025 on disposable plastic packaging for beverages and cleaning products containing less than 25% recycled material. Therefore, as of now, there are no imminent taxes on packaging or plastics in Switzerland (CMS, 2023n).

3.19 THE CZECH REPUBLIC

The general legal framework in the Czech Republic consists primarily of three key acts (CMS, 2023o): Act No. 477/2001 Coll. on Packaging, as amended (referred to as the "Packaging Act"), Act No. 541/2020 Coll. on Wastes, as amended (referred to as the "Wastes Act"), and Act No. 542/2020 Coll. on End-of-Life Products, as amended (referred to as the "End-of-Life Products Act"). The Packaging Act is primarily aimed at safeguarding the environment by minimizing waste generated from packaging. It focuses on reducing the weight, volume, and adverse environmental effects of packaging, including the regulation of chemical substances present in packaging materials. The Wastes Act establishes guidelines for waste prevention and management, emphasizing environmental protection, public health, sustainability, and responsible resource utilization. It aims to reduce the detrimental impact of natural resource usage and enhance resource efficiency. The recently enacted End-of-Life Products Act provides specific regulations for managing particular product types, such as electrical and electronic devices, batteries, accumulators, vehicles, and tyres. It addresses issues like hazardous substances in these products, waste prevention, product return and recycling procedures, and more (CMS, 2023o).

All three acts are aligned with relevant EU legislation, including the EU Circular Economy Package, and ensure compliance with these EU regulations. In February 2021, the Czech government proposed a new draft law aimed at reducing the environmental impact of specific plastic products. This legislation bans the use of certain single-use plastics, such as plastic straws and cutlery, starting from July 1, 2021. This legislative initiative aligns with Directive (EU) 2019/904 of the European Parliament and the Council, dated June 5, 2019, addressing the reduction of the environmental impact of select plastic products. Under the Packaging Act, entities introducing packaging or packaged products into the Czech Republic's market, referred to as "Introducing Entities," bear various obligations, including the following:

a) A general prevention obligation requires Introducing Entities to minimize packaging size and weight to comply with relevant regulations.
b) Introducing Entities must ensure the collection of such packaging or packaging waste, unless they can demonstrate that the packaging did not become waste.
c) Introducing Entities are responsible for ensuring the recycling of such packaging or packaging waste, if generated, in accordance with relevant rules.

These obligations can be fulfilled in several ways, including direct compliance by the Introducing Entities themselves, delegation of responsibilities to another entity along with ownership rights to the packaging, or engagement of an "Authorised Company," as defined in Section 16 of the Packaging Act, through a contractual arrangement. For Introducing Entities based outside the Czech Republic, some of these obligations may also be met through a designated responsible representative, a business entity located in the Czech Republic, which has entered into a contract with the Introducing Entity for this purpose (CMS, 2023o).

Under the Packaging Act, Introducing Entities have a responsibility to promote the repeated use of products that are returnable according to the Packaging Act. This may entail the imposition of a deposit for packaging that allows for repetitive use, with specific regulations, including the deposit amount, outlined in government decrees. As mentioned previously, the Packaging Act mandates that Introducing Entities ensure the collection of packaging or packaging waste unless they can demonstrate that the packaging did not become waste. This collection process must be provided free of charge, and there should be a sufficient number of collection points available. Introducing Entities can fulfil this obligation, along with others under the Packaging Act, by collaborating with an Authorized Company licensed for the collection of packaging and related waste (CMS, 2023o). The Wastes Act regulates the conditions for operating waste disposal sites, encompassing aspects such as waste sorting and storage in accordance with applicable regulations. It provides a framework for the proper management of waste. It's important to note that the return scheme established by the End-of-Life Products Act primarily applies to products governed by this specific legislation, as previously explained. In the Czech Republic, there are no dedicated taxes related to packaging and plastics. However, Act No. 235/2004 Coll., concerning value-added tax, specifies when introducing packaging and/or packaged products to the market is subject to VAT and outlines the tax base for calculating VAT. Additionally, Act No. 586/1992 Coll., on income tax, stipulates that contributions and subsidies given to meet obligations related to packaging and waste packaging return are exempt from income tax (CMS, 2023o).

Furthermore, Introducing Entities are obliged under the Packaging Act to register in a specific registry managed by the Ministry of the Environment. A registration fee of CZK 800 per year (approximately EUR 30) is applicable. The Wastes Act introduces a general obligation to pay for waste deposited at waste disposal sites by the waste's originator. The specific conditions for imposing this fee are defined in the Annexes to the Wastes Act, with the fee amount varying based on the waste type and weight. Municipalities are also authorized by the Wastes Act to levy waste fees for waste generated within their respective territories, referred to as "*poplatek za komunální odpad*" in Czech (CMS, 2023o). Amendments to the Packaging Act have introduced "eco-modulation," allowing Authorized Companies to charge lower fees to Introducing Entities that use more recyclable and reusable materials in their packaging production. A similar eco-modulation principle is incorporated into the End-of-Life Products Act concerning products and substances governed by this legislation. These regulations are primarily guided by EU legislation, including Regulation (EC) No. 282/2008 of March 27, 2008, on recycled plastic materials and articles intended for food contact and Commission Regulation (EU) No. 10/2011 of January 14, 2011, on plastic materials and articles intended for food contact. Czech law further outlines the powers of Czech authorities, particularly the Agriculture and Food Inspection Authority, to monitor compliance with these rules (CMS, 2023o).

3.20 DENMARK

In Denmark, the EU Directive 2019/904 of the European Parliament and of The Council of June 5, 2019 has authority in sections 9s and 9t of the Danish Environmental Protection Act. The following products are banned to be sold in Denmark with substitution options:

a) Cutlery (forks, knives, spoons, chopsticks).
b) Plates.
c) Straws and cotton bud sticks (with the exception of those that are used with active implantable medical devices or other medical devices).
d) Beverage stirrers.
e) Sticks to be attached to and to support balloons, except balloons for industrial or other professional uses and applications that are not distributed to consumers, including the mechanisms of such sticks.
f) Food containers made from expanded polystyrene (i.e., Styrofoam boxes with or without a lid) for fast food or other meal ready for immediate consumption that is typically consumed from the receptacle or is ready to be consumed without any further preparation.

g) Products manufactured from oxo-degradable plastic.
h) Beverage containers made of expanded polystyrene (Styrofoam), including their caps and lids.
i) Cups for beverages made of expanded polystyrene (Styrofoam), including their caps and lids.

Meanwhile, there are products which require specific attention, as shown in Table 3.4.

The directive also states that the marking of single-use plastic products is required to inform about the plastic content and correct disposal of the products to avoid incorrect disposal, causing environmental impacts. The pictograms need to be shown on the single-use products since July 3, 2021, as in Table 3.5:
Another significant Statutory Order, No. 1277, pertains to extended producer responsibility for tobacco product filters, which specifically focuses on the obligation to

TABLE 3.4
Single-Use Products Subjected to Specific Requirements (Dansk Producent Ansvar, 2023)

Restrictions on placing on the market/substitution/bans	Marking requirements	Producer responsibility
Until suitable substitution options are found, the aim is to attain a significantly reduce consumption and ultimately prohibit placing on the market.	Information to citizens about correct-suitable and incorrect-undesirable disposal options, the presence of plastics in products, and the environmental impacts from littering.	Costs for cleaning up litter/ public cleansing/collection. Data gathering/monitoring of quantities, awareness raising measures.
Entry into force: July 3, 2021	Entry into force: July 3, 2021	Entry into force no later than December 31, 2024
PRODUCT TYPE	**PRODUCT TYPE**	**PRODUCT TYPE**
Cutlery (incl. chopsticks)	Tobacco product filters	Tobacco product filters (responsibility for cleaning up litter January 2023)
Plates	Wet wipes	Wet wipes (no financing of cleaning up litter)
Straws	Sanitary towels	Balloons (no financing of cleaning up litter)
Cotton bud sticks	Cups for beverages+caps and lids	Cups for beverages+caps and lids (packaging)
Beverage stirrers		Lightweight plastic carrier bags (packaging)
Balloons and sticks for balloon		Food and beverage containers (packaging)
Styrofoam food boxes (EPS)		Plastic bottles (packaging)
Products made of oxo-degradable plastic (microplastics)		Plastic fishing gear

TABLE 3.5
Labelling of Single-Use Plastic Using Standard Pictogram according to Danish Directive

Single-use product	Label pictogram
Cigarette filters	
Sanitary towels, tampons, and related	
Wet wipes	
Beverage cups made partly from plastic	
Beverage cups made solely from plastics	
Beverage cups made solely from plastic must not be littered.	

address litter. Starting from January 5, 2023, producers bear the responsibility for tobacco product filters. Businesses falling within the defined scope can commence registration in the producer register, beginning May 1, 2023. Article 8 of the Single-Use Plastics Directive outlines the cleaning obligations as follows:

a) Covering the expenses related to waste collection for products discarded in public collection systems, which encompasses the infrastructure's establishment and operation, as well as the subsequent transportation and treatment of said waste.

b) These costs may also encompass the establishment of dedicated infrastructure for collecting this type of waste, such as suitable waste receptacles placed in commonly identified litter hotspots.
c) The expenses associated with litter cleanup should be restricted to activities carried out by public authorities or on their behalf.

3.21 ESTONIA

Estonia is considered to be one of the earliest countries in EU which has formed regulations to control substantial waste generation. The first act was the Packaging Act (2004). The legislation, initially enacted in 2004 and subsequently revised in 2021, establishes broad guidelines concerning packaging and its utilization, strategies for mitigating or lessening packaging waste generation, the establishment of a system to recover packaging and packaging waste, criteria for conducting audits and attaining recycling targets, as well as consequences for failing to adhere to these stipulations (Laspada and Karasik, 2022). This statute applies to all forms of packaging introduced into the market within the Republic of Estonia, regardless of its intended application (be it industrial, commercial, domestic, office, etc.). Packaging, as defined by this legislation, encompasses any item designed for housing, safeguarding, transporting, delivering, or showcasing goods at any point throughout its lifecycle. This encompasses plastic materials and plastic carrier bags, which are specifically addressed within this legislation. Plastic carrier bags, within the framework of this legislation, are characterized as bags made of plastic, with or without handles, dispensed to consumers at the point of sale of goods. These bags are classified into three categories: lightweight (with a wall thickness less than 50 microns), very lightweight (with a wall thickness less than 15 microns), and oxo-degradable bags (commonly known as "compostable plastic bags"). The legislation also sets forth objectives for reducing packaging waste and achieving recovery targets. According to this legislation, packaging must be designed, manufactured, and sold in a manner that facilitates its reuse or the recuperation of packaging waste, including recycling. Furthermore, it aims to prevent any detrimental environmental consequences when disposing of packaging waste or residues resulting from the processing of packaging waste. An amendment in 2018 prohibited the distribution of free, lightweight plastic carrier bags, while in 2019, the charge for heavier plastic bags was increased to one euro (Laspada and Karasik, 2022).

Additionally, Estonia implemented a Deposit Return System for beverage containers (such as bottles) in the early 2000s, aligning with the practices of numerous European Union Member States. This system covers cans, plastic bottles, and glass bottles. Under this scheme, a tax is imposed upon the purchase of a beverage container, which is subsequently refunded as a subsidy upon returning the container for collection and recycling. In many deposit return programmes, which often fall under the umbrella of extended producer responsibility, retailers and producers contribute to the funding required to sustain the infrastructure for collecting, sorting, and recycling returned beverage containers. In Estonia, the deposit for these containers amounts to 0.10 euros.

Meanwhile, the Packaging Excise Duty Act of 1996, which was recently updated in 2019, imposes an excise duty on packaging for goods introduced into the Estonian market or obtained from and imported from another European Union Member State.

Specifically, plastic packaging is subject to an excise duty rate of 2.5 euros per kilogram, making it one of the highest rates, along with metal packaging. This excise duty is the responsibility of the individuals who purchase the goods, those who utilize the packaging, or those who request the disposal of the packaging within Estonia, such as imported packaging waste. While this legislation provides a comprehensive framework, it also contains provisions for exemptions regarding plastic waste (Laspada and Karasik, 2022). These exemptions include plastics weighing less than 25 kilograms (approximately 55 pounds) per quarter and packaging made from alternative materials weighing less than 50 kilograms per quarter. The most recent waste management plan in Estonia, the Tallinn Waste Management Plan (2017–2021), includes provisions specifically targeting plastic waste. These provisions include replacing plastic cups with reusable ones at public events and enabling residents to sort plastic, glass, and metal waste before it is picked up by the city. Tallinn is also a partner in several regional associations that are working to reduce plastic waste. These associations include INTHERWASTE, which brings together European heritage cities to share experiences and lessons learned in municipal waste management, and Project BLASTIC, which is a project that aims to reduce the flow of plastic waste into the Baltic Sea (Laspada and Karasik, 2022).

3.22 FINLAND

Finland boasts a rich history of encouraging the return of refillable containers, and the deposit refund system is in place to sustain this tradition of promoting packaging reuse and recycling. To further motivate producers and importers to participate in this deposit refund system, a packaging tax on soft and alcoholic drinks packaging was introduced back in 1994. This tax was imposed when these products entered the Finnish market. Importantly, it provided a reduced tax rate for producers and importers who engaged in a government-registered deposit refund system. This tax incentive played a pivotal role in the establishment of *Suomen Palautuspakkaus Oy*, more commonly known as PALPA (PALPA, 2023).

The Finnish government has issued a decree outlining recycling goals for return systems and the minimum deposit values for various beverage containers. This decree applies to a return system for beverage containers as defined in section 68 of the Waste Act (646/2011) (Waste Act, 2022). In Section 2 of the decree, the minimum deposits for beverage containers included in the return system are specified as follows:

a) EUR 0.15 for metal containers.
b) EUR 0.20 for plastic containers larger than 0.35 litres but smaller than 1.0 litre.
c) EUR 0.40 for plastic containers of at least 1 litre.
d) EUR 0.10 for containers other than those described in paragraphs a)—c).

Moving on to Section 3 of the decree, it outlines the responsibilities related to reuse and recycling:

(1) The entity responsible for maintaining a return system for beverage containers must oversee the collection, reuse, and recycling of beverage containers

within the system. The objective is to achieve the following annually, in relation to the number of containers placed on the market by the members of the return system:

a) At least 90% by weight of reusable beverage containers should be reused.
b) At least 90% by weight of non-refillable beverage containers should be recycled.

(2) The entity taking on the role of the party responsible for maintaining a return system for beverage containers must fulfil its obligations, as outlined in subsection (1), within the third full calendar year after commencing operations.

In practice, most of beverage manufacturers and importers are members of return systems managed by PALPA. When the packaging tax was introduced, it was based on the volume of the packaging and set at FIM 4 (EUR 0.67) per litre for packaging, not in a deposit refund system; at FIM 1 (EUR 0.17) per litre for one-way packaging in a system; with refillable containers in the deposit refund system exempt from the tax entirely. The tax was revised in 2004. From 2005 to 2008, the tax on one-way packaging in deposit refund systems was halved to EUR 0.085 per litre. Since 2008, one-way packaging in deposit refund systems has been zero-rated. The rate is EUR 0.51 per litre for all drinks packaging not captured in a deposit refund system.

As previously mentioned, a tax of EUR 0.51 per litre is imposed on specific alcoholic beverages and soft drink packaging. However, becoming a member of an approved and operational return system or establishing a new return system exempts one from this tax. The deposits on these packages serve as an incentive for consumers to return empty beverage containers for recycling, preventing them from ending up in nature or mixed waste. These deposits effectively encourage recycling. Additional factors, such as the accessibility of the nearest return facility and the functionality of reverse vending machines, also influence people's willingness to recycle. Cultural habits and attitudes play a significant role: in Finland, the practice of returning bottles and cans is instilled in individuals during childhood and is considered a matter of importance. Consequently, Finland boasts some of the world's highest beverage packaging recycling rates. Efficient recycling of beverage containers enables the conservation of natural resources and contributes to a cleaner environment. For instance, manufacturing an aluminium can from recycled aluminium cans consumes only 5% of the energy required to produce a can from new materials. PET plastic bottles and glass bottles are recycled materials that find applications in various industries, including the manufacturing of new bottles and other products. Manufacturers and importers of beverage packaging pay membership fees and specific recycling fees, which are used to cover the expenses associated with return systems. These expenses encompass logistics, package transportation throughout the recycling process, compensation to return points, and material processing, among other things. PALPA, the administrator of the system, operates as a nonprofit organization.

The establishment of PALPA and the one-way can deposit refund system was primarily initiated by the brewing industry, although the shift from refillable glass bottles to cans was also partly influenced by retailers. However, the cost of the one-way packaging tax, even with the reduction for participation in a deposit refund scheme, initially hindered the adoption of one-way packaging. It was only after 2004, when the tax was halved and the intention to phase it out completely within four years was announced, that the transition to one-way packaging gained momentum. This, in turn, directly paved the way for the implementation of systems for one-way PET and glass bottles.

The fees contributed by producers who participate in the PALPA program serve as the financial backbone for various components of the system, including the upkeep of reverse vending machines, system administration, and the expenses related to material transportation and sorting. PALPA operates as a non-profit organization, which means that fees are established at a level that matches the operational costs of running the system. An estimate from 2011, which takes into account the fees paid by producers per can and the additional income generated from unclaimed deposits, suggests that the total operational expenses for the system amount to approximately EUR 0.019 per can (Ettlinger, 2016). Unclaimed deposits, a figure kept low due to the system's high return rate, are retained by PALPA and allocated towards the system's operational expenses, rather than being returned to retailers or producers, as seen in some other deposit refund schemes in Northern Europe.

3.23 GREECE

Greece implemented a tax on disposable plastic bags in 2018 (law 4496/2017), which was subsequently reviewed in 2020 (law 4685/2020) and more recently updated according to law 4819/2021 (Government of Greece, 2021). Consumers are required to pay an environmental fee of EUR 0.07 (plus VAT) per carrier bag, with the exception of biodegradable and compostable plastic bags. This fee serves a dual purpose, with the generated revenue considered public funds. The Independent Authority for Public Revenue collects this fee and then transfers it to EOAN. Additionally, in line with Law 4736/2020 (GG 200 A), which aligns with Directive (2019/904) on reducing the environmental impact of specific plastic products, an environmental levy of EUR 0.04 (plus VAT) per designated single-use plastic product category (beverage cups and food containers) is scheduled for implementation in early 2022. AADE will collect this fee and allocate it to the Green Fund (European Environment Agency, 2022). Furthermore, article 80 of law 4819/2021 introduces a specific provision imposing an environmental levy of EUR 0.08 (plus VAT) on products with packaging containing PVC, effective from June 1, 2022. However, it's worth noting that these taxes primarily pertain to a small portion of packaging, suggesting that Greece has limited packaging taxes in place.

Conversely, Greece operates four Producer Responsibility Organizations (PROs) for packaging, including two collective Alternative Management Systems known as SSEDs: Hellenic Recovery Recycling Corporation (HERRco) and Rewarding Packaging Recycling. There is also one individual Alternative Management System (ASED), called AB Vassilopoulos, along with PRO KEPED, responsible for lubricating oil packaging. The Extended Producer Responsibility (EPR) scheme encompasses both household and non-household sources and encompasses all packaging material categories (European Environment Agency, 2022). The recently adopted law 4819/2021 (Government of Greece, 2021) introduces an extension to the obligation for separate collection of non-household waste.

In Greece, the implementation of Extended Producer Responsibility (EPR) schemes includes a fee modulation system, often referred to as eco-modulation. This system involves applying varying fees based on different types of packaging materials and designs. While the common practice is to employ basic fee modulation, distinguishing between fees for primary material groups, more advanced fee modulation can create stronger incentives for packaging producers to design products with recycling in mind, thereby promoting higher recycling rates. The degree of advancement in fee modulation is assessed against four specific criteria considered benchmarks for a well-designed eco-modulated fee system:

a) Recyclability: This includes differentiation between materials such as PET and PS, distinguishing between various colours of PET, and recognizing distinctions like 100% cardboard boxes versus laminated beverage cartons.
b) Sortability and Disruptors: This involves considering penalties for the presence of labels, caps, sleeves, or other materials that are not compatible with the main packaging materials' recycling processes.
c) Recycled Content: This criterion evaluates the use of recycled materials in packaging.
d) Transparent Compliance Check: It assesses whether there is a clear and verifiable mechanism for ensuring that producers accurately report their contributions.

Greece has implemented advanced fee modulation, particularly for plastics, as outlined in law 4819/2021 (Government of Greece, 2021). According to this law, PRO (Producer Responsibility Organization) fees are modulated, taking into account factors such as durability, reparability, reusability, recyclability, and the presence of hazardous substances. The fee modulation criteria for plastic packaging include increasing fees for coloured PET bottles, multilayer plastic packaging, composite packaging, PVC, expanded polystyrene packaging, and PVC labels. Additionally, reduced fees are applied for packaging with recycled material content (a 50% reduction for plastic packaging consisting of at least 25% recycled material, except for plastic carrier bags, where the reduction is set at 30%). These measures align with the first three assessment criteria.

Furthermore, HERRCo (Hellenic Recovery Recycling Corporation) conducts audits of declarations submitted by contracted companies (producers). These audits are carried out by certified auditors selected either by HERRCo or the contracted companies themselves. The audits verify the total contributions paid to HERRCo by the producers. In 2018, audits covered 77% of the total contributions made to HERRCo that year. Additionally, minimum mandatory criteria for fee modulation for each type of packaging will be established in the near future.

Ensuring the timely incorporation of the Packaging and Packaging Waste Directive, as amended by Directive 2018/852, into national legislation is crucial to align a waste management system with EU standards. Greece has successfully transposed the amended Directive into its national legal framework through Law 4819/2021 (Government of Greece, 2021). Additionally, Article 89 of Law 4819/2021 (Government of Greece, 2021), which supersedes Article 8 of the former Law 2939/2001 on Packaging and recycling of packaging and other products, grants municipalities the autonomy to implement alternative packaging-waste management independently. However, to date, no municipalities have expressed an interest in pursuing this option (European Environment Agency, 2022). According to Law 4819/2021 (Government of Greece, 2021), which replaces Law 2939/2001, administrative penalties can be imposed on producers, Producer Responsibility Organizations (PROs), and municipalities for violations of the law's provisions. The approval decision of a PRO may be revoked if it substantially deviates from its business plan, approval conditions, or EOAN-set requirements on two separate occasions. The Joint Ministerial Decision outlining administrative penalties for first-degree local authorities in handling packaging waste responsibilities establishes a framework for determining fines imposed on municipalities failing to fulfil their obligations in this regard. It's noteworthy that, as of now, no penalties have been levied (European Environment Agency, 2022).

3.24 ITALY

Italy's plastic and environmental tax, initiated through the Budget Law 2020 (Law No. 160/2019), has experienced a series of delays in its implementation. Originally slated to commence in July 2020, its enforcement was postponed multiple times due to the disruptions caused by the Covid-19 pandemic. First, the implementation date was moved to January 1, 2021, followed by further delays to July 1, 2021, and then to January 1, 2022. Subsequently, the Budget Law 2022 (Law No. 234/2021) set a new effective date for the plastic tax in Italy, scheduled for January 2023, with no more anticipated deferrals until November 2022. However, the draft of the 2023 Budget Law, approved on November 22, 2022, postponed the Plastic Tax once more, this time to January 2024. This deferral was confirmed by the enactment of Law No. 197 on December 29, 2022, known as the Budget Law 2023. Presently, the legal framework for the plastic tax is delineated in Article 1, paragraphs 634 to 658, of Law No. 160/2019. These provisions establish subjective and objective requirements, define the tax's scope, and outline obligations and penalties. Additionally, a draft of the implementing measure by the Customs and Monopolies Agency has been published. While not definitive, it offers insights into practical aspects like the detailed content of quarterly declarations and reimbursement procedures.

The plastic tax in Italy serves as a measure aimed at discouraging the consumption of single-use plastic products known as "MACSI," an acronym for *Manufatti in plastica con singolo impiego*, or Single-Use Plastic Manufactures. Determining the entity liable to pay the plastic tax hinges on the production and consumption locations of MACSI as well as the entity's nature (whether it is a private consumer or not). Specifically, the responsible taxpayer could be the manufacturer for domestically produced MACSI, the purchasing business entity in Italy for MACSI originating from other EU countries, the supplier in cases of MACSI bought by private consumers (requiring the appointment of a fiscal representative), the importer for MACSI from third countries outside the European Union, or any business, whether resident or non-resident in Italy, that intends to sell MACSI obtained on its behalf in a production plant to another national entity. Notably, there is an option to request a refund for the tax paid on the purchase of MACSI that ultimately does not enter the domestic market. The plastic tax in Italy operates on a tax base of EUR 0.45 for each kilogram of plastic present in single-use products; specifically, those categorized as MACSI (*Manufatti in plastica con singolo impiego*, or Single-Use Plastic Manufactures) (Rodl and Partner, 2023a).

The scope of the tax is defined by Law 160/2019, which identifies MACSI as products meeting two key criteria: they incorporate plastics and organic synthetic polymers in forms such as sheets, films, or strips, and they are not designed for multiple transfers or reuse for their original purpose. This encompasses not only finished single-use products but also semi-finished items used in MACSI production, including preforms moulded from PET intended to become containers through a blowing process. Consequently, taxation applies at various stages of MACSI production, extending beyond consumption to cover semi-finished products and preforms. Law No. 160/2019 (Article 1, paragraph 644) permits the deduction of plastic quantities contained in MACSI used for creating other MACSI from the tax base, provided the tax has already been paid by other liable parties. The introduction of the plastic tax has generated mixed reactions (Rodl and Partner, 2023). Critics argue that such a substantial manoeuvre, particularly amid rising inflationary pressure, may result in increased prices and job losses, impacting both the plastics industry and the beverage sector. Notably, the Italian association of soft drinks producers, Assobibe, strongly protested against the tax. While consumption has shown improvement since 2020, it remains below pre-pandemic levels, with a 6% decrease compared to 2019. Projections for 2022 anticipate a slight consumption decline of -0.4%. However, if contagions resurge post-summer, necessitating new restrictions, the estimated sales drop could reach as much as 2.3%. Concerns arise from the potential sharp decline in the industry if the plastic tax takes effect in January 2024, as per the latest law update.

Exemptions and exclusions from the plastic tax are tightly regulated. Only MACSI certified as compostable, those constituting medical devices, and MACSI used to contain and protect medicinal preparations are excluded from the tax's purview. Additionally, there exists a minimum exemption threshold of EUR 25, below which neither declaration nor payment is required. Penalties for violations of the plastic tax depend on the specific circumstances. Non-payment of the tax can incur an administrative penalty ranging from double to five times the evaded tax, but not less than EUR 250. Late payment results in an administrative sanction of 25% of the tax due, with a minimum of EUR 150. Late submission of the declaration can lead to an administrative sanction ranging from EUR 250 to EUR 2,500 (Rodl and Partner, 2023a).

In terms of deadlines, Law No. 160/2019 mandates quarterly declaration submission and tax payment to the Italian customs and monopolies agency. The deadline for both filing and settling the tax due is the end of the month following the quarter's close. For non-resident entities, appointing a tax representative in Italy is mandatory to ensure proper declaration and tax payment. Drafted quarterly declaration forms encompass an extensive list of requested information, including details on purchases, quantities, types of plastic raw materials, and more. Regarding the refund procedure, insights are provided in the draft implementing document from the customs and monopolies agency. To claim a refund, the transferor or exporter must submit a formal request, attaching copies of purchase invoices proving tax payment and documentation substantiating the export or intra-EU transfer of MACSI (Rodl and Partner, 2023a).

3.25 LATVIA

The implementation of the plastic and environmental tax, known as the Natural Resources Tax (NRT), dates back to 1991. This tax system operates based on the quantity (in kilograms) of plastic source materials utilized by taxpayers. The NRT Law plays a pivotal role in governing the products subject to NRT, identifying eligible taxpayers, outlining potential exemptions, and specifying the tax rates applicable to various NRT objects (Rodl and Partner, 2023b). The responsibility for paying the NRT falls upon several categories of entities, including companies holding special permits and licenses for natural resource extraction, those involved in selling environmentally harmful goods or goods in packaging in Latvia, sellers of single-use tableware and plastic bags, entities utilizing radioactive substances leading to radioactive waste in Latvia, and specific scenarios related to vehicles. This tax applies both to Latvian and foreign companies when relevant conditions are met. The product scope encompassed by the NRT is extensive and covers natural resources, the collection of edible park snails, the use of natural gas in geological structures, waste disposal, goods detrimental to the environment (e.g., household and IT equipment, tyres, vending machines, light bulbs, medical devices), product packaging, disposable tableware, and accessories, radioactive substances, specific vehicles, and certain fuels such as coal, coke, and lignite.

Reporting of the NRT is conducted on a quarterly basis, with the option of annual reporting if the NRT liability remains below 142.29 euros. The primary goal of the NRT is to mitigate environmental pollution and curtail the production and sale of environmentally harmful materials. Businesses involved in polluting activities are obligated to pay the NRT or establish agreements with waste recycling companies, which often offer more favourable tax rates. Exemptions from the NRT are attainable under certain conditions, particularly when entities have entered into agreements with waste recycling companies for the recycling of packaging and disposable tableware, excluding those made from plastics and laminates. In the event of violations, sanctions can be imposed, including fines amounting to twice the unpaid NRT and late payment fees calculated at 0.05% per day. This regulatory framework aims to incentivize environmentally responsible practices and penalize non-compliance (Rodl and Partner, 2023b). See Table 3.6.

3.26 LITHUANIA

The implementation of plastic and environmental taxes in Lithuania has seen several significant developments over the years. The Law on Environmental Pollution Tax, which came into effect on January 1, 2000, marked the initial introduction of environmental taxation measures. Subsequently, on January 22, 2002, amendments to the Pollution Law introduced the taxation of plastic packaging. Notably, as of January 1, 2022, distinct tax rates have been established for recyclable and non-recyclable packaging types (Rodl and Partner, 2023c).

TABLE 3.6
Latvia's Natural Resources Tax (NRT) Rates

No.	Type of material of the packaging and disposable tableware	Rate (Euro) per 1 kg of material
1	Glass source materials	0.44
2	Plastic (polymer) source materials, except for bioplastic or polystyrene source materials	1.22
3	Metal source materials	1.10
4	Wood, paper, and cardboard or other natural fibre and bioplastic source materials	0.24
5	Polystyrene source materials	2.20

TABLE 3.7
Pollution Tax Rates and Tax Base of Lithuania

Type of packaging	Tariff for reusable packaging and recyclable disposable packaging, EUR/tonne	Tariff for non-recyclable disposable packaging, EUR/tonne
Plastic packaging	618	875
PET (polyethylene terephthalate) packaging	618	875
Combined packaging	900	1,200

As of the present, there is no ongoing legislative initiative aimed at taxing end-users purchasing plastic packaging or lightweight bags directly. Moreover, Lithuania has adopted the "Next Generation EU" decision from July 21, 2020, which includes provisions for a new tax based on the weight of non-recycled plastic packaging waste, set at EUR 0.80 per kilogram. Importantly, this tax is not collected directly from end-users or manufacturers/importers but is instead reimbursed by the Lithuanian state budget. Additionally, since 2016, Lithuania has implemented a deposit system for plastic bottles and various containers. Consumers are charged a EUR 0.10 deposit fee when purchasing these items, which is refundable upon returning the containers to designated collection points for recycling. However, for the purpose of this discussion, we will focus primarily on the Pollution Tax ash shown in Table 3.7, as it directly impacts manufacturers and importers, with the latest version coming into effect on January 1, 2021.

In the current landscape, various media and initiatives are actively promoting reduced plastic usage and discouraging the use of plastic bags in retail settings. Some supermarket chains have even transitioned away from providing single-use plastic bags. The legal basis for the Pollution Tax is firmly established, targeting products that have a detrimental impact on the environment. Regarding the nature of the tax itself, it is categorized as a tax, not a levy. Importantly, the responsibility for paying this tax lies with the manufacturer or importer supplying filled packaging to or within the

Lithuanian market upon their initial supply. However, there are provisions for potential exemptions from paying the pollution tax under certain circumstances. Exemptions may apply when the quantity of single-use packaging does not exceed 0.5 tonnes of the total filled packaging supplied in the Lithuanian market during a given tax period, when the quantity of filled packaging corresponds to the portion of the recovery or recycling task performed, or when reusable packaging is involved, provided that the complete collection and reuse process is executed (Rodl and Partner, 2023c).

3.27 MALTA

In Malta, the eco-contribution measure, initially introduced in 2005 under Chapter 473 of the Eco-Distribution Act, states that producers of plastic bags listed in the First and Second Columns of the First Schedule must adhere to certain obligations. These obligations are specified in the Fourth Schedule, and producers liable for the eco-contribution rate detailed in the Third Column of the First Schedule must comply with them.

However, it's worth noting that the First Schedule was removed and replaced by Schedule Five A of the Excise Duty Act through Act XVI. 2017.84. The new Schedule Five A outlines the eco-contribution for various plastic products, including plastic sacks, bags, cones, films, sheets, and tubes falling under HS Headings 3917 and 3920. The eco-contribution rate is €425 per 100 kgs, capped at a maximum of €290 per 10,000 units.

This contribution does not apply to:

a) Plastic films, sheets, or tubes that are an integral part of food packaging before retail or transfer.
b) Biodegradable sacks and bags compliant with specific standards (MSA EN 13432:2000, MSA EN14046:2003, MSA EN 14047:2003, and MSA EN14048:2003).
c) Printed plastic sacks and bags purchased for recycling waste collection by authorized waste management schemes.
d) Plastic sacks and bags acquired for non-recyclable waste collection.

Failure to comply with any obligations may result in penalties and legal consequences. Those found guilty of an offense could face fines as follows:

a) A fine of no less than €1,000 for the first conviction.
b) For a second conviction within six months of the previous one, a fine of no less than €1,500.
c) For a third conviction within 12 months of the first one, a fine of no less than €2,000.
d) For a fourth conviction within 24 months, a fine of no less than €2,500. Additionally, the court may suspend all licenses, permits, warrants, or other authorizations related to the economic activities associated with the offenses for a period ranging from one week to one month. This suspension is imposed at the court's discretion.

3.28 CONCLUSION

In conclusion, the European countries have worked thoroughly to establish a variety of policies to control the usage of plastics in the region in order to minimize the impacts to the environment. Many times, established policies in European countries have been a role model for other countries to initiate their own locally feasible policies. The European Union has formed diversified directives and frameworks, which applies to single-use plastics, plastic packaging, and microplastics to reduce the environment impacts of plastics through increasing the current recycling rates, while reducing over-dependency on any waste management like landfilling or incineration that can cause negative impacts to human health. Importantly, the plastic tax and the variety of mandatory monetary contributions as part of policy are not the main source of penalties on the usage of plastics, but such monies are used to further develop feasible infrastructure as well as campaign and teach the younger generations about the correct approach to using plastics responsibly and knowing how to manage wastes at the end of product life. Nevertheless, European countries are still on a journey to combat plastic issues, and the awareness and cooperation of citizens are of the utmost importance in determining the success of the implementation of policies and regulations on the continent.

REFERENCES

Card, D. (2016). Packaging taxes in Belgium. Institute for European Environmental Policy. Available at https://ieep.eu/wp-content/uploads/2022/12/BE-Packaging-Tax-final.pdf. Accessed on 20 August 2023.

CMS (2023a). Plastics and Packaging Law in Bulgaria. Available at https://cms.law/en/int/expert-guides/plastics-and-packaging-laws/bulgaria. Accessed on 10 September 2023.

CMS (2023b). Plastics and Packaging Law in Croatia. Available at https://cms.law/en/int/expert-guides/plastics-and-packaging-laws/croatia. Accessed on 10 September 2023.

CMS (2023c). Plastics and Packaging Law in France. Available at https://cms.law/en/int/expert-guides/plastics-and-packaging-laws/france. Accessed on 10 September 2023.

CMS (2023d). Plastics and Packaging Law in Germany. Available at https://cms.law/en/int/expert-guides/plastics-and-packaging-laws/germany. Accessed on 10 September 2023.

CMS (2023e). Plastics and Packaging Law in Hungary. Available at https://cms.law/en/int/expert-guides/plastics-and-packaging-laws/hungary. Accessed on 10 September 2023.

CMS (2023f). Plastics and Packaging Law in Luxembourg. Available at https://cms.law/en/int/expert-guides/plastics-and-packaging-laws/luxembourg. Accessed on 10 September 2023.

CMS (2023g). Plastics and Packaging Law in The Netherlands. Available at https://cms.law/en/int/expert-guides/plastics-and-packaging-laws/the-netherlands. Accessed on 10 September 2023.

CMS (2023h). Plastics and Packaging Law in Poland. Available at https://cms.law/en/int/expert-guides/plastics-and-packaging-laws/poland. Accessed on 10 September 2023.

CMS (2023i). Plastics and Packaging Law in Portugal. Available at https://cms.law/en/int/expert-guides/plastics-and-packaging-laws/portugal. Accessed on 10 September 2023.

CMS (2023j). Plastics and Packaging Law in Romania. Available at https://cms.law/en/int/expert-guides/plastics-and-packaging-laws/romania. Accessed on 10 September 2023.

CMS (2023k). Plastics and Packaging Law in Romania. Available at https://cms.law/en/int/expert-guides/plastics-and-packaging-laws/romania. Accessed on 10 September 2023.

CMS (2023l). Plastics and Packaging Law in Slovenia. Available at https://cms.law/en/int/expert-guides/plastics-and-packaging-laws/slovenia. Accessed on 10 September 2023.

CMS (2023m). Plastics and Packaging Law in Spain. Available at https://cms.law/en/int/expert-guides/plastics-and-packaging-laws/spain. Accessed on 11 September 2023.

CMS (2023n). Plastics and Packaging Law in Switzerland. Available at https://cms.law/en/int/expert-guides/plastics-and-packaging-laws/switzerland. Accessed on 11 September 2023.

CMS (2023o). Plastics and Packaging Law in Switzerland. Available at https://cms.law/en/int/expert-guides/plastics-and-packaging-laws/czech-republic. Accessed on 11 September 2023.

Commission to the European Parliament, The Council, The European Economic and Social Committee, and The Committee of The Regions (2018). Available at https://eur-lex.europa.eu/legal-content/EN/TXT/?uri=COM%3A2018%3A28%3AFIN. Accessed on 2 January 2024.

Dansk Producent Ansvar (2023). Single-use Plastic. Available at https://producentansvar.dk/en/products-and-responsibility/single-use-plastic/. Accessed on 2 September 2023

Ettlinger, S. (2016). Deposit Refund System (and Packaging Tax) in Finland. Institute for European Environmental Policy- Eunomia. Available at https://ieep.eu/wp-content/uploads/2022/12/FI-Deposit-Refund-Scheme-final.pdf. Accessed on 2 January 2024.

Ettlinger, S. and Ayesha, B. (2016). Landfill Tax, Incineration Tax and Landfill Ban in Austria. Institute for European Environmental Policy- Eunomia. Available at https://ieep.eu/wp-content/uploads/2022/12/AT-Landfill-Tax-final.pdf. Accessed on 10 September 2023.

European Commission (2018). ANNEXES to the Communication from the Commission to the European Parliament, the Council, the European Economic and Social Committee and the Committee of the Regions A European Strategy for Plastics in a Circular Economy. Available at https://eur-lex.europa.eu/resource.html?uri=cellar:2df5d1d2-fac7-11e7-b8f5-01aa75ed71a1.0001.02/DOC_2&format=PDF. Accessed on 10 September 2023.

European Environment Agency (2022). Early warning assessment related to the 2025 targets for municipal waste and packaging waste. Available at https://www.eea.europa.eu/publications/many-eu-member-states/early-warning-assessment-related-to. Accessed on 7 Feb 2024.

Government of Greece (2021). Law 4819/2021 on Integrated framework for waste management. Available at https://www.fao.org/faolex/results/details/ar/c/LEX-FAOC211896/. Accessed on 2 January 2024.

Keller & Heckman (2023). England to Expand Bank of Single-use Plastic Items. Available at www.packaginglaw.com/news/england-expand-ban-single-use-plastic-items. Accessed on 10 September 2023.

Laspada, C. and Karasik, R. (2022). Plastic Pollution Policy Country Profile: Estonia. Policy Brief. Duke Nicholas Institute.

PALPA (2023). Available at www.palpa.fi/beverage-container-recycling/deposit-refund-system/. Accessed on 11 September 2023.

Rodl & Partner (2023a). Plastic Tax: Italy. Available at www.roedl.com/insights/plastic-tax/italy-eu-green-deal. Accessed on 11 September 2023.

Rodl & Partner (2023b). Plastic Tax: Italy. Available at www.roedl.com/insights/plastic-tax/latvia-eu-green-deal. Accessed on 11 September 2023.

Rodl & Partner (2023c). Plastic Tax: Italy. Available at www.roedl.com/insights/plastic-tax/lithuania-eu-green-deal. Accessed on 11 September 2023.

Simon, J. (2018). Changing trends in plastic waste trade. [ebook] ZeroWasteEurope, p. 30. Available at https://zerowasteeurope.eu/wp-content/uploads/2019/11/zero_waste_europe_report_changing_trends_in_plastic_waste_en.pdf. Accessed on 8 February 2021.

Waste Act (2022). 646/2011; Amendments up to 494/2022 Included. Available at https://finlex.fi/en/laki/kaannokset/2011/en20110646.pdf. Accessed on 11 September 2023.

4 Policies of Plastic Use in American Countries

4.1 INTRODUCTION

Nowadays, plastic pollution has become a serious threat to the environment and humans due to the high demand for plastic materials in our daily lives. Plastic pollution has caused various detrimental problems to our environment such as the pollution of the land, impacts on the food chain, jeopardization of marine life, and other impacts. The overuse of plastic materials in human life has become the major contributing factor to plastic pollution. According to a study conducted by Condor Ferries Ltd (2021), a total of approximately 500 billion plastic bags were found to be used annually around the globe, while approximately 381 million tonnes of plastic waste was found to be generated annually around the globe. The United States has generated 38 million tonnes of plastic waste annually, which is approximately 10% of global plastic waste generation globally.

The detrimental effects of plastic waste pollution on the environment have currently gained much attention from the governments of various countries in the world. Plastic wastes need a very long time to degrade and decompose, and the decomposition time required is enormous. As an example, the disintegration of a plastic bottle requires approximately 450 years. The governments of various countries have implemented a diverse range of policies such as banning of plastic, pricing mechanisms on plastic usage, relevant campaigns, etc., in order to reduce the usage of plastic materials and thus reduce the plastic pollution. The implementation of these policies is essential in reducing the impact of plastic pollution. The banning of plastic bags and the levying of taxes on plastic bags usage can limit or reduce the usage frequency of plastic bags. Also, the production and manufacturing of plastic materials or products causes air pollution and the greenhouse phenomenon. This is mainly attributed to the production and manufacturing process of plastic contributing to the emission of carbon dioxide gas. In current technology, plastic is a derivative product from the processing of fossil fuels such as petroleum and natural gas. According to Wright (2019), the release of carbon dioxide gas during the plastic manufacturing process was found to constitute approximately 3.8% of the greenhouse gas that is discharged worldwide. This indicates that the production and manufacturing of plastic materials exacerbate the release of greenhouse gas and air pollution.

In this chapter, the implementation of relevant policies in American countries that consist of three continents, which are North America, Central America, and South America, are summarized in Table 4.1.

DOI: 10.1201/9781003387862-4

TABLE 4.1
The Countries That Were Selected for the Implementation of Plastics Policy Discussion

American continents:	Countries involved:
North America	Antigua and Barbuda; the Bahamas; Canada; United States (California, Hawaii, and Chicago); Jamaica; Saint Vincent and the Grenadines.
Central America	Belize; Costa Rica; Guatemala; Honduras; Panama.
South America	Argentina; Brazil; Chile; Colombia; Peru; Uruguay.

TABLE 4.2
The Year and Detailed Description of the Implementation of Plastic Ban Policy in a Total of Eight Different Locations in North America

Country/States	Year	Type of policy	Description
Antigua and Barbuda	2016	Ban	• Plastic bags are banned. • Styrofoam products are also banned.
Bahamas	2020	Ban	• The use of disposable (single use) plastic bags is banned.
Canada	2018	Ban	• The plastic bags with thickness less than 50 microns are prohibited. • Single use disposable plastic bags are banned.
Jamaica	2018	Ban	• The manufacturing, distribution, import, and use of disposable plastic bags are banned.
Saint Vincent and Grenadines	2017, 2019	Ban	• Styrofoam and expanded polystyrene.
United States (California)	2016	Ban and pricing mechanism	• The use of plastic bags is prohibited. • Implementation of the price of $0.10 per single-use plastic bag.
United States (Hawaii)	2015	Ban and pricing mechanism	• The selling of disposable, single-use plastic bags and Styrofoam dishware are banned.
United States (Chicago)	2014, 2017	Ban and pricing mechanism	• The use of disposable, single-use plastic bags is banned. • Apply a checkout tax on paper bags or plastic checkout-bags of $0.07.

4.2 POLICIES IN NORTH AMERICA

The implementation of policies on plastic usage in a total of eight locations in North American is clearly summarized in Table 4.2. These eight locations include Antigua and Barbuda, the Bahamas, Canada, Jamaica, the United Sates (California), the United States (Chicago), the United States (Hawaii), and Saint Vincent and the Grenadines.

4.2.1 Antigua and Barbuda

4.2.1.1 The External Trade (Shopping Bags Prohibition) Order, 2017

Antigua and Barbuda, an island country in North America, has implemented a ban on the use of plastic shopping bags. The government of the nation has initiated this ban to reduce plastic pollution. The plastic ban law was introduced by the government of Antigua and Barbuda under the Department of Environment of Antigua and Barbuda in 2016. This plastic ban law is known as "The External Trade (Shopping Bags Prohibition) Order. 2017" (Department of Environment Antigua and Barbuda, 2016). According to this policy, shopping bags that are made from polyethylene or petroleum are prohibited. Plastic shopping bags are also known as "T-shirt" plastic bags due to the appearance of these plastic bags. There is a strict prohibition on the importation, distribution, trade, and usage of shopping bags in this policy, starting from June 2016. Nevertheless, there are six categories of plastic bags that are exempt from this policy and are not prohibited, as listed in Table 4.3.

The execution of banning the usage of plastic shopping bags was started in the big grocery stores in the country. After that, this banning action was further implemented in smaller convenience stores and shops after being fully executed in big grocery stores. The banning of plastic bags has caused supermarkets are to compulsorily supply paper bags made from recycled paper. Also, supermarkets can provide reusable bags to the customers for purchase. The government of Antigua and Barbuda has implemented and separated these banning activities in three phases,

TABLE 4.3
Six Categories of Plastic Bags That Are Not Prohibited in the External Trade (Shopping Bags Prohibition) Order, 2017

Categories of plastic bags:

1. Plastic bread wrap
2. Plastic packaging for fresh products (whether repacked or not) such as:
   - fresh meat and fresh meat products,
   - fresh fish and fresh fish products; and
   - fresh poultry and fresh poultry products.
3. Plastic packaging for the products of:
   - Nuts and fruits,
   - Dairy products,
   - Seeds,
   - Cooked food (including hot and cold),
   - Bakery products,
   - Frozen products, liquid products,
   - Pharmaceutical products,
   - Veterinary products,
   - Minor hardware items.
4. Seedling tubes made of polythene
5. Dry cleaning bags used for laundry purposes
6. Trash storage and plastic bags for waste disposal purposes
   - Liners for dustbin, bin refuse

which are Phase 1, Phase 2, and Phase 3. Phase 1 started with the cessation of the import plastic bags and is estimated to be fully implemented within six months. Phase 2 is due to be implemented once Phase 1 is fully completed. Phase 2 is implemented to eliminate the usage of plastic bags in big grocery stores. Phase 3 is due to be initiated after the completion of Phase 2 in all big grocery stores. Phase 3 is implemented to target the elimination of plastic bags usage in smaller convenience stores. The implementation period for Phase 3 was given a longer period due to small businesses' scale and lack of flexibility. The implementation of the plastic bags ban is important and helpful in reducing the negative impact on the environment. The shopping plastic bag prohibition act has helped in reducing the disposal of plastic bags in landfills. This is further proven by the fact that the disposal of plastic bag waste has been significantly reduced by 15.1% after implementing the plastic bags ban for one year. The implementation of the plastic bags ban has played an important role in encouraging the introduction of more relevant policies to reduce plastic pollution (United Nations Environment Programme, 2021).

4.2.1.2 The External Trade (Import Prohibition) Order, 2017

This policy has been added to encourage the usage of biodegradable and reusable materials and ban the usage of expanded polystyrene in making the food service containers such bowls, plates, beverages cups, etc. The prohibition of the import of certain goods made of expanded polystyrene has resulted in a good impact by further banning the plastic usage. The implementation of Styrofoam prohibition was classified into three schedules (Antigua Nice Ltd, 2017).

In the first schedule of prohibition, the importation and usage of food service containers that were made of expanded polystyrene were prohibited. The examples of the food service containers were clamshell, hinged containers, all other containers made of expanded polystyrene, Styrofoam such as plates, bowls, cups for both hot and cold drinks, etc. This stage was started after July 1, 2017. The second schedule of this prohibition was started on January 1, 2018. In the second schedule, plastic utensils such as straws, forks, knives, spoon, etc., trays for fruit and vegetables, meat trays, and egg cartons were prohibited from usage in terms of importation and application. The third schedule of this prohibition policy was started in July 1, 2018. In this schedule, naked polystyrene coolers were prohibited from importation and usage.

Merchants must monitor regularly after the depletion of the expanded polystyrene usage in their inventory. Shopping malls and grocery stores are considered to be in the food services sector and are also required to observe this Styrofoam ban. The catering sector is also categorized in the food services sector and needs to implement the Styrofoam ban. However, there are some exemptions to this import prohibition order. The exemption items in this prohibition include airline carriers, airline private charters, and passenger cruise vessels. The government has provided a list of alternatives to replace the usage of Styrofoam after the banning of the Styrofoam in food service containers. The government also provided for the tax-free usage of these alternatives, listed as follows:

1. Sugar cane/bagasse.
2. Wheat straw.
3. Bamboo.

4. Potato starch.
5. Areca palm.
6. Carboard/paper.
7. PLA corn-starch (non-GMO).

4.2.1.3 Litter Control and Prevention Act, 2019

The implementation of this act was to control and impede acts of littering in any form. In the policy of Litter Control and Prevention Act, 2019, a littering action in any public place is considered as an offence. In this Act, only a site established by the minister or any person obtaining approval from the Minister are eligible to handle the site for litter-disposal purposes. By referring to the Department of Environment Antigua and Barbuda, 2019, the offenders will be penalized if they breach the following Act, as summarized in Table 4.4. The offensive actions under the Litter Control and Prevention Act, 2019, are divided into three different categories, which are dumping of trash in a public area, dumping of waste from motor vehicle, illegal dumping, and dumping of trash from construction sites.

According to Table 4.4, any person caught in the act of "dumping of trash in a public area" will be subjected to fine of up to $3,000 or one year of imprisonment for an individual fine, while the fine can reach up to $15,000 for a corporate body. The action of improper waste disposal by simply throwing waste away in a public area will be subjected to punishment, as mentioned earlier. For the act of "dumping of waste from motor vehicle or trailer," if trash or any items that are carried by a person in a vehicle falls out from the vehicle, this person will be considered to have offended against this act and will be subjected to punishment, as listed in Table 4.4. Any person will be punished if they throw away rubbish or waste on the property of someone else due to the act of "illegal dumping." If a person receives several penalties due to the "illegal dumping offence," the vehicle of this person will be confiscated, and their driver's licence will also be suspended. Further punishments of persons and corporate bodies violating the "illegal dumping act" are summarized in Table 4.4. Lastly, if the construction site's proprietor is found to dispose the waste at a place which is not designed for construction waste disposal purposes, the proprietor is considered to be in violation of the Act of "Dumping of Trash from Construction Sites." The proprietor who violates this Act will be subjected to a fine of $15,000. The proprietor of a construction site must obtain a permit from the Central Board of Health for proper disposal of construction waste.

In conclusion, the introduction of the plastic ban policy by the government of Antigua and Barbuda has been effectively implemented in this country, with significant results. After the implementation, the amount of plastic litter found in landfills was observed to be significantly reduced by 15.1%. Also, some small shops and small-scale supermarkets started the implementation of the plastics ban earlier, although they are only required to implement the plastics ban after the large supermarkets have completely implemented and adopted the policy (United Nations Environment Programme, 2021). In practising the plastics ban, their government also encountered a challenge in getting the media to play a jingle on their platforms without payment. The playing of the jingle is mainly important to the awareness-raising campaign. However, the playing of the jingle is continuing on the state-run media (Hill, 2016).

TABLE 4.4
The Components, Description, and Penalties of the Litter Control and Prevention Act, 2019

Dumping action that is in breach of the act:	Descriptions:	Penalties:
a) Dumping of trash in a public area	Any person who does not dispose of waste properly in an authorized area but simply disposes of the waste in a public area will be subjected to punishment.	a) Individual fine: Fine up to $3,000 or subject to one year imprisonment. b) Corporate body: Fine up to $15,000.
b) Dumping of trash/ waste from motor vehicle or trailer (car, trailer, etc.)	Any person is considered as an littering offender if the item or materials in or on a motor vehicle along any road or street falls from the motor vehicle due to: i. Not fastened securely enough to keep it from falling out of the motor vehicle. ii. Not sufficiently or well enough covered to keep it from falling from the motor vehicle.	a) Individual fine: Fine of up to $5,000 or subjected to one year of incarceration. Suspension of driving license for six months. b) Corporate body: Fine up to $15,000.
c) Illegal dumping	Any person who throws or abandons rubbish or waste on the property or occupied space of another person will have committed an infraction.	a) Individual fine: Fine up to $3,000 and may also be subjected to clean up of the waste or one year of imprisonment. b) Corporate body: Fine up to $15,000.
d) Dumping of trash from construction sites	a) Waste disposal by the proprietor of a construction site in a place that is not designed for waste disposal will be charged for violation of the act and will be subjected to punishment. b) For proper waste disposal, a permit is necessary to be obtained from the Central Board of Health.	a) The proprietor that violates this act will be subjected to a fine of $15,000.

4.2.2 Bahamas

Plastic pollution awareness in the Bahamas was initially promoted by the non-profit organization the Bahamas Plastic Movement. The Bahamas Plastic Movement is a group organized to increase awareness of plastic pollution and reduce plastic pollution in the Bahamas. The organization of Bahamas Plastic Movement was founded by Kristal Ambrose in 2014. The launching of the organization the Bahamas Plastic Movement is to promote public awareness of the plastic pollution phenomenon and obtain a suitable solution (Bahamas Plastic Movement, 2014). The goal of the Bahamas Plastic Movement is to create a community that has advocacy skills in supporting plastic pollution reduction. The organization aims to develop and build the plastic-ban mindsets of Bahamians. They wish to trigger these mindsets in the cultural habits of all Bahamians by implementing policy changes in the Bahamas. The development of a plastic-ban mindset in Bahamians can be done by encouraging Bahamians to participate in citizen science and environmental leadership on a hand-on basis. The organizer of the Bahamas Plastic Movement, Kristal Ambrose, and her students from the Bahamas Plastic Movement have travelled to Nassau, Bahamas, to conduct a meeting with the Minister of Environment and Housing, Romauld Ferreira, to discuss the plastic pollution issues in Bahamas by banning plastic bags in Bahamas. They have discussed the severe, detrimental effects of the continuing plastic pollution problem on the Bahamian nation. Before the discussion meeting, Kristal Ambrose even drafted a bill about the plastic bag ban with the assistance of a lawyer to share with the government. The Bahamas Plastic Movement claimed that the increase of plastic pollution on the beaches would cause a significant loss in tourism in the Bahamas. Eventually, the Minister of Environment and Housing made an official announcement of the plastic bag ban on single-use plastic at a nation level on January 1, 2020, after the discussion meeting (Goldman Environmental Prize, 2020). The main purpose of the implementation of a single-use plastic ban is to effectively reduce daily plastic consumption in order to decrease the plastic-pollution phenomenon in the country. This could help to reduce plastic proliferation in the community of the island, and thus, could create a better, healthier environment. The plastic ban also could prevent massive landfill fires and further reduce litter on the beaches and streets. The implementation of plastic ban was observed to improve the general health of the nation.

4.2.2.1 Environmental Protection (Control of Plastic Pollution) Act, 2019

As mentioned earlier, the Minister of Environment and Housing officially announced that the plastic bag ban on single-use plastic is effective from January 1, 2020 onwards (Goldman Environmental Prize, 2020). In this plastic ban, four single-use plastic products are banned, which are single-use plastic bags, plastic utensils, plastic straws, and Styrofoam containers and cups. This policy mainly emphasizes the following three actions (The Government of Bahamas, 2019):

1) Forbidden: the importation, distribution, manufacturing, selling, supplying, or using disposable, single-use plastic dishware and non-biodegradable, oxo-biodegradable, and biodegradable, single-use plastic bags.
2) Forbidden: the release of balloons into the air.
3) Control the usage and application of compostable, single-use plastic bags and other related matters.

The Environment Protection Act, 2019, was officially approved by the legislative assembly of the Bahamas on December 19, 2019. This act was further enforced on the January 1, 2020. However, manufacturers and exporters of expanded polystyrene in the Bahamas were exempted from this act. The details of this environmental act will be discussed by dividing it into four main parts, as listed in Table 4.5. In this first part of this ban, all single-use plastic dishware such as cups, plates, and other ware for food-serving purpose that is made of polystyrene and cutlery sets that made of

TABLE 4.5
The Details and Information in the Environment Protection Act (2019)

Plastic items	Detail information:
Prohibition of single-use plastic dishware	• The importation, distribution, owning, production, trading, supplying, and use of single-use plastic dishware is banned in the Bahamas. • Single-use plastic dishware that is forbidden: polystyrene cups, plates, and wares to serve food that are made of polystyrene, and plastic knives, forks, spoons, and straws. • Plastic dishware that is excluded from this ban: reusable plastic dishware, compostable plastic dishware, and plastic packaging components used in sealing food and drinks for delivery purposes.
Prohibition of the release of balloons	• No one is allowed to release balloons loaded with gas (to enable the balloons to move higher) into the atmosphere. • Penalties: a) First offence: subjected to fine of not more than $2,000. b) Second and subsequent: subjected to fine of not more than $3,000. c) Corporate bodies proved to commit an offence: subjected to a fine of not more than $5,000.
Prohibition on permitting or causing the release of balloons	• No person, either one person or more, shall cause or permit the releasing of balloons which are loaded with gas that causes them rise in the atmosphere. • However, three cases of releasing balloons are excluded from this ban: inadvertent or careless release within a building when the balloon does not escape to an outdoor area and weather forecasting, research purposes. • Penalties: a) First offence: subjected to fine of not more than $2,000. b) Second and subsequent: subjected to fine of not more than $4,000. c) Corporate body proved to commit an offence: subjected to a fine of not more than $5,000.
Prohibition on non-biodegradable, single-use plastic bags	The importation, distribution, owning, production, trading, supplying, and usage of oxo-biodegradable, non-biodegradable, or biodegradable, single-use plastic bags are not allowed in Bahamas. However, this plastic ban does not cover the following items: a) Compostable, single-use plastic bags. b) Plastic bag designed to fully or partially contain food ready for eating such as fruits, nuts, ground coffee, vegetables, cereals, or candy. c) Plastic bags that are designed to hold fresh products such as raw fish and poultry products or raw meat. d) A specially designed bag to contain unwrapped seeds, corns, bulbs, flowers, rhizomes, and soil-contaminated commodities. e) Bags that are used to pack sand, pea rock, bulk fasteners, stone, etc. f) Bags that used for the purpose of administering pharmaceuticals or other medical supplies. g) Bags that are used for dry cleaning. h) Bags used to hold the newspapers during delivery. i) Bags used to dump the rubbish from homes, public spaces, commercial establishments, or other locations.

(Continued)

TABLE 4.5
Continued

Plastic items	Detail information:
	j) Bag that are designed primarily for the purpose of containing living aquatic animals in water. k) Bags primarily designed to pack hardware goods. l) Bags exclusively used to transport or store agricultural goods. m) Bags that are designed as an element of packaging and are used to seal a product before the delivery process. n) Bags that are designed and used to store ice for retail purposes. o) Bags that are designed specifically for damp umbrellas. p) Bags designed to be used as party bags. • Penalties: 1st offence: Subjected to a fine of not more than \$2,000; also, subjected to a fine of \$500 per each day or part of a day for a continuing offence; 2nd or succeeding offence: Subjected to a fine of maximum \$3,000; Also, subjected to a fine of \$700 per each day or part of a day for a continuing offence.
Selling of single-use, compostable plastic bags	The trader can sell compostable plastic bags at a cost of a minimum of 25 cents and not more than 1 dollar per a plastic bag without the charge of Value Added Tax (VAT) on the day this plastic ban act has been enforced. The cost of the selling the plastic bags must be stated independently as "checkout bag fee" on the receipt of the customer.

plastic are forbidden from usage. However, this ban is not applied to plastic dishware that is reusable, compostable, and the components of which are used as packaging to seal food and drinks for delivery purposes. The minister might announce the latest altered ban list of single-use plastic dishware in the future. Concerning the forbidding of the releasing of balloons, the releasing of balloons that are loaded with gas that enables them to travel higher in the atmosphere is prohibited under this plastic ban. The person who is found to offend against this act will be subjected to penalties as listed in Table 4.5. Part of the prohibition on permitting or causing the release of balloons mentioned that no person (either one or more than one person) shall permit or cause the release of balloons that are inflated with gas and cause these balloons to rise in the atmosphere. This act is not applied to conditions such as the unintentional release of balloons due to careless release within a building that does not cause the balloons to escape into the atmosphere, and lastly, scientific research using balloons for weather forecasting and meteorological purposes. Concerning the forbidding of non-biodegradable, single-use plastic bags, the import, distribution, production, trading, owning, supply, and use of non-biodegradable, oxo-biodegradable, or biodegradable, single-use plastic bags are prohibited in the Bahamas. However, this ban does not apply to the plastic items that are listed in Table 4.5. Concerning the selling of single-use, compostable plastic bags, the merchant can sell a compostable plastic bag with a cost between 25 cents and 1 dollar, and it must be stated as a checkout-bag fee on the receipt.

4.2.3 Canada

The generation of large quantities of plastic waste in Canada is estimated to reach up to approximately 3.3 million tonnes per annum. According to research conducted by Martinko (2020), the generation of huge plastic waste amounts was mainly contributed by the high daily consumption of Canadians in using approximately 15 billion plastic bags and 57 million straws per day. The high daily consumption of plastic bags and plastic straws has further worsened the plastic pollution in Canada. In fact, the implementation of a plastic ban was conducted by the Canadian government as early as 2007, as mentioned by Martinko (2020).

4.2.3.1 By Law 462

A town known as Leaf Rapid, which is located in Northern Manitoba, Canada, introduced a plastic ban law in 2007. This is the first community in North America that implemented a plastic bag ban law. Initially, Leaf Rapids authorities discovered that the usage and improper disposal of plastic bags had severely harmed the environment and caused plastic bag pollution. Plastic pollution caused the government to allocate a budget of $5,000 per year to tidy up the disposed plastic bags that were scattered across the town. The Municipal Council legalized a plastic bag ban – namely, By-Law 462 – on March 22, 2007 (Leaf Rapids Council, 2006). According to By Law 462, the supply or selling of single-use plastic shopping bags to the customers by the merchants in Leaf Rapids is prohibited. Merchants that breach this By-Law will be subjected to a fine of up to $1,000 per day. If the merchants persist in the violation for more than one day, the merchants will be charged as guilty of a second offence for each day the violation persists. In this By-Law, plastic bags used to store non-packaged items as listed later are exempted from the plastic bags ban law. The non-packaged items include dairy products, fresh fruit, vegetables, nuts, confectionery, cooked foods (hot or cold), packaging plastic bags for raw fish, raw meat, candy, and poultry. Supplying plastic bags that cost more than $1.50 is also exempted from this law.

4.2.3.2 Plastic Bag Reduction Act

In the year of 2019, the Plastic Bag Reduction Act has been introduced in Prince Edward Island, a province of Eastern Canada. According to Government of Prince Edward Island (2019), the Plastic Bag Reduction At was enforced on July 1, 2019. The targets of the implementation of this act are listed as follows:

1. To minimize the usage of single-use plastic checkout bags in businesses.
2. To minimize waste generation and damaging impact on the environment.
3. To promote ethical and sustainable, prioritized business practices in Prince Edward Island.

After the implementation of this act, the merchant shall not refuse or discourage a customer from using his or her own reusable bag to carry the goods that her or she purchased or received in a business. On the other hand, the merchant shall not sell or provide a single-use plastic bag to a customer. The merchant also shall not provide a free checkout bag to a customer, but they can only provide a checkout bag to the customer under the following conditions:

1. The customer has initialled their request for a checkout bag, if they need a checkout bag, and acknowledges that they do.
2. A paper bag or reusable checkout bag is offered to customers.
3. The customer will be charged with a minimum of (a) $0.15 per paper bag or $1.00 per reusable checkout bag.

However, there are no limitations or prohibitions on the sale of bags that are provided in packages of numerous bags in this act. Some consumers are intentionally requested to purchase more checkout bags and to use these bags at their home or workplace. This prohibition act does not apply to the following usage of checkout bags:

a) Paper bags of small size.
b) Bags that are used to pack food products such as grains, fruit, nuts, sweets, vegetables, etc.
c) Plastic bags that are used to pack small hardware products such as bolts, nails, etc.
d) Plastic bags that are uses to plant potted plants and bundle flowers.
e) Plastic bags that are used to pack and enclose the fish, poultry, and frozen foods.
f) Plastic bags that are used to pack the prescription pharmaceuticals obtained from pharmacies.
g) Plastic bags that are used to convey live fish.
h) Bags that are used to pack ready meals that are available for immediate consumption or bakery products that are not pre-packed.
i) Bags that are used to protect the bedding, linens, or other bulky goods that are unable to be put in a reusable bag.
j) Plastic bags that are used to enclose newspapers or printed information and are to be placed in the house or workplace of the customers.
k) Plastic bags that are used to pack and secure already-laundered or dry-cleaned clothes.
l) Plastic bags that are used to pack medical equipment and medical products, which equipment and products are used under the provision of health services that are provided by a member of the Pharmacy Act, R.S.P.E.I. 1988, Cap. P-6.1.
m) Bags that are used to secure tyres which are too large to be put inside a reusable bag.
n) Transparent plastic bags utilized to pack foods which contain liquid and are estimated to have a leakage possibility during shipment.
o) A bag that has been made from a particular type of material and is designed for a specific purpose, as prescribed by the prohibition act.

4.2.3.3 Montreal By-Law 16–051

Montreal, the second largest city in Canada, officially announced the implementation of By-Law 16–051 in 2016. The main purpose of implementing this by-law was to forbid the supply of certain plastic shopping bags in retail outlets. Plastic shopping bags, in this by-law, are the plastic bags that are made with traditional, non-degradable, oxo-degradable, or biodegradable plastic. The city council aims to encourage a change in mindset regarding the application of plastic bags and decrease the environmental impact of plastic bags by enforcing the implementation of this by-law. According to By-Law 16–051, the following subsequent stated actions are prohibited (Nicholas Institute for Environmental Policy Solutions, 2016):

a) Provide or supply traditional plastic shopping bags less than 50 microns thick to customers in retail outlets, irrespective of whether free of charge or requiring payment.
b) Provide or supply to customers with oxo-degradable, oxo-fragmentable, or biodegradable plastic bags of any thickness.

Besides, the usage of plastics bags for the following applications are not subjected to By-Law 16–051:

a) Plastic bags that are used to pack or contain food products such as nuts, fruits, meat, fish, vegetables, and daily products and are paid for at the cashier counter.
b) Plastic bags that are used to ensure hygiene by separating the food product from other non-food products.

Persons who violate By-Law 16–051 may be subjected to the punishment. For individuals, the person who violates the by-law for the first time will be considered as first-time offender and will be subjected to a fine of $200 to $1,000. The person who is charged for a subsequent offence in violating the By-Law 16–051 will be subjected to $300 to $2,000. For corporations, a fine of $400 to $2,000 will be issued for the first offence, while a fine of $500 to $4,000 will be issued for subsequent offences.

After two years of implementation, By-Law 16–051 has been found unable to achieve the desired outcome. Valerie Plante, the Mayor of Montreal City, announced that the city of Montreal will prohibit all plastic bags, regardless of thickness. Previously, all the stores were restricted from selling plastics bags that were 50 microns in thickness. However, they are prohibited from selling all types of plastic bags, regardless of thickness. The introduction and implementation of the Plastic Bag Reduction Act in Prince Edward Island has been found to be a remarkable success. The Canadian maritime provinces were collectively found to dispose of approximately 15 to 16 million plastic bags per annum. However, this issue has been completely eradicated after the implementation of this by-law, with no plastic bags in sight. Retailers in Prince Edward Island have offered paper or reusable bags since the implementation of this plastic bag ban. Notably, there were no penalties imposed in the first year of the Act's enforcement. For the person who violate the Act, merchants and their customers can be fined $10,000 and $500, respectively. This has demonstrated the full cooperation of residents in the ongoing effort to combat plastic pollution (Martinko, 2020).

4.2.4 California

4.2.4.1 Senate Bill 270

California has passed Proposition 67 through a referendum in 2016 by initiating the implementation of Senate Bill 270. The inception of this Senate Bill 270 was primarily moved by the severely incremental problem of plastic pollution. This is underscored by the fact that Californians have consumed an estimated 15 billion single-use plastic bags per annum, which is equal to approximately 400 bags per Californian. Senate Bill 270 has prohibited stores or retailers from providing single-use plastic bags to their customers (Legislative Analyst's Office, 2016). Senate Bill 270 has comprised three critical provisions, which are the prohibition of single-use plastic bags and creating new standards for reusable carryout bags.

(a) The prohibition of single-use plastic bags.

According to Senate Bill 270, stores such as convenience shops, liquor stores, large pharmacies, and supermarkets in California are prohibited from providing or giving plastic bags to customers. However, plastic bags can be provided by these stores for some specific purposes such as unwrapped, raw food (fresh meat, fish, poultry), un-packed nuts, fresh-baked buns, etc.

(b) Creating new standards for reusable plastic carryout bags.

This provision is primarily targeted on establishing a standard concerning the material composition and durability of both reusable bags and recycled paper bags. The California Department of Resources Recovery and Recycling (CalRecycle) will monitor the involved bag manufacturers to ensure that the produced bags achieve all the specified requirements. The stipulated requirements for the reusable plastic carryout bags are listed as follows:

(1) These bags must qualify as a recycled paper bags.

Under this requirement, the materials that are used to produce reusable carryout bags must consist of at least 40% post-consumer, recycled materials. All the information such as the name of the manufacturer, location, or country of the manufacturing site and the total percentage of post-consumer recycled materials must be printed on the reusable carryout bags.

(2) These bags must qualify as reusable grocery bags.

The qualified reusable grocery bag is designed to provide at least 125 uses and has a minimum of 15 litres of volumetric capacity. Besides, the reusable grocery bag must be able to be washed or disinfected by using a washing machine. The materials used to fabricate the reusable grocery bags must avoid using toxic materials. Reusable bags produced from plastic film must consist of a minimum of 40% post-consumer materials. On the other hand, reusable bags produced from fabric must be sewn to meet the requirement of being able to carry a least 22 pounds of goods for a travelling distance of 175 feet for 125 times. Reusable bags must be printed with information such as the manufacturer's name and country, the recycling instructions, and use statement. The use statement must describe the bag as a reusable bag able to be used at least 125 times. Reusable bags must be manufactured by the companies in the list of CalRecycle's Certified Reusable Grocery Bag Producer.

(3) Impose charges on the other carryout bags.

Stores can provide reusable grocery bags or recycled paper bags to customers with a charge amount of not less than $0.10. However, plastic bags that are used for specific purposes such as prescription medicines, raw meat, fish, etc., are excluded from the charge. On the other hand, free, reusable grocery bags or recycled paper bags will be provided to certain low-income customers or clients who have paid by using EBT cards or WIC cards or vouchers.

The implementation of a plastic bag ban in California has successfully reduced plastic bag consumption. The consumption of plastic bags was found to be highly reduced by 71.5%. This action also could keep the recycling system from using plastic bags which can cause the machinery to be clogged. The clogging of plastic bags in machinery has significantly caused a rise in costs. However, the prohibition of plastic bags in California has also rapidly increased the demand for the usage of paper bags and caused some of the customers to purchase thicker plastic bags. High-thickness plastic bags were used to replace and substitute the secondary usage of plastic bags for the purpose of trash-can liners or pet-waste-collection purposes. According to HITE (2020), 28.5% of plastic bag usages are considered to be offset by other bag usages. Besides, the plastic industry has paid to petitioners to put a bill on the ballot through the voting process after it was approved by legislature after Senate Bill 270. The banning of plastic bags will affect business. In conclusion, the banning of plastic bags has been successfully implemented and generated a great outcome.

4.2.5 Hawaii

In 2011, the government of Hawaii officially implemented a plastic bag ban, known as Disposable Foodware Ordinance (Ordinance 19–30/Bi40l1 40). Some ordinances have been introduced in Hawaii for the past few years with the purpose to regulate the plastic bag ban in Hawaii. The Mayor of Honolulu, Oahu in Hawaii approved and signed Bill 40 in 2019. The objective of Bill 40 was to clarify the regulations that govern the use of plastic bags and single-use plastic products. According to City and County of Honolulu (2019), there are several sections in Bill 40, and the details of this Bill 40 are listed as follows:

Section 1: Amendments on the Revised Ordinance of Honolulu 1990

(1) Plastic checkout bag: Plastic bag that is provided by the supermarkets, stores, etc., to a customer for the purpose of transporting the grocery items or other retail items. This plastic bag is not specially produced and designed for the reuse purpose.

(2) Plastic: Materials that are derived from fossil fuel or are produced from the petrochemical polymeric compounds through the reaction process. The plastic materials can be further moulded by streams.
(3) Plastic film bag: The plastic film bag is produced from blow-moulded, thin-elastic film sheets that have a thickness less than 10 mils, where mils is equivalent to 0.0254 mm.

However, all the previously mentioned bags do not include the bags that used for the following purposes:

(a) Bags without handles that are used to pack items such as vegetables, grains, nuts, fruits, candles, or small hardware parts.
(b) Bags (without handle) that are used to pack frozen food, raw meat, fresh fish, potted plants, or other items that contain moisture.
(c) Bags that are used to place newspaper for delivery purposes.
(d) Bags used to pack cleaned clothes and garments after laundry or dry-cleaning purposes.
(e) Bags that are used to transport chemical pesticides, drain cleaners, or other caustic chemical which are available at retail. However, this exemption is only applied to one bag per client.
(f) Bags that are used to transport live animals such as fish or insects which are purchased from pet shops.
(g) Bags that are used to pack garbage, pet waste, and yard waste. Bags that used for this purpose can be purchased with multiple bags in one package.

Section 2: Article 27 in Chapter 41 of the Revised Ordinance of Honolulu 1990 is abrogated.

Section 3: Introduction of New Article 27 with title of "Polystyrene Foam and Disposable Food Service Ware" to replace the previous one.

(1) Regulation on the use of polystyrene foam dish ware, disposable plastic service ware, and plastic food ware. In this regulation, polystyrene foam food ware, disposable plastic service ware, and disposable plastic food ware are prohibited from being sold to customers or being used to serve prepared food.
(2) Exemptions are applied on the application as follows:

(a) Exemption will be provided when receiving an application accompanied by credible supporting documentation.
(b) Exemption for industry application.
(c) Disposable plastic straws can be exempted and provided to the customers if the customer is either physically or medically unable to use non-fossil-fuel-based straws.
(d) Plastic packaging that is used to pack unprepared products such as raw meat, fresh fish, fresh seafood, and poultry products.
(e) Packaging for pre-prepared food, shelf-stable food, also known as non-perishable food items, and prepared meals or food.

(f) Plastic packaging that requires immediate action to safeguard life, health, property, safety, or essential public infrastructure in any situation deemed an emergency by the city.

(3) Polystyrene foam food ware, disposable plastic service ware, and food ware are prohibited from being sold to customers:

- Polystyrene foam food ware, disposable plastic service ware, and disposable plastic food ware are prohibited from being sold in any stores in Honolulu. However, an exemption is applied for the following applications:

  (a) Plastic bags that are used to pack unprepared food products such as raw meat, fresh fish, seafood, and poultry products.
  (b) Packaging for pre-packaged food and shelf-stable food.
  (c) Packaging for products that do not meet compliance standards and are sold to a food vendor with an exemption granted.

- Stores can apply for an exemption in order to adhere to the prohibition.

(4) Disposable service ware can be supplied to customers by the food vendor upon request by customers. The food vendor can offer disposable service ware for food and beverages only upon request or inquiry from a customer or person receiving the food or drink or in a self-service site.
(5) Penalties involved:

  (a) The Department of Environmental Services is the charged and responsible department to enforce and administrate this provision.
  (b) Local restaurants or businesses that have breached the regulation or rule implemented in this article will be subjected to the following penalties:

- The restaurant or business will be warned to stop distributing or selling the prohibited items in this article such as polystyrene foam food ware, disposable plastic service ware, and disposable plastic food ware to customers.
- If the restaurant or business is found to continue to provide these items to customers, the restaurant or business will be subjected to a fine of between $100 and $1,000.

4.2.6 Jamaica

4.2.6.1 The Trade Act

(a) The Trade Act (Plastic Packaging Materials Prohibition Order, 2018) (Government of Jamaica, 2018b)

The Government of Jamaica introduced a plastic ban policy, "The Trade Act," in the year 2018. The Trade Act of Jamaica is also known as the "Plastic Packaging

Materials Prohibition Order, 2018." According to the policy of The Trade Act, all single-use plastic is prohibited from importation and distribution in large quantities in Jamaica, effective January 1, 2019. In this policy, there are two different sets of guidelines on the dimensions of single-use plastics, where each of the guidelines will be effective at different points in time. Single-use plastic products with dimensions of 610 mm × 610 mm × 0.03 mm is banned starting from January 1, 2019 and expected to be fully executed before January 1, 2021. On the other hand, single-use plastic products with dimensions of 610 mm × 610 mm × 0.06 mm are prohibited starting January 1, 2021. However, there are some single-use plastic products that are exempted from this prohibition, which are mentioned as follows:

(1) Single-use plastics that were imported into and distributed in Jamaica before January 1, 2019.
(2) Single-use plastic products that are used to preserve public health and comply with the food safety standard. These single-use plastic that are provided by the retailers to contain eggs, flour, rice, baked goods, and fresh products such as fresh fish, raw meat, etc., for hygiene and safety purposes.
(3) Single-use plastics that are imported and distributed by the Ministry of Health, which is responsible for health and plastic use in dentistry practises and other medical fields. Also, single-use plastics that are used to carry medicines.
(4) Single-use plastic products that are used to carry or contain personal belongings that are placed inside the luggage of a person entering or leaving Jamaica.
(5) Single-use plastic drinking straws that are provided to a person with disabilities are exempted from this act. These plastic straws are imported and distributed by a body that represents a person with disabilities who is recognized by the government of Jamaica.
(6) Single-use drinking straws made of polyethylene or polypropylene that are attached on the boxes of juice or drink will be valid until January 1, 2021.

This act has prohibited the application of all types of single-use plastic products, which includes plastic bags that are identified as photodegradable, oxo-degradable, biodegradable, degradable, or compostable. The person who has violated this Act will be subjected to a fine with not more than $2 million or will be subjected to maximum two years' incarceration.

(b) The Natural Resources Conservation Authority (Plastic Packaging Materials Prohibition Order, 2018) (Government of Jamaica, 2018a)

This Order is similar to the Trade Act that was discussed earlier. However, this Order has prohibited the commercial use and manufacturing of single-use plastic bags. In this order, the following terms are defined:

(1) An approved person is referred to as a person that is designated by the Minister to have the exemption to manufacture or use single-use plastics.

(2) A person with disabilities or person with a disability is defined as any person with a permanent physical, mental, intellectual, or sensory impairment that might impede their complete and effective engagement in society, at the same level as other persons.
(3) Single-use plastics:

- Single-use plastic bags.
- Packaging that is fully or partially made of expanded polystyrene foam.
- Drinking straws that are fully or partially made of polyethylene or polypropylene and manufactured for single-use purposes.

(4) Single-use plastic bags:

- Any bag that is fully or partially made of polyethylene or polypropylene in the applicable dimensions.

(5) According to this Order, a quantity is considered commercial if it is reasonably intended for wholesale or retail sale to gain economic benefit, either in cash or kind. Starting from January 1, 2019, this order is prohibited for all the manufacturing or use of single-use plastic in commercial quantities.

The exemptions on single-use plastic bags for this order are slightly different from the exemptions in the Trade Act listed as follows:

(1) Single-use plastics that are used to maintain public health and food hygiene and safety purposes which include packaging to contain and distribute flour, rice, eggs, baked goods, and fresh products such as raw meat, fresh fish, etc.
(2) The manufacturing and usage of single-use plastic that has been approved by the Minister of Job Creation and Economic Growth.
(3) Single-use plastics that are used to pack personal belongings and are placed inside the luggage of a person when travelling into or out from Jamaica.
(4) Drinking straws that are manufactured for the use of persons with disabilities by organizations with granted approval from the Minister.
(5) Single-use plastics that are used to contain food or beverages will only remain valid until January 1, 2020.
(6) Drinking straws (which are made of polyethylene or polypropylene) that are manufactured for one-time use and attached to the packaging of juice boxes is only valid until January 1, 2021.

The order has prohibited the use of all types of single-use plastics, which include plastic bags that are termed as oxo-degradable, photodegradable, degradable, bio-degradable, or compostable plastics. The person who has violated this order will be subjected to a fine of not more than $50,000 or subjected to maximum two years of imprisonment. The implementation of this single-use plastic ban order is the first phase of the plastic ban in Jamaica.

Besides the implementation of the single-use plastic ban, the government of Jamaica has also considered banning the use of Styrofoam and plastic straws in Jamaica. The implementation of the plastic ban in Jamaica was divided into two phases: the first phase of the single-use plastic prohibition started on January 1, 2019, while the second phase of the plastic ban started on January 1, 2020. In the first phase of the single-use plastic ban, the importation of expanded polystyrene foam into Jamaica has been restricted. In the second phase of the plastic ban, the manufacturing and distribution of plastic products that are made of polystyrene foam are prohibited in Jamaica. The measure on the plastic straws ban is divided into phase 1 and phase 3. In the phase 1, the importation of plastic straws from other countries into Jamaica are strictly prohibited, while the manufacturing of plastic straws is also banned in Jamaica. In phase 3, the importation of the plastic straws that are affixed on boxes of juice into Jamaica from other countries is restricted. A grace period of 6 months is provided to relevant businesses to transition away from plastic straws to other alternative-material straws.

The implementation of the plastic ban policies in Jamaica is observed to be progressing well and has received a positive response from the public. The government of Jamaica has received approximately 90% support from the consumers and stakeholders. The effectiveness of the plastic ban is yet to be identified, as the implementation of the single-use plastic ban has just been underway for approximately one to two years. However, businesses found to violate this single-use plastic ban order are subjected to the fining action.

4.2.7 Chicago

4.2.7.1 Addition of Article XXIII into Municipal Code Chapter 11–4

Plastic pollution in Chicago is increasingly getting worse due to the high usage amounts of plastic bags. The aldermen in Chicago, Illinois authorized an ordinance in an effort to curb the plastic pollution problem in April 2014. Aldermen are elected members in a municipal council in local governments which include cities and towns. The city council of Chicago is expected to eliminate the disposal of plastic bags from garbage streams. According to NBCChicago (2014), the daily used plastic bags amount in Chicago is estimated to be approximately 3.7 million plastic bags. Approximately 3% to 5% of the daily used plastic bags are disposed of as waste. This is one of the reasons that has led to the approval of this ordinance. According to this ordinance, all stores are prohibited from offering plastic bags to customers to contain and transfer purchased goods. This ordinance was effective starting from August 1, 2015, and this ordinance is only applicable for retail stores that occupy more than 10,000 square feet. According to Chicago City Council (2014), this implementation of this ordinance for retail stores with land area less than 10,000 square feet will be effective starting from August 1, 2016.

In order to promote the prohibition on the usage of plastic bags, retail stores can offer reusable bags or recyclable paper bags to customers to contain and transfer purchased goods to a car or home. In this ordinance, reusable bags are referred to as washable bags that can be repeatedly used for more than 125 times, contain at least 15 litres, and are at least 2.2 mils in thickness if these bags are made of plastic materials. According to this ordinance, the compostable type of bag can also be provided to the customers of retail stores at the payment counter for the customers to contain and transport their purchased goods to a car or home. Retail stores can also prefer to provide or not provide any bags to their customers. However, this ordinance does not provide the authority to the retail stores to prohibit customers from using their own shopping bags when they are shopping in the retail stores. Under this ordinance, any person who has breached this ordinance will be subjected to a fine of between $300 and $500. If the person has continued to violate this ordinance, this person will be subjected to another fine, as the offence against this ordinance is considered to be a distinct one each day.

The implementation of this ordinance has not been very effective, as the usage of plastic bags is observed unable to be fully prohibited. According to this ordinance, plastic bags with a thickness of more than 2.25 mils are considered to be reusable bags. Although retail stores have been prohibited from providing customers with plastic bags, these retail stores have replaced the offering of plastic bags to customers with paper bags or thicker plastic bags that are four times thicker than normal plastic bags. However, replacing normal plastic bags with paper bags is observed to cause higher energy consumption. This is mainly because the manufacturing of paper bags requires higher energy consumption compared to the manufacturing of plastic bags. According to the Illinois Retail Merchants Association, a retail store in Chicago has provided an approximate amount of 20,000 plastic bags to their customers weekly before the implementation of the ordinance. However, this retail store has offered a total of approximately 10,500 plastic bags to customers after the implementation of the ordinance. Although the number of plastic bags used has been significantly reduced after the implementation of this ordinance, the overall usage of plastic bags is still considered to be at least two times higher than plastic bag usage before the implementation of the ordinance. This is because the plastic bags provided after the implementation of the ordinance are at least four times thicker than the plastic bags before the implementation of the ordinance. Furthermore, thicker plastic bags require more time for the decomposition process when compared to thin plastics bags. Thus, the implementation of this ordinance was found unable to reduce the plastic pollution problem and may cause a detrimental impact on plastic pollution (SCHEIBE, 2016).

4.2.7.1.1 Chicago Checkout Bag Tax Ordinance

The implementation of the ordinance in 2015, it was observed, did not achieve the desired outcome by reducing plastic waste amounts. This has further caused the city council to introduce and implement another relevant policy in order to reduce the plastic pollution problem. This newly implemented policy is called as the Chicago Checkout Bag Tax Ordinance, while the levy is known as the Checkout Bag Tax. The implementation of this tax is intended for the widespread distribution of checkout bags throughout the city of Chicago. Upon the request of customers for checkout bags, the customers will be charged $0.07 for the purchase of checkout bags. Checkout bags are referred to as paper bags or reusable plastic carry bags. However, this tax is not categorized as inflicted on businesses or stores (Chicago City Council, 2016).

When customers request to purchase checkout bags during their payment for goods, the retail store must make sure that the fee charged for the checkout bags are stated separately on the receipt as independent items and indicated as the "checkout bag tax." If the retail stores fail to follow the terms stated in this ordinance, they will be considered to have breached this ordinance. However, retail stores have two choices on the sale of checkout bags at payment counters, which include:

1. Charge tax on customers who request checkout bags, and state the charge fee independently as "checkout bag tax" on the receipt.
2. Retail stores cannot charge customers with the "checkout bag tax" if they do not list the "checkout bag tax" independently on the receipt.

If retail stores fail to separate the checkout bag tax independently from the customers, the retail stores do not bear the responsibility to pay the checkout bag tax for the customers. The customers must pay the "checkout bag tax" to the respective Department. After the retail stores solicit the checkout bag tax from the customer, they will transfer the collected checkout bag tax to the wholesalers that sold the checkout bags to the stores. When the wholesalers pay the collected checkout bag tax to the Department, they need to prepare the details, which are the tax and the filed returns. The tax payment can be done in two ways, which are the payment of actual tax liabilities and the payment of predicted taxes.

Retail stores that separate the checkout bag tax of $0.07 independently in the receipt of customers are given the right to keep an amount of $0.02 per checkout bag. This also indicates that retail stores are required to pay a total of $0.05 per checkout bag. The wholesalers that collect the checkout bag tax from the retail stores must pay the exact amount of $0.05 per checkout bag to the respective Department, and they are not allowed to keep any amount. However, if the wholesalers sell the checkout plastic bags to other purchasers or individuals, the wholesalers can also keep $0.02 from the tax collected for every checkout bag. If the wholesalers do not collect the checkout tax from the retail stores, the retail stores have to pay the collected checkout bag tax to the respective Department themselves. If the retail stores are unable to pay the collected checkout bag tax to the Department by the due date, the retail stores will be subjected to a fine of $100. The fine can be abandoned if the retail stores are able to show a proof of late payment.

According to a study conducted by GIANGRECO (2019), they found that the number of plastic shopping bags used after one month of implementation of the Checkout Bag Tax Ordinance dropped by 42%. Before implementation of this ordinance, the people in Chicago are estimated to have carried an average of 2.3 plastic bags per person when they went grocery shopping. After the implementation of the Checkout Bag Tax Ordinance, the number of plastic bags per person significantly dropped to approximately 1.8 plastic bags per person. On the other hand, the number of people who use or carry their own reusable bag during their grocery shopping was observed to increase by approximately 20%. From this observation, the levy-type policy is found to be more effective than the ban-type policy in Chicago. This is because most people are more concerned with spending their money on used checkout bags than the spirit of saving the environment. Bringing their own reusable bags during their grocery shopping can prevent them from having to spend extra money for a plastic checkout bag.

After the success of the implementation of the Checkout Bag Tax Ordinance in Chicago, the aldermen in Chicago have mentioned that they will introduce a new ordinance prohibiting the use of polystyrene and Styrofoam dishware in the year of 2021. However, the implementation of the new ordinance in Chicago requires the approval of the Chicago City Council. And until now, this proposed ordinance has yet to be implemented. Some plastic products such as plastic spoons, plates, forks, etc., will be banned in restaurants if this proposed ordinance is approved by the Chicago City Council. However, restaurants can still provide recyclable and compostable plastic to customers, and these are subjected to a full or half waiver in case they are unable to wash the plates. Overall, the implemented policies and the policies that are going to be implemented are found effective in curbing the plastic pollution problem in Chicago (Bauer and Alami, 2020).

4.2.8 Saint Vincent and the Grenadines

4.2.8.1 Environmental Health (Expanded Polystyrene Ban) Regulations, 2017

The government of the Saint Vincent and Grenadines (SVG) has introduced and implemented a new policy which is known as the Environmental Health (Expanded Polystyrene Ban) Regulations, 2017 in 2017. The purpose of the implementation of this policy is to help address the plastic pollution problem in the country, and this policy was officially effective starting May 1, 2017 (Food and Agriculture Organization (FAO), 2017). There are three main objectives in this regulation, as follows:

1. The importing, production, and selling of disposable dishware made of expanded polystyrene are prohibited in Saint Vincent and the Grenadines.
2. Dishware that is made from expanded polystyrene is prohibited from being used in Saint Vincent and the Grenadines.
3. The replacement of containers or food packaging that are made of expanded polystyrene with containers or food packaging that are made from biodegradable, recyclable, or other environmentally friendly materials are highly encouraged.

According to the first objective, the importing, production, and selling of any dishware products that are made of expanded polystyrene are strictly prohibited in Saint Vincent and the Grenadines. Any person that has breached this policy will be subjected to a fine of $5,000 or face incarceration of 12 months or both. Restaurants are also prohibited from serving food using dishware made of expanded polystyrene (Styrofoam). Dishware is referred to as containers, plates, cups, trays, and other goods that are used to contain or serve food to customers. When the Chief Environmental Health Officer suspects that a person has violated the policy, the person will be advised to comply with the regulations. If the person is still found to have violated the policy continuously, the person will be considered as an offender against the policy. This person will be subjected to a penalty of $5,000 or incarceration for 12 months. If the person has purchased dishware products made of expanded polystyrene before the kick-off period of the regulations, the person will be exempted from this regulation. However, the person must provide the evidence to customs. With the presenting of evidence, a period of 9 months will be given for stocks to reach Saint Vincent and the Grenadines and another 9-month transition period will be given for the selling, using, and serving of dishware products made of expanded polystyrene after the kick-off of the Regulations. The purpose of giving the transition period is to provide enough time for people in Saint Vincent and the Grenadines to adapt to the implemented Regulation.

Other than the Regulations as mentioned previously, the government of Saint Vincent and the Grenadines also waived the Value Added Tax (VAT) in 2017. The purpose of the waiver of the VAT is to promote the replacement of packaging and dishware that is made of expanded polystyrene dishware with biodegradable packaging and dishware. The implementation of this repeal will be effective starting from May 1, 2017 (Jamaica Observer Ltd, 2017).

4.2.8.2 Environmental Health Control of Disposable Plastics Regulations, 2019

After the implementation the Environmental Health (Expanded Polystyrene Ban) Regulations, 2017, for two years, the Minister of Health Senator officially declared the implementation of new regulations in Saint Vincent and the Grenadines in November 2019. The purpose of implementing this regulation was to ban the usage of single-use plastic bags (SearchLight, 2020). There are four main concerns throughout this policy which are stated as follows:

1. All the disposable plastic bags are prohibited from being imported into Saint Vincent and the Grenadines, starting from March 1, 2020.
2. All single-use plastic dishware or containers are banned from being imported into Saint Vincent and the Grenadines, starting from August 1, 2020.
3. All single-use plastic bags are prohibited from being used, sold, and distributed, starting from August 1, 2020.
4. All plastic food containers are prohibited from being used, sold, and distributed, starting January 1, 2021.

There are some exemptions for using plastic bags, as mentioned in the Regulations. Plastic bags or containers exempted from this Regulation are as follows:

a) Plastic bags used to contain breads.
b) Plastic bags used to contain fresh products such as meat and fish.
c) Plastic bag or containers used to contain prepared food.
d) Plastic bags used to contain medicine.
e) Plastic bags used to contain fruits and nuts, dairy products, and confectionary products.
f) Plastic bags used to contain tiny hardware goods.
g) Plastic bags or containers used to contain veterinary products.

h) Plastic bags used to contain seedlings.
i) Plastic bags used to serve as trash and laundry bags.
j) Plastic bags or containers used for export purposes.
k) Single-use bags that are made of biodegradable materials.

If a person has breached this regulation, the person will be subjected to a penalty of $5,000. If the person persists in the violation action against the regulation despite the previous given warnings or advice, an additional penalty will be issued to the person every day until the violation action stops. With the implementation of this heavy penalty action, the Minister has expected to significantly reduce the plastic pollution problem within a five-month period. Other than the policy implemented by the government to ease the pollution problem in Saint Vincent and the Grenadines, Massy Stores (a local chain store in Saint Vincent and Grenadines) also took their own initiative to help curb the plastic pollution problem in Saint Vincent and the Grenadines. Massy Stores has officially announced that they will start to implement a charge on every plastic bag that is requested by customers during checkout from their three supermarkets. The implementation of the charge on plastic bag could encourage customers to use reusable bags instead of using the single-use plastic bags. Massy Stores has monitored the usage of plastic bags in these three supermarkets for more than one year of implementation. Massy Stores has found that plastic bags usage has rapidly decreased by 67%, which is equivalent to reducing usage to 480,000 disposable plastic bags. This observation indicated that the action conducted by Massy Stores to implement a charge on plastic bags is a great success. This also shows that the reduction in the plastic pollution problem does not necessarily have to be initiated by and depend on government action.

4.3 SOUTH AMERICA

Policies on plastic bag or container usage have been implemented in several countries in South America such as Argentina, Brazil, Chile, Colombia, Peru, and Uruguay. The details of these policies will be clearly discussed in detail in the next section.

4.3.1 Argentina

4.3.1.1 Law 13868

Severe plastic pollution in Argentina has become a critical problem in the country. In order to reduce the plastic pollution problem in Argentina, the government of Argentina has initiated some actions to curb the plastic pollution problem. The governor of the Buenos Aires Province has introduced Law 13868 through Decree 1521. According to this law, all the business premises and shops in the province are prohibited from offering non-biodegradable plastic bags to customers. The term "non-biodegradable bags" refers to the plastic bags that are made of non-biodegradable plastic such as polyethylene, polypropylene, etc. Non-biodegradable plastic bags must be replaced with the plastic bags that are made of biodegradable plastics to reduce the environmental impact. According to a report of Argentina Ambiental (2008), this law is officially effective starting from September 29, 2008.

According to this law, businesses are categorized under three main sections. These businesses will be given 12 months by the government to gradually replace non-biodegradable plastic bags with bags that are more environmentally friendly. Business is categorized as follows:

1. The sale of food and beverages in a hypermarket.
2. Retailing supermarkets that focus on food and beverages.
3. Small retail markets that focus on food and beverages.

For businesses that are not categorized in the previous list, a period of 24 months is given to these businesses to select for replacement of non-biodegradable plastic bags. The manufacturers of the plastic bags should convert their technology to produce biodegradable plastic bags. They can supply biodegradable plastic bags to retail stores in order to replace the non-biodegradable plastic bags. However, there is an exemption in this law for plastic bags that are used to pack food and wet, pre-processed and processed food. This is because biodegradable plastic bags are not suitable to be used in packing food and wet, processed food and processed food due to hygienic concerns. The contact between water and biodegradable plastic bags might initiate the hydrolysis process; thus, the replacement of non-biodegradable plastic bags with biodegradable plastic bags for wet, processed food or processed food is not applicable. In this case, non-biodegradable plastic bags are still allowed to be used to contain food and wet, processed food and processed food. If retail businesses are found to breach this law, they will be subjected to the following penalties:

1. Warning will be given to the offender for a first-time violation.
2. A fine between ten (10) and up to one thousand (1,000) basic salaries of the Entering Category of the Administrative Group – class 4 – or the one that replaces it in the future, of the salary scale of Law No. 10,430 (Text Ordered by Decree No. 1,869/96 and its amendments), with a regime of thirty (30) hours of work per week.
3. Confiscation of non-biodegradable plastic bags will result in penalties as outlined in subsections of 1, 2, or 3.
4. The premises of the business will be temporarily closed for less than one month.
5. The premises of the business will be closed permanently.

4.3.1.2 Law No. 3147

After the implementation of Law 13868 in year 2008, another law was introduced in 2009 to mitigate the plastic pollution issues in the country. This law is known as Law No. 3147. The purpose of the implementation of this law is to promote the development of the manufacturing of biodegradable plastic bags. This law is also intended to gradually decrease the usage of non-biodegradation plastic bags in retail businesses before the complete prohibition on providing non-biodegradable plastic bags to customers becomes effective. Lastly, this law is also expected to substitute the non-biodegradable plastic bags with the biodegradable plastic bags. The objective of implementing this law is to reduce the generation of waste and the disposed amount of non-biodegradation plastic waste to the environment. This law is officially effective starting on September 17, 2009. The enforcement authority should develop a suitable scheme for the "Reduction of Non-Biodegradable Bags" and "the Replacement of Non-Biodegradable Bags." The development of this scheme must be conducted within 180 days after the law takes effect. According to The Legislature of the Autonomous City of Buenos Aires (2009), this scheme must contain the following criteria:

1. Policy that coordinates with the retail businesses such as retail chains, hypermarkets, and supermarkets with the aim to reduce the number of non-biodegradable plastics bags provided to customers.
2. Policy that coordinates with the manufacturing sector (preferably those manufacturers that are located in Buenos Aires City) regarding the replacement of non-biodegradable plastic bags.
3. Create financial incentives to stimulate technology development, especially in the SMEs. This can further help ease the related manufacturing sector in the production of biodegradable plastic bags to substitute for non-biodegradable plastic bags.
4. Create and formulate a time schedule for the involved parties in the substitution of non-biodegradable plastic bags with the biodegradable plastic bags.
5. Organize awareness campaigns promoting the importance and advantages of using their own shopping bags in reducing plastic pollution issues and thus helping to conserve the environment.
6. Provide technical assistance and training involved.

In this law, substitution for non-biodegradable plastic bags must follow the following schedule:

a. Businesses for non-biodegradable plastic bags are given a two-year period to complete the replacement of non-biodegradable plastic bags after enforcement of this law.
b. For the usage of non-biodegradable plastic bags:

   (i). In supermarkets and hypermarkets for selling food and beverages are given four years for the replacement of non-biodegradable plastic bags.
   (ii). Retail businesses that are not mentioned in (i) are given five years for the replacement of non-biodegradable plastic bags with other substitutes.

If the retail businesses or owners are still using non-biodegradable plastic bags and containers and not practising the replacement of non-biodegradable plastic bags with biodegradable bags, the businesses' owners or the person will be subjected to a fine. Businesses or persons found to breach this law will be subjected to a penalty of ARS 1,000 and ARS 100,000 and the non-biodegradable plastic bags will be confiscated (ARS: Argentina Peso).

4.3.1.3 Law 3847-D-2018

Plastic bags and other plastic products are not the key source of plastic pollution; microplastics are also one of the main sources which have contributed to plastic pollution. It is estimated that there are approximately 0.95 million tonnes of microplastics, on average, that are being disposed of in the ocean per annum. Microplastics mostly originate from cosmetics and personal care products. These products containing microbeads are the main contributor of solid, primary microplastics. Plastic microbeads discharged from wastewater treatment plants flow into the ocean. These microplastics can absorb the organic pollutants and industrial chemicals that flow into the ocean from industrial areas. This has further caused the entry of microplastics into the food chain system, as fish or other aquatic animals might consume microplastics and bring them into the food chain. In order to overcome this problem, the government of Argentina introduced and implemented Law 3847-D-2018 in 2020. The purpose of implementing this law was to ban the manufacturing, importation, and selling of cosmetics, personal care products, and dental hygiene products consisting of plastic microbeads. In this law, the details of the law and the definitions for cosmetics, personal care products, and dental hygiene products are listed as follows (Honourable Chamber of Deputies of the Argentine Nation, 2019):

1. Cosmetic products are referred to as synthetic substances that are applied on different parts of the human body such as face skin, body skin, lips, etc., to maintain the external parts of human body in good condition, provide sanitization, and alter body scent.
2. Dental hygiene products are referred to as products such as toothpastes and mouthwashes that are applied on the teeth and oral mucosa to clean them and keep the teeth in good condition.
3. Plastic microbeads are referred to as solid substances that are derived from petroleum in microparticle form with a particle size equal to or less than 5 millimetres. Plastic microbeads are insoluble in water and difficult to degrade.

Persons who have breached this law will be subjected to the penalty of Law 16,463, also known as the Law on Medicines. According to Mosteirin (2019), the person who violates this law can be subjected to the following penalties:

1. Warning will be given to the violator.
2. Subjection to a fine range of ARS 1,000 to ARS 1,000,000 (ARS: Argentina Peso).
3. The subjection of penalty depends on the severity of the infraction and the premises where the violation may partial, complete, or temporary.
4. If a business has violated this law, the business license will be suspended for three years. If the violation of the business is serious, the licence of the business will be suspended for a maximum of three years.
5. Detention of infringing products.
6. The licenses of the retail stores which are selling and manufacturing non-biodegradable plastic bags will be revoked.

4.3.2 Brazil

4.3.2.1 Law No.8006

Rio de Janeiro, a state in Southeast Brazil, officially enforced Law No. 8006 in 2018. The purpose of implementing this law was to ban trading, usage, and distribution of plastic bags in Rio de Janeiro, Brazil. Law No.8006 is an amendment of the existing Law No.5.502. Law No. 5.502 pertains to the substitution and accumulation of plastic bags in commercial establishments.

According to this law, the corporations and merchants in the state of Rio de Janeiro are prohibited from providing customers with single-use bags or bags that are made of polyethylene, polypropylene, etc. Corporations and merchants are restricted from providing single-use plastic bags, regardless of whether the plastic bag is costless or with cost. The merchants will be given 18 months to find a substitution for the plastic bags with reusable bags after the enforcement of the law. The merchants are referred to as the companies that are categorized as microenterprises, also known as small businesses. The companies which are not included in the category of microenterprises will be given only 12 months to find a replacement for single-use plastic bags.

The reusable bags that are used to replace the single-use plastic bags must be able to withstand 4, 7, or 10 kilogrammes of goods. Also, the materials used to manufacture the reusable bags must be at least equal to or more than 51% renewable sources. According to the law, the colour of the reusable bags should be two colours: green and grey. The green colour represents recyclable waste, while the grey colour represents the other tailings. The colour labelling of the reusable bag can help consumers to separate waste and preform the action of waste collection easily. Reusable bags will be offered to customers, and these bags can cost equal to the cost of the bags or below (Assembleia Legislativa do Estado do Rio de Janeiro, 2018).

4.3.2.2 Decree No.55,827

Sao Paulo, a city in Brazil, introduced a policy known as Decree No. 55,827 in 2015. Decree No.55,827 is a revised version of Law No. 15,374, which was implemented in 2011. The objective of implementing Decree No.55,827 was to prohibit merchants or businesses from providing free or selling plastic bags to customers in Sao Paulo of Brazil. As announced by Civil House of The Mayor's Office (2015), this decree was officially effective starting from February 5, 2015, and some of the highlights of this decree are listed as follows:

1. The action of providing plastic bags (either free of charge or with charge) to customers for them to pack and carry their purchased goods is prohibited in Sao Paulo.
2. Businesses or corporations must support and encourage customers to bring and use their own reusable bags to pack and carry their purchased goods. The reusable bags should be produced from resistant materials.
3. Reusable bags are referred to as bags that comply with the standard of the Municipal Authority for Urban cleaning purposes. These reusable bags can also be used to collect dry, solid household litter.
4. Reusable bags are prohibited from being used to collect the common, unsorted litter.

According to National Congress of Brazil (2008), any person that does not separate solid waste accordingly will be subjected to the following punishment:

a) Consumers that do not follow their commitments under the logistics system will be given a warning.
b) For subsequent offence cases, the person will be subjected a fine between 50 to 500 reals (also equal to USD$7.20 to USD$87.20).

For persons who are involved in activities such as production, storing, using, etc., of plastic products that can cause harmful effects on human health or the environment, they will be subjected to the following penalty:

a. The person will be subjected to a fine of between 500 reals and 2,000,000 reals (or equivalent to USD 87.20 to USD 348,802).
b. The fine amount will be increased by five times if the product is found to be reactive.

4.3.2.3 Law No.17,261

The City Council of Sao Paulo announced the implementation of Law No. 17,261 in 2020. The objective of implementing this law was to prohibit the distribution of single-use or disposable plastic products. This law was officially effective starting from January 1, 2021. The Civil House of the Mayor's Office (2020) has highlighted the following main points of this law:

a. Businesses such as restaurants, bakeries, hotels, and other related shops are strictly banned from providing disposable cups, dishware, drinks stirrers, and cutlery that made of plastic to customers.
b. Paper plates and reusable plastic cups can be used to substitute for disposable or single-use plastic products.
c. Disposable or single-use plastic products can be replaced with products that are biodegradable or reusable so that these products can be recycled.

Any person that has breached this will be subjected to the following penalties:

a. A warning and summons will be issued to the person for a first violation.
b. Another summons and a penalty of 1,000 reals, equivalent to USD 176, will be issued to the person for a second violation.
c. Another summons and a penalty of 2,000 reals, equivalent to USD 353, will be given to the person for a third violation.
d. For fourth and fifth violations, another summons and a fine of 4,000 reals, equivalent to USD 706, will be issued to the person.
e. A penalty of 8,000 reals, equivalent to USD 1,413, will be given to the person for a sixth violation, together with administrative closure.
f. If the person who has been issued with administration closure is still violating the law, then a police investigation will be conducted.

4.3.3 Chile

4.3.3.1 Law No.21,100

The government of Chile implemented a plastic ban policy, which is known as Law No 21,100, in 2018. According to EcoWatch (2018), the people in Chile were found to use an approximately 3.4 billion plastic bags per year. The high consumption of plastic bags in this country has contributed to the generation of high plastic bag waste. Furthermore, the improper management of the plastic bag waste and the long degradation period of plastic bags has caused 90% of the plastic bags wastes to be disposed of in landfill, and some of the plastic bags waste was further flushed into the ocean. Due to plastic pollution issues in Chile, Law No. 21,100 was introduced by the Chile government to mitigate the plastic pollution problem. In this law, the usage of plastic bags is strictly banned in the whole nation of Chile. According to Chile Ministry of Environment (2018), the following is the key information of this law:

1. The objective of this law is to ban the usage of plastic bags to protect the environment.
2. Plastic bags are prohibited from being distributed on business premises.
3. Plastic bags that are used as the primary food packaging are exempted from this law for hygiene purposes.
4. The violation and fine for people who have violated this law:

   a) Any person who has violated this law will be to a maximum penalty of five monthly tax units. This mainly depends on every plastic bag that has been used or distributed.
   b) The confirmation of the penalty will be contingent upon the following conditions:

      (i). The distribution or usage amount of the plastic bag.
      (ii). The previous behaviour of the violating person.
      (iii). The financial capability of the violating person.

5. Perform environmental education. The Ministry of Environment will initiate and organize programmes related to environmental education.
6. Validity of the enforcement for this law. The law will be officially effective and require full compliance from all commercial establishments within a period of six months after the announcement of this law on August 3, 2018. However, this law will only take effect after two years from the publication of the law for the small, micro, and medium sized businesses.
7. Business establishments or business premises can still give plastic bags to their customers within a period of six months (large business) or two years (small, micro, and medium business) from publication. However, the giving of plastic bags to customer is limited to only two plastic bags per customer.

After one year of the implementation of Law No. 21,100, the Ministry of Environment has claimed that the use of plastic bags after implementing this law has been highly reduced, with a total amount of 16,000 tonnes of plastic bags. This has also led to a high reduction in the total amount of plastic bags (approximately 2.2 billion tonnes) produced in Chile (The Working Forest Staff, 2019).

4.3.4 Colombia

4.3.4.1 Resolution No, 0669

The government of Colombia has organized a programme to provide and discuss the rational usage of plastic bags through the implementation of this resolution. The objectives of this resolution are:

1. To reduce the production of plastic waste caused by the disposed plastic bags.
2. To alter consumer behaviour regarding the use of plastic products in order to conserve the environment as well as human health.

The implemented programme is known as "Rational Use of Plastic Bags." According to Ministry of Environment and Sustainable Development of Colombia (2016), the purpose of this programme is to provide proper guidelines on reuse and recycling and to help create awareness of the use of plastic bags on business premises. The distributors of plastic bags must take charge of the development, execution, and management of this programme with the presentation of an annual report on their commitment on this programme. The distributors are referred to as the chain stores, shops, huge business premises, and chain pharmacies. The distributors are required to fulfil the following responsibilities:

1. Developing, executing, and managing the programme.
2. Need to do annual report submission.
3. The plastic bags need to show the information listed below to inform consumers:
    a. Message reminding of the rational usage of plastic bags (this message must occupy at least 10% of the surface area of the bags).
    b. The loading capability of the plastic bags (in unit kilograms).
    c. Quality of the plastic bags.
    d. Suggestion of methods in reusing the plastic bags.
    e. Offer other alternatives bags for the customers such as the reusable bags to carry and transport their purchased goods.

The distributors must establish the following targets according to two groups:

1. The existing distributors:
    a. The distributors must make sure that the total plastic bags offered to the customers is decreased by 10% after one year of implementation of the programme.

   b. The distributors must ensure the amount of plastic bags need is decreased by at least 5% annually in the following year until the number of plastic bags has been reduced by 60% when compared to the amount of plastic bags in year 2016 (the base year for reference).

2. For the new distributors:

   a. Start to reduce the number of plastic bags that are provided to the consumers.

The implementation of this resolution has led to a significant decrease in the usage of plastic bags. The allocation of tax was $3.6 million. This was a good outcome, and it has provided confidence to the government on further implementation of more policies to curb plastic pollution.

4.3.4.2 Law No.1819

According to the government of Columbia, a person in Columbia was found to use an average of 288 plastic bags per year. The high consumption of plastic bags in Columbia might be one of the main contributing reasons for the severe plastic pollution condition in the country. The worsening effect of plastic pollution has caused the government of Columbia to take some approaches to mitigate plastic pollution in the country. In order to overcome this problem, the government of Colombia introduced and implemented a law known as Law No. 1819 in 2017. The objective of the implementation of this law was to regulate and control the usage of plastic bags by forcing the imposition of tax on every plastic bag purchased by customers (ElTiempo, 2017).

The elements in the Law No. 1819 are listed as follows:

1. The usage of plastic bags with dimension of 30 cm × 30 cm or smaller is illegal.
2. Consumers that purchase disposable plastic bags will be charged for COP$20, also known as 20 Colombian Pesos (COP).
3. The tax on the usage of disposable plastic bags will increase COP$10 per year until 2020, when the tax reaches COP$50.
4. Plastic bags that are used to pack the raw foods such as fishes, meats, etc., are exempted from this tax payment.
5. The usage of reusable and biodegradable bags is exempted from this taxation.

The implementation of the Law No. 1819 was found to be a great success, as the usage of plastic bags was significantly reduced by 27%, as reported by the companies that produced the plastic bags. Besides, the tax collected from the purchase of plastic bags in the first year of implementation of this law was COP$10,404 million, which was equivalent to USD$3 million. The total sales of plastic bags were found to be approximately COP$475,000 million, equivalent to USD$118 million. On the other hand, the supplying of plastic bags in business premises was found to have decreased by 59.4% from 2016 to 2019. After one year of implementation, the usage of plastic bags has showed a significant decrease, with an approximate number of 575 million. The government of Colombia has also organized an event known as "*Reembolsale al Planeta*," meaning reimbursing the planet. The objective of organizing this event is to publicize the disadvantages of the using of plastic products to our environment (United Nations Environment Programme, 2021).

4.3.4.3 Law No. 1973

The implementation of Law No. 1819 has significantly reduced the consumption of plastic bags in Colombia. Due to this success, the government of Colombia continued to take initiative to implement another law, which is known as Law No. 1973, to further reduce the plastic pollution in Colombia in 2019. According to this law, the importation, trading, and usage of plastic bags as well as the other plastic products are prohibited from Archipelago of San Andrés, Providencia, and San Catalina. The aim of the implementation of this law is to reduce the environmental impact caused by the improper disposal of plastic wastes (Republic of Colombia-National Government, 2018). The details of this law are listed as follows:

1. Plastic bags are strictly prohibited from the importation, distribution, and usage in San Andrés, Providencia, and Santa Catalina.
2. The usage of plastic products such as plastic straws, plates, and cups that are either made of plastic or polystyrene are also banned.
3. Merchant vessels that arrive to the department are prohibited from using plastic bags for disposing of waste. They must dispose of the waste when they reach the port.
4. The merchants can provide reusable bag or paper bags to customers, subject to a charge.
5. Organizing pedagogical campaigns to develop awareness among local citizens and tourists on the environmental impact of using plastic bags.

There are some exemptions stated in this law:

1. Plates, straws, cups, and bags that are made from recyclable materials or biodegradable materials are exempted from this law.

4.3.5 Peru

According to UNEP (2018a), there are approximately 18,000 tonnes of waste that are generated daily in Peru, of which 10% is plastic waste. The amount of plastic waste that proceeds to recycling processes is also very minimal. The huge plastic wastes generated in Peru has further caused plastic pollution to become a critical problem in Peru. The occurrence of severe plastic pollution in Peru has alerted the government of Peru to investigate this issue seriously. The government of Peru has implemented several policies to decrease the generation of plastic wastes by reducing the consumption of plastic bags in order to protect the environment.

4.3.5.1 Supreme Decree No.013–2018-MINAM

The government of Peru introduced and implemented a policy called Supreme Decree No. 013–2018-MINAM in 2018. The objective of this implemented policy was to encourage the citizens of the country to use plastic products more responsibly by reducing the consumption of the disposable plastic products. This policy also targets the replacement of plastic bags with biodegradable plastic bags. Biodegradable plastics are referred as bags that can be reused and materials that can be bio-degraded into harmless substances which do not contaminate the ground. This is mainly because non-biodegradable plastic degrades into small fragments, which is known as microplastics, and some might release dangerous substances during the degradation process. The release of microplastic and dangerous substances into the environment could cause harmful effects to the environment and require a longer time to be fully degraded (Ministry of the Environment, 2018). According to Ministry of Environment (2018), the details discussed in the supreme decree are listed as follows:

1. The purchasing, importation, and consumption of single-use disposable plastic bags, plastic straws, plastic containers, and Styrofoam containers are prohibited from Peru. The following plastic products that are banned from purchase within 180 business days after the implementation of the decree:

    a. Disposable single-use plastic bags.
    b. Plastic bags used to contain harmful chemical substances and additives which can cause the contamination.
    c. Plastic straws.
    d. Styrofoam containers or containers that made of expanded polystyrene which are used to contain food and beverages for consumption.

2. Single-use, disposable plastic bags, straws, containers, and the other plastic products are prohibited from being brought in or used in the following places within 30 business days after the implementation of this decree:

    a. Natural areas which are under protection.
    b. Places that are claimed as Cultural Heritage.
    c. Museums that are under the charge of Executive Branch agencies.

3. Plastic bags for the following application are exempted from this decree:

    a. Plastic bags that used to pack foods for asepsis and safety purpose such as raw food (fish, poultry, etc.) and processed foods.
    b. Plastic bags which are used as garbage disposal.
    c. Plastic bags that are used for the purpose of ensuring health safety and hygiene effects.
    d. Plastic straws used for medical treatment reasons on medical premises and those persons with temporary or permanent disabilities and older folks.

According to Travelling & Living in Peru (2019), the implementation of this decree was found to be successful, as the consumption of plastic bags has significantly decreased by 30%. The reduction of 30% in consumption is equivalent to approximately 1 billion plastic bags. This decrease in plastic bags consumption is expected to be increased annually, as the government of Peru has imposed a tax on every plastic bag purchased by customers.

4.3.5.2 Law No. 30884

The government of Peru has continued to take initiative to curb plastic pollution issues by introducing a new law, known as Law No. 30884, in 2019. The objective of this law is to regulate the usage of single-use plastics products, including plastic bags and disposable containers. According to Ministry of Environment (2019), the details of this law are further discussed, as listed:

1. The purposes of this law:
    a. To create a regulatory framework for single-use plastic products, where single-use plastic products are referred to as non-reusable, and containers that are made of expanded polystyrene for food and beverage packages are used for consumption purposes.
    b. To educate people to realize their right to live in a balanced and adequate environment by reducing the negative effects caused by the littering of plastic products in the environment and marine environment.

2. Gradually reducing the usage of plastic bags:
    a. Commercial establishments such as retail stores, supermarkets, etc., are required to replace single-use and non-biodegradable polymer-based bags with reusable and biodegradable bags within 36 months after the law takes effect. Reusable and biodegradable bags are referred to as bags that can be reused and will not release harmful contaminants to the environment during the degradation process.
    b. Merchants must charge customers for every plastic bag that they purchased. The charging amount must be at least equivalent to the market price.

3. Prohibition of the disposable plastic products such as bags, straws, and containers:
    a. 120 days after the implementation of this law:
        i. The purchasing, using, and selling of polymer-based bags, straws, containers, and the containers that are made of expanded polystyrene are prohibited in the country.
        ii. The plastic bags or wrappers used to contain the newspaper, magazines, or any other written materials are prohibited.
    b. 12 months after the implementation of this law:
        i. The importation, production, trading, usage, delivery, and marketing of polymer-based plastic bags are strictly not allowed. Polymer-based

bags are referred to as bags with an area less than 900 cm^2 and thickness less than 50 microns.

   ii. The importation, production, trading, usage, delivery, and marketing of polymer-based straws are prohibited.
   iii. The importation, production, trading, usage, delivery, and marketing of polymer-based plastic bags.

c. 36 months after the implementation of this law:

   i. The importation, production, distribution, delivery, and usage of non-reusable, polymer-based plastic bags and plastic bags that will break down and result in contamination by releasing microplastics or other harmful compounds are prohibited.
   ii. The importation, production, distribution, delivery, and usage of non-reusable, polymer-based utensils such as plastic plates and glasses which will breakdown and release contaminants such as microplastics or harmful compounds are prohibited.
   iii. The importation, production, distribution, delivery, and usage of expanded Styrofoam containers are prohibited.

4. Plastic products that are used in the following applications are exempted from this law:

   a. Plastic bags that are utilized to pack and contain foods such as the raw food (fish and poultry) and processed food for hygiene and safety purposes.
   b. Plastic bags that are utilized to ensure health or hygiene purposes.
   c. Plastic straws that are required to be used in medical establishments for medical purposes, especially for persons with temporary and permanent disabilities, and older folks.

5. Educate the citizens on the environmental commitment:

   a. The government of Peru will organize some education programmes, activities, and training, which are aimed at:

      i. Creating awareness among citizens, children, and teenagers regarding to the negative environmental impacts of the usage of the polymer-based bags or goods. It is necessary to replace the usage of single-use polymer-based bags or goods with reusable bags/goods or bags/goods that do not cause the contamination of microplastics or the release of other harmful substances during the degradation process.
      ii. Establishing an environmental obligation that encompasses importers, manufacturers, and dealers of polymer-based items such as plastic bags, containers, etc. This regulation aims to encourage the adoption of innovative technologies to enable them to produce eco-friendly products.

   b. The government of Peru has declared July 3 as "International Plastic Bag Free Day" and Wednesdays as "Plastic Recycling Day" to encourage the reduction of plastic waste.

6. The person who violates this law will be subjected to penalties as follows (Congress of the Republic, 2005):
   a. Subjected a fine of 10,000 Tax Units (UIT), where the tax rate depends on the payment date.
   b. Confiscation of items or substances involved in the violation of the regulation, whether temporarily or permanently.
   c. Suspension or limitation of the action that has violated the regulation.
   d. Suspension or revocation of a licence, concession, or any other form of authorization.
   e. Temporary or permanent shutdown of location or establishment where the violation occurred, partly or wholly.
   f. If the violations persist after paying the fine, a new fine with amount not more than 100 UIT will be imposed for each month until the violations stop.

7. Enforce the usage of recycled materials in manufacturing polyethylene terephthalate (PET) bottles:
   a. The manufacturers of PET bottles must ensure the usage of at least 15% of recycled PET material (PET-PCR) in the manufacturing process.

8. Subjection to a levy tax on the purchased plastic bags:
   a. The purchase of plastic bags will be subjected to a tax for the purpose of impeding the use of plastic bags and reducing the plastic waste released to the environment.
   b. The application of tax on any kind of plastic bags that are used to carry or transport goods.
   c. The tax levied on the purchase of plastic bags will be gradually increased. It was started at $0.10 per bag in 2019 and will reach $ 0.50 per bag in 2023. The amount of the tax will be maintained at $0.50 per bag in the subsequent years ($ is referred to Peru Sol currency).
   d. Merchants must include the purchasing number of plastic bags and the amount of tax applied on the plastic bags inside the receipts for acknowledgement purposes.
   e. The imposition of the tax on the plastic bags will be officially effective starting from August 1, 2019.
   f. The tax imposed on the purchase of plastic bags will be collected and considered as income of the public treasury.

4.3.6 Uruguay

4.3.6.1 Law No. 19655

The government of Uruguay initiated the implementation of Law No. 19655 in Uruguay in 2018 to curb the plastic pollution problem. The aim of implementing this law was to reduce the environmental impact by decreasing the usage of plastic bags. According to the National Registry of Laws and Decree (2018), this law has discouraged the usage of single-use, disposable plastic bags and encouraged the usage of reusable and recyclable plastic bags. The following details are the contents of this law:

1. All the plastic bags that are provided to the customers by the merchants to carry purchased goods are covered in this law.
2. The importation, production, distribution, trading, and delivery of non-compostable and non-biodegradable plastic bags are prohibited.
3. The plastic bags that are exempted from this law:

    a. Usage of plastic bags to pack or contain fish, raw meats, and poultry for hygiene and safety purposes.
    b. Plastic bags that can be reused many times.

4. Responsibilities of the merchants who provide plastic bags to customers:

    a. Organize an awareness campaign that can provide environmental education to citizens, adults, and children to use the plastic bags rationally and responsibly. This awareness campaign can also educate on the impact of plastic waste to the environment.
    b. The regulation on the plastic bags can be included in logo form or inscription form.
    c. Set up a plastic bag waste reception system.
    d. Operate the reception system in accordance with the regulation.
    e. Provide and encourage the purchasing of reusable bags.
    f. Establish methods that can reduce the use of plastic bags.

5. The usage of plastic wrappers in the delivery of any written papers such as newspaper and magazines is prohibited. This regulation will be effective after the implementation of this law.

4.4 CENTRAL AMERICA

4.4.1 Belize

4.4.1.1 Environmental Protection (Pollution from Plastics) Regulations, 2020

The country of Belize has imported approximately 200 million plastic bags and 52 million pieces of containers which are made of Styrofoam and plastic per annum. According to Department of Environment (2020b), the people in Belize have consumed 11 plastic bags and three Styrofoam containers every week. The government of Belize implemented this regulation in 2020 in order to protect and conserve the environment. The objective of this regulation is aimed to prohibit the usage of plastic and Styrofoam products at the national level. However, the production and importation of the products that are made of plastic and Styrofoam are still allowed in Belize, but it is necessary to obtain a permit or licence. According to Department of Environment (2020a), the regulation consists of several parts, and the details of this regulations are summarized in three main parts, as follows:

Part 1: The importation of banned products

1. A permit is required to be obtained from the Department to perform the importation of the banned products as listed in Table 4.6.
2. The application of permit must be conducted by:
   a. Using the form provided by the Department.
   b. A fee of $25, which is not refundable.
3. If the banned products are categorized as biodegradable or compostable, the following documents listed shall be submitted:
   a. The safety data sheet and the certificate of the product.
   b. The third-party certification to prove that the product has been evaluated at an independent laboratory for a conformity assessment.
4. If the banned products have fulfilled the following criteria listed, a permit will be granted:
   a. The banned products are not made of plastic defined in the Table 4.7.
   b. Restricted products that are used for the purposes of medical and pharmaceutical fields or function as a barrier bag despite listed in Table 4.7.

TABLE 4.6
The Banned Products Listed by Department of Environment (2020a)

Description of Banned Products
A propylene polymer or other primary form olefins.
Example: Polypropylene.
Primary form of styrene polymers.
Example: Expansible polystyrene.
Plates, sheets, plastics, foils, strips, films, non-cellular and not reinforced, laminated, supported, or similarly merged with the other materials from: • Ethylene polymer. • Propylene polymer. • Styrene polymer.
Packaging of goods and or closures: • Boxes, cases, and crates such as egg boxes. • Sacks and bags that made of ethylene polymer or other plastic. • Closures such as caps, stoppers, and lids. • Trays, plates, and cups.
Tableware and kitchenware: • Straws. • Spoons, forks, knives, tumblers.

TABLE 4.7
The Description of the Prohibited Products as Listed in Department of Environment (2020a)

Description	
Single-use Styrofoam clamshell	Single-use plastic bags (also known as T-shirt bags)
Single-use Styrofoam food and soup containers	Single-use plastic cups and lids
Single-use Styrofoam cups and lids	Single-use plates and bowls
Single-use Styrofoam and plastic bags	Single-use plastic and Styrofoam food containers
Single-use drinking straws	Single-use fork, spoons, eating utensils

c. After the non-refundable fee is paid.
d. The application form has been completed correctly.
e. The information of the imported products is correct and accurate.
f. The other necessary information is clearly provided.

5. The application for a permit will be approved if the banned products are made of biodegradable or compostable materials and fulfil the following criteria:

    a. Meet the required national standard.
    b. Obtain the third-party certification from an independent laboratory.
    c. Categorized as a commercial type of biodegradable plastic which is defined as bio-based, biodegradable, compostable, or environmentally degradable plastic.
    d. Has been subjected to a conformity assessment, as stated in criterion 6.

6. A conformity assessment is necessary if:

    a. The banned products are categorized as a commercial type of biodegradable plastic.
    b. The banned products are imported for the first time.
    c. The banned product is not certified by a third-party, independent certification or the third-party certification is more than five years old.
    d. The Department believes that it is necessary to guarantee the products are biodegradable or compostable to meet the national requirement standard.

7. The Department may conduct a random conformity assessment to evaluate the biodegradability of the products.
8. The importer or producer bears the cost of the conformity assessment.
9. The permit is granted for:

    a. Only valid for 180 days and single time importation.
    b. Not transferrable and renewable.

10. The permit will not be granted if:

    a. The banned products that are listed in Table 4.7 are not used for medical and pharmaceutical purposes.

b. The information provided is incorrect.
c. The imported products do not fulfil the requirements as commercial-type biodegradable plastic and do not meet the national standards.
d. The importer has violated a permit condition.
e. The imported product is found to be hazardous or harmful to the health of humans and the environment.

11. If any person imports the products listed in Table 4.7, the person will be subjected to the following penalties, as listed:

    a. Penalty between $10,000 and $20,000.
    b. Fine with three times the products' value but not exceeding $20,000.

12. A person can bring a maximum of 10 plastic products, as defined in Table 4.6, when travelling into or out from the country. If the person is found to have more than 10 plastic products, the additional products will be retained.

Part II: Production of banned products

1. A licence must be obtained from the Department by a person before starting to produce the restricted products.
2. A licence is required to be obtained for each manufacturing facility.
3. The application of licence must be:

    a. Using the form provided by the Department.
    b. Paid with $500, which is non-refundable.

4. If a person wants to produce banned products which are considered as biodegradable or compostable, they are required to submit the following documents:

    a. The safety data sheet and the certificate of the product.
    b. A third-party certification which is issued by an independent laboratory after being subjected to a conformity assessment.
    c. Documents that certify the products are categorized as bio-based, biodegradable, compostable, or environmentally degradable plastic.

5. A permit will be granted to the applicant if the banned products have fulfilled the following criteria:

    a. The banned products are used in the application of medical and pharmaceutical or as barrier bags, despite being listed in Table 4.7.
    b. Paid the fee, which is non-refundable.
    c. The other necessary information is provided.
    d. The applicant has obtained environmental clearance by signing the environmental compliance plan.
    e. The licence application form is completed, with accurate information.

f. The applicant has environmental clearance and environmental compliance plan for the renewal applications.
g. The applicant has fulfilled the requirements for the importation and manufacturing of the banned products.

6. The following criteria are necessary to be met in order to be granted the licence:

   a. The banned products have met the requirement of national standards.
   b. Able to provide the evidence that the banned products are biodegradable or compostable.
   c. The manufacturing equipment has complied with the environmental laws and environmental clearance.
   d. The applicant has signed the environmental plan to obtain environmental clearance.
   e. The obtained licence must be placed at the apparent location of the facility.
   f. A report with the following details must be submitted by the licence holder:

      i. The name of product.
      ii. Indication of the "Harmonized System Code" for the products which are for exportation purposes.
      iii. The type of product and the composition of the products.
      iv. The amount and volume of the products.
      v. The purpose and sector that is intended for the products.
      vi. Product description.
      vii. Provide information such as details of usages, details of delivery, and suitability to be used as barrier bags.
      viii. Indication whether the products are made of biodegradable material or non-biodegradable material.
      ix. If the product is indicated as biodegradable, the type of material must be indicated as bio-based plastic or biodegradable plastic.
      x. The registration number of the products which are biodegradable.
      xi. The manufacturing reports must be in a particular format and submitted to the Department within 15 working days at the end of every quarter.

7. The permit will not be granted if:

   a. The banned products that to be produced are defined in Table 4.7 and these products are not applied for medical and pharmaceutical purposes.
   b. The provided information is incorrect.
   c. The applicant does not meet the requirements of environmental clearance and the compliance plan.
   d. The importer has violated the permit requirements.
   e. The imported product is found to be hazardous or harmful to the health of humans and the environment.

8. If a person is found to produce the banned products without obtaining a licence, the person will be penalized as follows:

    a. A fine between $10,000 and $20,000.
    b. Maximum six months of imprisonment.
    c. The facility is forced to shut down.
    d. Any combination of the penalties a to c.

9. If the licensee is unable to submit the report as listed in 6(f), the licensee will be subjected to one of the following fines, as listed:

    a. Subjected to fine of $2,500 for a first offence.
    b. Subjected to fine of $5,000 for a second offence.
    c. Subjected to fine of $10,000 and the cancellation of licence.

10. The licensee will be subjected to a penalty of $10,000 if the licensee submits a report with the wrong information.

Part III: The prohibited products.

1. The banned product (as listed in Table 4.7) is prohibited from importation unless it is:

    a. Used in the application of medical and pharmaceutical purposes or functions as a barrier plastic bag.
    b. There is no suitable substitute on the market.

2. Any person who produces the banned products will be considered to breach the regulation and the person will be subjected to the following punishment:

    a. A fine of within $10,000 and $20,000.
    b. A fine with three times the value of the banned goods.
    c. A maximum six months of incarceration.
    d. Combination of fine and incarceration together with the dumping cost.

3. Any person found to have sold the prohibited products with no legal reason as stated in (1) will be subjected to the following penalties:

    a. A fine of between $10,000 and $20,000.
    b. A fine of three times the value of the prohibited goods.
    c. A maximum of six months' incarceration.
    d. Combination of fine and incarceration with the dumping cost.

4. The ownership requirements of the prohibited products:

    a. A person is only allowed to own a maximum of 10 or less prohibited products.
    b. If a person is found to own more than 10 but less than 100 prohibited products, the extra prohibited products will be seized.

c. If a person is found to own more than 100 prohibited products, the act of the person will be considered as an offence to this law.
d. If a person is found to own more than 100 prohibited products but less than 500 prohibited products, the person will be subjected to the following penalties:

    i. A fine of three times the value of the prohibited products or $5,000, depending on whichever value is greater.
    ii. The dumping cost of the prohibited products in accordance with the instructions of the Department.

e. If a person is found to own more than 500 prohibited products, the person will be subjected to the following penalties:

    i. A fine with three times of the value of the prohibited products or $5,000, whichever value is greater.
    ii. The cost of dumping the prohibited products, in accordance with the instructions of the Department.

5. However, all the regulations in this law will be ineffective in the condition of:

    a. The country has officially announced a disaster emergency in accordance to Disaster Preparedness and Response Act.
    b. The country has announced the emergency state and national security.

6. A person must apply and obtain a permit to conduct the importation of the raw materials for the banned goods. A permit is also required to produce the banned goods with the imported raw materials.

4.4.2 Costa Rica

4.4.2.1 Law No. 9703

According to Costa Rica Horizons Nature Tours (2019), approximately 550 tonnes of plastic wastes are generated per day. Approximately 80% of the plastic wastes are discharged into the ocean and only 9% of the plastic wastes are recycled. The generation of large amounts of plastic wastes has severely caused plastic pollution in Costa Rica and further affected the economy of Costa Rica. In order to overcome this problem, the government of Costa Rica has introduced and implemented Law No.9703, which was implemented to prohibit the usage of expanded polystyrene. According to the Legislative Assembly of The Republic of Costa Rica (2019a), the details of this law are summarized as follows:

1. The objective of this law: To prohibit the importation, sale, and distribution of containers that are made of expanded polystyrene in all business establishments in Costa Rica.
2. The exemptions of this law:

    a. If the alternative materials are not environmentally feasible due to conservation or product protection concerns.

   b. Packaging that used for household appliances and the other similar items.
   c. Industrial applications.

3. Violation on the law according to TCRN Staff (2020):

   a. The person who performs the importation or distribution of containers made of expanded polystyrene in commercial establishments is considered as a minor offender. The person will be subjected to penalties as follows:

      i. A fine with base wage (CRC450). (CRC is referred to Costa Rican colon, or Costa Rica currency.)
      ii. Perform compensation and remediation for the harm done to the environment.

4. The prohibition of the containers made of expanded polystyrene will be initiated 24 months after the implementation of this law.
5. In the period with the 24 months after the enforcement of this law:

   a. The government promotes and support the productive transformation of industries which are involved in importing and producing containers made of expanded polystyrene.
   b. The government highly encourages the development of containers that are made of environmentally friendly materials.
   c. The involved industries can obtain financial credit for the development of environmentally friendly containers.

6. The Ministry of Health (MoH) is required to formulate a plan to obtain the replacement of containers made expanded polystyrene. The incentives and the environmental education on the negative impacts of the expanded polystyrene on the environment should be included in this plan.

4.4.2.2 Law No. 9786

The government of Costa Rica has introduced and implemented another new policy: Law No.9786. This new policy is introduced after a few months of the implementation of Law No.9703, which further prohibits the usage of plastic straws, bags, and bottles in Costa Rica. The aim of this law is to mitigate the plastic pollution problem and protect the environment from plastic pollution. According to The Legislative Assembly of The Republic of Costa Rica (2019b), the details of this law are summarized as follows:

1. The prohibition of plastic straws:

   a. The distribution and commercialization of plastic straws is prohibited in the Costa Rica.
   b. Plastic straws that are attached to the packaging of the products are given exemption from this law for three years.

2. The prohibition of plastic bags:

   a. Plastic bags which are used to pack and transfer goods are prohibited from distribution and selling in all the business premises and supermarkets in Costa Rica.
   b. Exemption is given by this law to the plastic bags that can be re-used, pose minor environmental impact (or environmentally friendly), and fulfil the following requirements listed:

      i. Small plastic bags with dimensions of 45 cm in width × 60 cm in length and thickness 0.75 inches and above, and these plastic bags must be made of material that consist of a minimum 50% recycled material.
      ii. Medium-sized plastic bags with dimensions of 52 cm in width × 68 cm in length and a thickness of 0.88 inches and above. These plastic bags must be made of a minimum of 50% recycled materials.
      iii. The usage of biodegradable bags is exempted from this law.
      iv. Materials that are categorized as having minor environmental impact must be certified by the respective organizations. These organizations are authorized by Costa Rican Accreditation Entity.

3. The prohibition of plastic bottles: The importation, manufacturing, distribution, and marketing of plastic bottles must fulfil at least one of the following criteria listed:

   a. The polymer resins used to fabricate plastic bottles must contain recycled resin. The amount of recycled resin used in fabricating the plastic bottles is used to determine several conditions such as the availability in the local market, asepsis conditions, public health and safety, hygiene, technology availability and accessibility, and type of product.
   b. Organization of a valorised program such as the reuse and recycling of the wastes generated from the usage of plastic bottles.
   c. Involvement in the waste management programme.
   d. Perform cooperation with at least one municipality to improve the waste collection and management system.
   e. Manufacturing and using the products that can reduce the generation of waste or the disposal of the waste generated contributes a less harmful effect to the environment.

4. Procurement and purchasing:

   a. The purchasing of single-use plastic products such as cutlery sets (knives, spoons, forks), plates, and other plastic products that are used for food consumption in public institutions, companies, and municipalities are prohibited.
   b. Plastic products that can be reused and recycled are permitted to be purchased.

5. The person who violates this law is considered a minor offender, and the person will be subjected to the following penalties:

    a. A fine of 1–10 base wage (CRC450). CRC is defined as Costa Rican currency.
    b. Perform compensation and remediation for the harm done to the environment.

4.4.3 Guatemala

4.4.3.1 Government Agreement No. 189–2019

The government of Guatemala introduced and implemented a law known as Government Agreement No. 189–2019 in 2019. The initiative of the government of Guatemala to implement this law is to curb the plastic pollution issue in Guatemala. According to Ministry of Environment and natural Resources Guatemala (2019), the following are the summarized details of this agreement:

1. The distribution and usage of single-use plastic bags, plastic cups and plates, plastic stirrers, and containers that are made of expanded polystyrene or plastic are prohibited in Guatemala. These products are independent of shapes and designs.
2. The residents of Guatemala are given a period of two years, starting from the enforcement date of this regulation, to substitute the specified products with items made from biodegradable or compostable materials. The materials used must fulfil the requirements of UNE-EN 13432, a European Union Standard. However, the selection of materials is also subjected to the availability in the local market.
3. Exemptions in this law:

    a. Products that produced to be used for medical and therapeutic purposes.
    b. Plastic or expanded polystyrene bags used to pack and seal imported goods during the packaging process.

However, the new government proclaimed that they planned to abolish this policy after two years of implementation. Currently, there is still no news disclosing the result of this plan. According to Briz, Volpicella and Gramajo (2021), the plastic prohibition in this country is most likely to remain in effect, regardless of whether the national prohibition is retained, and more local governments in Guatemala may choose to pass similar regulations.

4.4.4 Honduras

Since the government of Honduras has not implemented any policies, the municipality in Honduras has enforced a prohibition on the usage of plastic bag at the local level. This plastic prohibition has been implemented in three main islands of Honduras, which are Roatán, Utila, and Guanaja. This plastic ban was announced one year prior to its implementation, giving the citizens of Honduras sufficient time to prepare for this new law. The governing body of the municipality has carried out an awareness campaign and activities by providing two reusable bags that were made of canvas and leaflets to each household. The purpose of organizing this campaign is to inform and induce the awareness of the citizens about the reason of implementing the plastic ban in the country. Events were also organized in schools to provide awareness to the students on the environmental problem that is caused by the massive disposal of plastic wastes. Besides, there are approximately 5 million plastic bottles that are imported into Utila island per annum. The huge importation of plastic bottles into Utila island could contribute to the generation of large amounts of plastic wastes on Utila island. The mayor of Utila island has decided to add and mix the plastic bottle wastes with cement and use the mixed materials to pave the streets. The purpose of this action was to help to reduce plastic waste on Utila island. At the same time, roads built with the materials are more sustainable and cheaper than roads built with conventional materials. According to UNEP (2018b), the implementation of the plastic ban has highly reduced the usage of plastic bags in Guanaja, Utila, and Roatan by reducing 100%, 80% and 50%, respectively.

4.4.5 Panama

4.4.5.1 Law No.1

Plastic waste in Panama is comprised of 19% of the total waste in the capital vicinity. In 2019, the government of Panama introduced and implemented Law No. 1 to control the plastic pollution issue. The aim of this law was to encourage the usage of reusable bags in all business establishments. According to The Legislative Assembly of Panama (2022), the details of this law are listed as follows:

1. The usage of polyethylene bags in carrying goods is prohibited in business establishments such as supermarkets, warehouses, and self-service stores.
2. The term periods that are given to business establishments to obtain a suitable substitution in replacing the polyethylene bags are summarized as follows:

    a. 18 months after the implementation and enforcement of the law: This is applicable to shop retailers, supermarkets, and pharmacies.
    b. 24 months after the implementation and enforcement of the law: This is applicable to the warehouses and wholesalers.

3. The exemption of this law:

    a. The application of polyethylene bags to pack prepared or pre-prepared foods and wet supplies for hygiene purposes is exempted from this law.
    b. The goods which the compatible substitutes are not feasible.

4. The responsibility of the Ministry of Environment:

    a. Establish nationwide educational campaigns to promote responsible usage of non-degradable or non-biodegradable materials and highlight

the environmental advantages of opting for reusable bags and eco-friendly materials.
b. Implementing programmes aimed at incorporating companies not included in this law, which are engaged in product marketing to ensure their compliance with the requirements of law.

5. Merchants can decide to impose a charge on the purchasing of reusable bags. If the merchants decide to charge for the provided reusable bags, the price that they decide to charge must be equal to the cost of the reusable bags. The charge or tax collected by the merchants must be submitted to the respective authority at the beginning of each year.
6. The tax collected and submitted by the merchants to the authority are used to fund the organization of the relevant programmes such as the importance of recycling and education on the pollutants' materials.

4.4.5.2 Law No. 187

The government of Panama has continued to take initiative to control the plastic pollution issue in the following year. Since 2020, the government of Panama has implemented and enforced Law No. 187, establishing a national framework to regulate single-use products (Panama, 2020). According to Panama (2020), the details of this law are summarized as follows:

1. The purpose of implementing this law:
   a. To promote and intensify the long-term development of the country in reducing and replacing the use of single-use plastic items with sustainable items which have lower environmental and health impacts.
2. The responsibility of citizens:
   a. Encourage a culture of alertness, education, and discussion regarding the adverse effects of using single-use plastics on the environment.
   b. Acquire and encourage the use of the environmentally friendly alternatives to replace the use of non-biodegradable, single-use plastic products. Promote and encourage reuse and recycling practises.
   c. The purchasing of packaging and single-use plastics products is gradually eliminated in Panama.
   d. Developing personal and family practises to reduce the usage of single-use plastics.
   e. Shared responsibility in the disposal of litter generated by the consumption of single-use plastic products.
3. The responsibility of a person who is involved in trading:
   a. Develop a strategy to reduce the use of single-use plastics and design a plan that consists of corporate social responsibility.
   b. Gradually eliminate the usage of single-use plastic products and packaging.

c. Adopt the usage of reusable packaging.
d. Share responsibility for products and anticipate a similar commitment from vendors.
e. Conduct evaluations of progress in reducing the usage of plastic and generation of garbage. Authorize inspection for the purpose validating said progress.
f. Encourage the organization of internal and external programmes to increase the awareness, education, and communication of citizens on the consequences of using single-use plastic.

4. The gradual substitution of single-use plastic items with:

   a. The gradual replacement of the single-use products with products made of reusable, biodegradable, and compostable materials.
   b. The replacement of single-use plastic products with plastic-type materials that are degradable, bioplastics, biodegradable, or any disposable single-use plastic is prohibited.

5. The regulation on the demand and consumption of single-use plastic. The goal of demand and consumption management is to gradually substitute single-use plastic products with reusable, recyclable, or biodegradable products. The types of products that are the focus of this regulation are listed as follows:

   a. Plastic ear swabs.
   b. Plastic covers used for laundry clothes.
   c. Plastic packages used to contain eggs.
   d. Disposable plastic stirrers.
   e. The plastic rods used to hold balloons.
   f. Plastic toothpicks.
   g. Plastic reeds.
   h. Plastic cocktail sticks.
   i. Candy sticks made of plastic.
   j. The rings attached on cans.
   k. Disposable plastic plates.
   l. Single-use plastic items labelled as degradable plastic.

6. Prohibition of single-use plastic products on the national level. The products that listed in item (5) will be prohibited on the national level based on the following time schedule:

   a. The prohibition of single-use products as listed in criteria (7), parts a, b, e, f, h, i, and j, starts from July 1, 2021.
   b. The prohibition of single-use products as listed in criteria (7), part c, d and i, starts from July 1, 2022.
   c. The prohibition of single-use products as listed in criterion (7), part g, is starts from December 31, 2023.

7. The exemption in this regulation:

    a. Plastic reeds used in medical application and for people who are suffering from chronic and involutional diseases, disabilities, and old folks are exempted from this law.
    b. Items that can only be used with the medical prescription in pharmacies and medical premises are exempted from this law.

8. The responsibilities of public institutions:

    a. Take responsibility in using the single-use plastic products.
    b. Eliminate the use of single-use plastic gradually and invest in the substitution of materials that are environmentally friendly.
    c. Execute the National Strategic Plan for Single-Use Plastic Reduction and track the progress progressively.
    d. Organize public awareness campaigns to enhance and educate on the advantages of using renewable alternatives that are compostable which can be long-lasting in replacing the manufacturing and purchasing of single-use plastic products.

9. This law will be regulated by the Minister of Environment in a maximum of three months, starting from the date of the implementation of this law.

REFERENCES

Antigua Nice Ltd, 2017. Antigua News: Styrofoam Ban in Antigua: Stages and Implementation. www.antiguanice.com/v2/client.php?id=806&news=10298

Argentina Ambiental, 2008. Law 13868—Use of Polyethylene Bags. Prohibition—Environmental Argentina. https://argentinambiental.com/legislacion/buenos-aires/ley-13868-uso-bolsas-polietileno-prohibicion

Assembléia Legislativa do Estado do Rio de Janeiro, 2018. Ordinary Laws. https://oglobo.globo.com/rio/pezao-sanciona-lei-que-proibe-sacolas-plasticas-em-supermercados-do-rio-22823112

Bahamas Plastic Movement, 2014. About | BPM NEW. www.bahamasplasticmovement.org/about

Bauer, K. and Alani, H., 2020. Chicago Moves To Ban All Styrofoam, Single-Use Plastics. https://blockclubchicago.org/2020/01/15/chicago-moves-to-ban-all-Styrofoam-single-use-plastics

Briz, R., Volpicella, K. and Gramajo, J. P., 2021. Environmental Law and Practice in Guatemala: Overview | Practical Law. https://uk.practicallaw.thomsonreuters.com/w-013-2794?transitionType=Default&contextData=(sc.Default)&firstPage=true

Chicago City Council, 2014. Office of the City Clerk—Record #: SO2014–1521. https://chicago.legistar.com/LegislationDetail.aspx?ID=1676473&GUID=E704F6F5-3960-4327-BC60-07B8766C35FA&Options=Advanced&Search

Chicago City Council, 2016. Chicago Checkout Bag Tax. www.chicago.gov/content/dam/city/depts/rev/supp_info/TaxPublicationsandReports/3-50ChicagoCheckoutBagTaxOrdinance.pdf

Chile Ministry of Environment, 2018. Ley-21100 03-Ago-2018 Ministerio Del Medio Ambiente—Ley Chile—Biblioteca del Congreso Nacional. www.bcn.cl/leychile/navegar?idNorma=1121380

City and County of Honolulu, 2019. A Bill for an Ordinance. http://www4.honolulu.gov/docushare/dsweb/Get/Document-248953/ORD19-030.pdf

Civil House of the Mayor's Office, 2015. Decree No. 55,827 of January 6, 2015 Catalog of Municipal Legislation. http://legislacao.prefeitura.sp.gov.br/leis/decreto-55827-de-06-de-janeiro-de-2015/

Civil House of the Mayor's Office, 2020. Law No. 17.261 of January 13, 2020 Catalog of Municipal Legislation. https://legislacao.prefeitura.sp.gov.br/leis/lei-17261-de-13-de-janeiro-de-2020

Condor Ferries Ltd., 2021. 100+ Plastic in the Ocean Statistics & Facts (2020–2021). www.condorferries.co.uk/plastic-in-the-ocean-statistics

Congress of the Republic, 2005. LAW No. 28611. http://icsidfiles.worldbank.org/icsid/icsidblobs/onlineawards/C3004/C-184_Eng.pdf

Costa Rica Horizons Nature Tours, 2019. Costa Rica Is Striving to Become the First Country to Ban Single-use Plastics—Horizontes.com. www.horizontes.com/blog/costa-rica-is-striving-to-become-the-first-country-to-ban-single-use-plastics-by-2021

Department of Environment, 2020a. Environmental Protection Regulations 2020. https://doe.gov.bz/wp-content/uploads/2020/03/Environmental-Protection-Regulations-2020.pdf

Department of Environment, 2020b. Single Use Plastic—Department of the Environment. https://doe.gov.bz/single-use-plastics

Department of Environment Antigua and Barbuda, 2016. The External Trade (Shopping Plastic Bags Prohibition) Order. https://elaw.org/system/files/attachments/publicresource/External_Trade_Prohibition_of_Plastic_Bags_Order_2017.pdf

Department of Environment Antigua and Barbuda, 2019. Litter Control and Prevention Act, 2019. https://elaw.org/system/files/attachments/publicresource/No.-3-of-2019-LITTER-CONTROL-AND-PREVENTION-ACT-2019-No.-3-of-2019.pdf

EcoWatch, 2018. Chile to Become First Country in the Americas to Ban Plastic Bags—EcoWatch. www.ecowatch.com/chile-plastic-ban-2573881713-2573881713.html

ElTiempo, 2017. Know the Details About the Tax that Plastic Bags Will Charge in Colombi—if—LTIEMPO.COM. www.eltiempo.com/vida/conozca-los-detalles-sobre-el-impuesto-que-cobrara-las-bolsas-plasticas-en-colombia-104296

Food and Agriculture Organization (FAO), 2017. Environmental Health (Expanded Polystyrene Ban) Regulations 2017. https://nicholasinstitute.duke.edu/sites/default/files/plastics-policies/2343_N_2017_Environmental_Health_Polystyrene.pdf

Giangreco, L., 2019. How Behavioral Science Solved Chicago's Plastic Bag Problem—POLITICO. www.politico.com/news/magazine/2019/11/21/plastic-bag-environment-policy-067879

Goldman Environmental Prize, 2020. Kristal Ambrose—Goldman Environmental Foundation : Goldman Environmental Foundation. www.goldmanprize.org/recipient/kristal-ambrose

Government of Jamaica, 2018a. The Natural Resources Conservation Authority Act. www.nepa.gov.jm/sites/default/files/2019-11/Proc_2_Plastic_Packaging.pdf

Government of Jamaica, 2018b. The Trade Act. www.nepa.gov.jm/sites/default/files/2019-11/Proc_1_Trade_Act.pdf

Government of Prince Edward Island, 2019. RSPEI 1988, c P-9.2 | Plastic Bag Reduction Act | CanLII. www.canlii.org/en/pe/laws/stat/rspei-1988-c-p-9.2/latest/rspei-1988-c-p-9.2.html

Hill, A., 2016. Synopsis: Implementation of the Plastic Bag Ban in Antigua and Barbuda. https://environment.gov.ag/assets/uploads/attachments/eab93-synopsis_plastic_bag_ban_-anb.pdf

HITE, J., 2020. The Truth about Plastic Bag Bans | Conservation Law Foundation. www.clf.org/blog/the-truth-about-plastic-bag-bans

Honorable Chamber of Deputies of the Argentine Nation, 2019. 3847-D-2018. www.hcdn.gob.ar/proyectos/textoCompleto.jsp?exp=3847-D-2018&tipo=LEY

Jamaica Observer Ltd, 2017. St Vincent Bans Styrofoam Products. www.jamaicaobserver.com/news/St-Vincent-bans-Styrofoam-products&template=MobileArticle

Leaf Rapids Council, 2006. Green Initiatives. https://web.archive.org/web/20120719073146/www.townofleafrapids.ca/green_initiaties.htm#Plastic Bag Ban By-law

Legislative Analyst's Office, 2016. Referendum to Overturn Ban on Single-Use Plastic Bags. https://lao.ca.gov/ballot/2016/Prop67-110816.pdf

Martinko, K., 2020. Prince Edward Island Has Removed Millions of Plastic Bags From Its Waste Stream. www.treehugger.com/prince-edward-island-has-eliminated-plastic-bags-from-its-waste-stream-5071927

Ministry of the Environment, 2018. El Peruano—Decreto Supremo que aprueba la reducción del plástico de un solo uso y promueve el consumo responsable del plástico en las entidades del Poder Ejecutivo-decreto supremo—N° 013–2018-minam—poder ejecutivo—ambiente. https://busquedas.elperuano.pe/normaslegales/decreto-supremo-que-aprueba-la-reduccion-del-plastico-de-un-decreto-supremo-n-013–2018-minam-1708562–2/

Ministry of the Environment, 2019. El Peruano—Law that Regulates Single-use Plastic and Disposable Containers or Containers—Law—N° 30884—Legislative Branch—Congress of the Republic. https://busquedas.elperuano.pe/normaslegales/ley-que-regula-el-plastico-de-un-solo-uso-y-los-recipientes-ley-n-30884–1724734–1/

Ministry of Environment and Natural Resources Guatemala, 2019. Acuerdo gubernativo número 189–2019. https://sgp.gob.gt/wp-content/uploads/2019/09/AG-189-2019.pdf

Ministry of Environment and Sustainable Development of Colombia, 2016. Resolution 668 of 2016. [online] https://nicholasinstitute.duke.edu/sites/default/files/plastics-policies/2516_N_Diario_Oficial_Final.pdf [Accessed 20 Apr. 2022].

Mosteirin, M., 2019. Distribution and Marketing of Drugs in Argentina: Overview | Practical Law. https://uk.practicallaw.thomsonreuters.com/w-014-7135?transitionType=Default&contextData=(sc.Default)&firstPage=true

National Congress of Brazil, 2008. Decree No. 6514. www.planalto.gov.br/ccivil_03/_ato2007-2010/2008/decreto/d6514.htm

NBCChicago, 2014. City Council Passes Ban on Plastic Bags-NBC Chicago. www.nbcchicago.com/news/local/chicago-city-council-plastic-bag-ban-ordinance/71347/

NEWS784, 2020. Massy Stores (SVG) Ltd Reports A 67% Decrease In Plastic Bag Usage—EWS784. https://news784.com/2020/03/05/massy-stores-svg-ltd-reports-a-67-decrease-in-plastic-bag-usage/

Nicholas Institute for Environmental Policy Solutions, 2016. Montreal Bylaw 16–051 Prohibiting the Distribution of Certain Shopping Bags in Retail Stores. http://ville.montreal.qc.ca/sel/sypre-consultation/afficherpdf?idDoc=27530&typeDoc=1

Panama, T.L.A. of 2020. LEY 187. http://extwprlegs1.fao.org/docs/pdf/pan199562.pdf.

Republic of Colombia—National Government, 2018. LAW 1973 OF 2019. https://www.suin-juriscol.gov.co/viewDocument.asp?ruta=Leyes/30036668, por medio de la cual se regula y prohíbe el,y se dictan otras disposiciones

Scheibe, T., 2016. Has Chicago's plastic bag ban helped?—Chicago Magazine. www.chicagomag.com/Chicago-Magazine/August-2016/Plastic-Bag-Ban/

SearchLight, 2020. Ban on use of disposable plastic bags suspended—PM—Searchlight. https://searchlight.vc/searchlight/breaking-news/2020/08/15/ban-on-use-of-disposable-plastic-bags-suspended-pm

TCRN Staff, 2020. The Costa Rica News Businesses in Costa Rica Have to Adapt to the Laws. https://thecostaricanews.com/businesses-in-costa-rica-must-adapt-to-the-new-laws-that-regulate-single-use-plastic-and-polystyrene

The Government of Bahamas, 2019. Environmental Protection (Control of Plastic Pollution) Act, 2019, Government Notices. www.bahamas.gov.bs/wps/portal/public/gov/government/notices/environmentalprotection(controlofplasticpollution)act2019/

The Legislative Assembly of Panama, 2022. LEY I. http://extwprlegs1.fao.org/docs/pdf/pan178033.pdf.

The Legislative Assembly of The Republic of Costa Rica, 2019a. Costa Rican Legal Information System. www.pgrweb.go.cr/scij/Busqueda/Normativa/Normas/nrm_texto_completo.aspx?param1=NRTC&nValor1=1&nValor2=89355

The Legislative Assembly of The Republic of Costa Rica, 2019b. Law NO.9786. www.pgrweb.go.cr/TextoCompleto/NORMAS/1/VIGENTE/L/2010-2019/2015-2019/2019/1604B/90187_118676-1.html

The Legislature of the Autonomous City of Buenos Aires, 2009. Law 3147 —Promote the Development of the Production of Biodegradable Bags. https://www.c40.org/case-studies/buenos-aires-reduces-single-use-plastics/

The National Registry of Laws and Decrees, 2018. Law No. 19655. https://www-impo-com-uy.translate.goog/bases/leyes/19655-2018?_x_tr_sl=es&_x_tr_tl=en&_x_tr_hl=en&_x_tr_pto=sc

The Working Forest Staff, 2019. Plastic Bag Usage in Chile Drops by 16,000 Tonnes in One Year | The Working Forest. www.workingforest.com/plastic-bag-usage-in-chile-drops-by-16000-tonnes-in-one-year/

Traveling & Living in Peru, 2019. Peru Reduces Use of Plastic Bags by 30% in 2019. www.livinginperu.com/peru-reduces-use-of-plastic-bags-by-30-in-2019/

UNEP, 2018a. Backpack Ponchos: Peru's Solution to Plastic Pollution. www.unep.org/news-and-stories/story/backpack-ponchos-perus-solution-plastic-pollution

UNEP, 2018b. Single-use Plastics: A Roadmap for Sustainability | UNEP—UN Environment Programme. www.unep.org/resources/report/single-use-plastics-roadmap-sustainability

United Nations Environment Programme, 2021. Policies, Regulations and Strategies in Latin America and the Caribbean to Prevent Marine Litter and Plastic Waste. https://wedocs.unep.org/bitstream/handle/20.500.11822/34931/Marine_EN.pdf?sequence=1&isAllowed=y

Wright, L., 2019. Plastic Warms the Planet Twice as Much as Aviatio—ere's How to Make It Climate-friendly. https://theconversation.com/plastic-warms-the-planet-twice-as-much-as-aviation-heres-how-to-make-it-climate-friendly-116376

5 Policies of Plastic Use in African Countries

5.1 INTRODUCTION

Concerns about impacts of plastics to the environment were raised in the 1960s. Environmental issues caused by exploitation of plastic usage were being widely posted in the media and made known to Americans in the early days. In the beginning, plastic wastes have ended up in landfill, while, after a long time of accumulation with low-degradability plastics, the landfills were full. Many landfills are forced to close, and local authorities have to reclaim hills or forests to handle such a huge volume of plastic waste daily. Those plastic pollution episodes in American and European continents are currently on-going in the African region, including the piling up of plastic wastes on the beaches of Africa countries, endangering the marine life of Africa countries, and the accumulation of plastic wastes on the ground in African countries. Plastic pollution also threatens the economy of African countries, which highly depend on the fishing industry, tourism industry, and agricultural industry. The severity of plastic pollution can be visualized through the data provided by Sambyal (2018), where it is estimated that 4.8 million tonnes of plastic waste was mismanaged during the year of 2010, and it was predicted that the amount would reach 11.5 million tonnes in 2025. Realizing this problem, most African countries have acted swiftly to implement their respective plastic usage policies and measures to combat plastic pollution. The first plastic usage policy observed in Africa was implemented as early as 2002, where South Africa introduced a ban on plastic bags below the thickness of 30 μm and placed a levy on plastic bags which are thicker than 30 μm. The awareness of African countries' governments is observed to have risen around the year 2010 as an increasing trend of African countries have implemented their respective plastic usage policies in their countries. As African countries are generally recognized as countries which have developed relatively late, keen attention is paid on the efforts, methods, and success of African countries in combatting plastic pollution. African countries are currently experiencing rapid population growth and rapid urbanization, causing Africa's middle class to grow, and informal shops are being replaced by supermarkets. Jambeck et al. (2018) stated that this has promoted growth of consumer markets for plastic goods, hence increasing the usage of plastics in the region. With the increase of population density and the big per capita consumption, the ability of underdeveloped waste-management infrastructure of the countries to cope with the waste generation of the countries turns up as a serious concern. The occurrence of the flood incident in Ghana is an example of the issue where irresponsible plastic dumping has caused the blockage or even damage to the local drainage system (Mensah and Ahadzie, 2020). According to United Nations Environmental Programmes UNEP (2018a), Africa is progressing

DOI: 10.1201/9781003387862-5

to eradicate plastics by employing plastic usage measures such as prohibiting the production and distribution of single-use plastic and employing a full ban on production and usage of plastic bags. However, it was also reported that plastics are being smuggled and sold in some African countries which have employed the plastic bans (UNEP, 2018b). This has raised concerns, as the issue of plastic pollution, which is intended to be reduced or solved through the implementation of the plastic usage policy, will not be effective. Additionally, plastic bag smuggling is concerning, as this indicates that the actual usage of plastics in the countries might differ significantly from the statistics which were provided by the official bodies of government, hence causing the governmental bodies to underestimate the severity of the issue in the region.

5.2 SOUTHERN AFRICA

Southern Africa consists of 16 countries, which are Angola, Botswana Comoros, Democratic Republic of Congo, Eswatini, Lesotho, Madagascar, Malawi, Mauritius, Mozambique, Namibia, Seychelles, South Africa, Tanzania, Zambia and Zimbabwe. According to Bezerra et al. (2021), all countries in Southern Africa have introduced plastic bag reduction policies. However, it was observed that only seven out of 13 of the countries are implementing plastic-reduction policies. Others are still in the progress of reviewing or have revoked policies. The most common plastic usage policy observed in Southern Africa is to implement legislation to ban plastic bags and products. However, there are two types of bans which are implemented in the countries of Southern Africa: (a) ban of production, importation, distribution, selling, and usage of plastic bags or non-biodegradable plastic bags; (b) ban of plastic bags below specified thickness, respectively. Besides of the banning of plastic bags which have less thickness than the specified thickness, some of the countries are found to be also enforcing levies on plastic bags which are thicker than the specified, banned thickness.

Overall, the policies of the South African countries are summarized in Table 5.1 for South Africa, as reported by Dikgang et al. (2010), who studied the results of the plastic levy implemented since year 2003 with five years of observation. The research was done by conducting analysis on the data, which are retrieved from the local retailers. In the research, it is observed that plastic usage in South Africa decreased by approximately 44% after implementation of levy. In fact, this is a significant amount since the implementation of a plastics levy. However, the decrease in usage of plastic carry bags, which was observed in low-income retailers, was relatively low, at 50%, when compared to high-income retailer where the reduction was observed at 57%. Dikgang et al. (2010) stated in a report that the decline in plastic bag usages are reported at 62% and 19% for upper-middle income retailers and low-income retailers, respectively, after the implementation of the plastic usage levy. However, the plastic bag levy is charged at different rates by each retailer three months after the implementation of the levy, and the plastic bag levy price collected by the retailers has been reduced due to pressure applied by local plastic bag manufacturers.

TABLE 5.1a
Details of Plastic Usage Policy of Countries in Southern Africa

Country	Policy	Year of implementation	Scope	Details
Angola	Levy	N/A	National	Announced, Reviewing [*a]
Botswana	Hybrid of ban and levy	2006	National	• Bans importation, production, and usage of bags < 24 μm.[*a *b] • Levy is placed on plastic bags >24 μm.[*a] • In case of conviction, sentence to three years of jail and a fine of BWP 5 000[*a]
Comoros	Ban Ban	2016 (Jan) 2018 (April)	Local capital National	• Bans importation, manufacture, commercialization, and distribution of non-biodegradable plastic bags.[*a] • In case of disobedience, fine range from 100,000 to 10 million francs and/or a period of three months to five years imprisonment.[*a]
Democratic Republic of the Congo	Ban	2018 (July)	National	Bans importation, manufacture, selling, and usage of non-biodegradable plastic bags and packaging[*a]
Lesotho	Levy	N/A	National	Announced, reviewing[*a]
Madagascar	Partial ban	2015	National	Ban plastic bags < 50 μm[*a] (implemented, not enforced)
		N/A		Ban plastic bags < 50 μm[*a] (not implemented)

Note:
[*a] (Bezerra et al., 2021)
[*b] (Excell et al., 2018)

The reduction of the plastic levy collected by the retailers has then led to the revival of plastic bag usage amongst the upper-middle income class a few months after the levy price reduction (Dikgang et al., 2010). However, despite the revival of the plastic bag usage demand, Dikgang et al. (2010) found that plastic carrier bags were being used by the middle-income class to carry more goods. They found that the number of plastic bags issued per 1,000 South Africa Rand (R) of real retail purchases is decreased and it is believed that the phenomena are caused by 2 factors which are the levy price collected by the retailers which are still relatively high and the customers are already normalised to pay for the plastic bags. The value of goods carried per bag is double when compared to the value of goods carried per bag of upper-middle-income-class consumers (Dikgang et al., 2010). It is stated that the

TABLE 5.1b
Details of Plastic Usage Policy of Countries in Southern Africa (Cont.)

Country	Policy	Year of implementation	Scope	Details
Malawi	Ban	2012	National	Acts of producing, distributing, selling, and using of plastic bags are prohibited (revoked) [*a]
	Ban	2015 (June)		Acts of producing, distributing, selling, and using of plastic bags are prohibited (revoked) [*a]
	Ban	2018 (June)		Acts of producing, distributing, selling, and using of plastic bags are prohibited (revoked) [*a]
	Ban	2019 (July)		Acts of producing, distributing, selling, and using of thin plastic bags are prohibited [*d]
Mauritius	Ban	2016 (Jan)	National	• Acts of producing and selling of non-biodegradable plastic bags are prohibited [*a] • In case of disobedience, fine of 10,000 rupees[*a]
	Ban	2021 (January)		Acts of importing, producing, owning, selling, supplying and, using of several non-biodegradable single-use plastic products[*c]
	Ban	2021 (March)		Ban on the importation, manufacturing, commercialization, and supply of new types of non-biodegradable plastic bags.[*c]

Note:
[*a] (Bezerra et al., 2021)
[*b] (Excell et al., 2018)
[*c] (Richard and Ribet, 2021)
[*d] (Princewill, 2021)

TABLE 5.1c
Details of Plastic Usage Policy of Countries in Southern Africa (Cont.)

Country	Policy	Year of implementation	Scope	Details
Mozambique	Hybrid of ban and levy	2016 (February)	National	• Partial ban of importation, manufacturing, and commercialization of plastic bags < 30 μm.[*a] • Introduced levy on other plastic bags. In case of disobedience, fine between 30 to 80 minimum wages[*a]
	Ban	N/A		Ban on usage of plastic bags.[*b] (not implemented)

(Continued)

TABLE 5.1c
Continued

Country	Policy	Year of implementation	Scope	Details
Namibia	Ban	2018 (November)	Protected areas	• Visitors visiting protected areas are prohibited from bringing plastic bags.*[a] • In case of disobedience, fine of maximum N$ 500 or subject to imprisonment maximum period of six months.*[a]
	Levy		National	• Approved, not implemented. • Levy of N$0,50 per bag on plastic carrier bags.*[c]

Notes:
*[a] (Bezerra et al., 2021)
*[b] (Magoum, 2020)
*[c] (Thikusho, 2019)

TABLE 5.1d
Details of Plastic Usage Policy of Countries in Southern Africa (Cont.)

Country	Policy	Year of implementation	Scope	Details
Seychelles	Ban	2017 (Jan)	National	• Acts of importing, manufacturing, distributing, and selling of plastic bags, plastic utensils, and polystyrene boxes.*[a] *[b] • In case of disobedience, maximum fine of SCR 20,000 or imprisonment for up to 1 year or both*[a] *[b] • 2017 (Jan), ban on importation*[a] • 2017 (Jul), ban on distribution.*[a]
	Levy	2020	National	• Levy of SCR 1.00/bottle on plastic bottles.*[d]
South Africa	Hybrid of ban and levy	2003 (May)	National	• Ban of lightweight bags < 30 μm*[c] • Levy of plastic bags of 25 cents per plastic bag which are thicker than 30 μm.*[c] • In case of disobedience, fine of maximum R10 0000, or three times the commercial value of which offense is committed, or imprisonment up to 10 years, or both fine and imprisonment.*[c]
Swaziland	Levy	2015	National	• Levy of 35 cents per bag on plastic carrier bags*[a] • Revoked 2015*[a]

Notes:
*[a] (Bezerra et al., 2021)
*[b] (Goitom, 2017)
*[c] (South Africa Government Gazette 7348 Vol 443 (2002)
*[d] (ENVIRONMENT PROTECTION ACT, 2016 (Act 18 of 2016))

TABLE 5.1e
Details of Plastic Usage Policy of Countries in Southern Africa (Cont.)

Country	Policy	Year of implementation	Scope	Details
United Republic of Tanzania	Partial ban	2006 (November)	Zanzibar, Tanzania	Ban on plastic bags < 30 μm*a
	Ban	2019 (June)	National	• Total ban on importations, exportation, producing, selling, possessing, supplying, and usage of plastic bags. *b • Visitors not encouraged for carrying plastic bags, except for ziploc plastic bags. Visitors, however, are not allowed to dispose of them in Tanzania. *b
Zambia	Partial ban	N/A	National	• Ban on plastic bags < 30 μm and related packaging*c • Announced, not implemented.
Zimbabwe	Partial ban	2010 (December)	National	• Partial ban on importing, manufacturing, and distributing of plastic bags < 30 μm. • In case of disobedience, fine USD$ 5,000 or one-year jail sentences. • Extended Producer Responsibility (EPR) is proposed.

Notes:
*a (Bezerra et al., 2021)
*b (US Embassy in Tanzania, 2019)
*c (Excell et al., 2018)

low-income class havs the highest consumption of plastic bag per R1,000 of purchases because of the lower amount of levy collection by low-income-class retailers when compared to the amount of levy collected by other income classes of retailers. Dikgang et al. (2010) also mentioned that the consumers of lowest income consume the highest amount of plastic bags, hence making the plastic bag levy regressive.

Dikgang et al. (2010) also stated that the consumption of plastic bags was increased and the value of goods in bag per R1,000 purchased was decreased over time. This can be due to two factors: firstly, the amount of levy collected has been decreased by the retailers, causing the increment of plastic bag consumption; secondly, South African consumers have become accustomed to payming for plastic bags, hence increasing the consumption of plastic bag per R1,000 of purchase, although the levy on plastic bags is set to be stabilized after a period of time. Therefore, though the usage of plastic bag per R1,000 of purchase has decreased by a significant number since the implementation of plastic bag levy, the consumption of plastic bag per R1,000 of purchase has steadily increased since the decline of levy amount collected by the retailers; hence, it is expected that the consumption behaviour of plastic bag usage has not been significantly changed, and the problem of plastic pollution in South Africa will continue to persist.

5.2.1 Case Study on Mauritius

Mauritius is considered to be one of the countries in Southern Africa which has the highest awareness and has a plastic reducing policy which consists of the majority of daily-use plastic products. Mauritius was found to impose a policy on polyethylene terephathalate (PET) plastic bottles in 2001 (ENVIRONMENT PROTECTION ACT, 2016 [Act 18 of 2016]) and plastic bag usage policy as early as 2004 by legislation. The legislation is known as the Environment Protection (Plastic Carry Bag) Regulations, 2004. However, no related information is found on the legislation act on plastic bag usage policy. The act was amended and cited as the Environment Protection (Banning of Plastic Bags) Regulations, 2015, in 2015 and further amended in 2020 and cited as the Environment Protection (Control of Single Use Plastic Products) Regulations, 2020. In the Environment Protection (Banning of Plastic Bags) Regulations, 2015, importation, manufacturing, selling, and supply of plastic bags by any individual and businesses, except businesses which were registered as a manufacturer or an importer of exempted plastic bags before October 31, 2015, are prohibited and are liable to a fine not exceeding 10,000 rupees in the case of breaching the act (Government of Mauritius, 2015). It is also stated in the gazette by the Government of Mauritius (2015) that the list of exceptions contains plastic bags which are:

i. Transparent roll-on bags solely used to contain fresh, chilled, or frozen food, expect for canned food or eggs.
ii. Bags which are used for waste disposal.
iii. Agricultural purposes bags.
iv. Sampling or analysis bags.
v. Packaging bags which seal goods before selling or exported.
vi. Transparent pocket-type bags smaller than 300 square centimetres.
vii. Transparent, resealable bags with security tamper.
viii. Bags which are used to carry personal belongings.
ix. Bags produced to be exported.

The Environment Protection (Control of Single Use Plastic Protection Control of Single Use Plastic Products) Regulation, 2020, was then introduced and executed in 2021. The execution of the act is divided into two parts and was executed on two different dates: January 15, 2021 and April 15, 2021, respectively. In the gazette, it was stated that the First Schedule, which involved biodegradable single-use plastic products, and Part II of the Second Schedule, which involves non-biodegradable single-use products, were executed on January 15, 2021, while Part II of Second Schedule was executed on April 15, 2021. The items stated in the gazette are tabulated and shown in Table 5.2: list of items in First Schedule and Second Schedule of Environment Protection (Control of Single Use Plastic Products) Regulations, 2020.

TABLE 5.2
List of Items in First Schedule and Second Schedule of Environment Protection (Control of Single Use Plastic Products) Regulations, 2020

Biodegradable single-use products	Non-biodegradable Single-use products	
	Part I	Part II
Plate	Plastic forks	Plastic tray
Cup	Plastic knives	Plastic hinged container
Bowl	Plastic spoons	Sealed plastic straw
Tray	Plastic chopsticks	–
Straw	Plastic plate	–
Beverage stirrer	Plastic cup	–
Hinged container	Plastic bowl	–
Cup lid	Plastic tray	–
Receptacles	Plastic straw	–
Forks	Plastic beverage stirrer	–
Knives	Plastic hinged container	–
Spoons	Plastic lid for single-use plastic products	–
Chopsticks	Receptacles	–

The legislation prohibits the public and businesses from importing and manufacturing non-biodegradable single-use plastic products listed in Second Schedule, Part I of the act and also prohibits a list of biodegradable single-use products which are listed in First Schedule of the act to be imported or manufactured, except for businesses which are registered and approved by the Director-General of the Mauritius Revenue Authority (Government of Mauritius, 2020). Under the legislation, businesses are required to apply for a certificate which is valid for three years in order to allow themselves to import or manufacture biodegradable single-use products. The application requires businesses to pay 10,000 rupees, provide valid reasons with support of documentation when making such application, and is subjected to the Director-General's approval.

The consequence of breaching the act is clearly stated in the act. Individuals possessing or using non-biodegradable, single-use plastic products listed in the Second Schedule of the act, except for the purpose of trade, are subjected to a fine not exceeding 2,000 rupees on the first conviction and subjected to fine maximum of 5,000 rupees on the second or subsequent conviction (Government of Mauritius, 2020). Government of Mauritius (2020) also stated that individuals who are found to possess or to use single-use plastic listed in the Second Schedule of the act for trade are subjected to a fine maximum of 20,000 rupees on the first conviction. The second and subsequent conviction, however, might subject the individual to a fine maximum 100,000 rupees and to imprisonment for a maximum of three months. It is also stated that importers of non-biodegradable, single-use plastic products stated in Second Schedule are subjected to fine not exceeding 50,000 rupees on the first conviction. The second or subsequent conviction can be subjected to fine of maximum 10,000 rupees and to imprisonment for maximum period of two years. Manufacturers of the non-biodegradable, single-use plastic products stated in Second Schedule, however, are subjected to a fine of maximum 100,000 rupees on the first conviction and are subjected to both fine of maximum 250,000 rupees and imprisonment for a maximum period of two years for the second and subsequent convictions.

As there is no study found to be conducted on the effects of the current plastic usage policy, which was implemented in 2021; studies of the effect of plastic usage policy in Mauritius are conducted by referring to studies which are conducted on the effects of previous plastic usage policy implemented during 2016. Therefore, the effect of the plastic ban in Mauritius in this study was referred to the findings by Foolmaun et al. (2021) on the effect of the ban of plastic bags in Mauritius, starting in 2016, inasmuch as it prohibits the public and businesses from importing, manufacturing, selling, and distributing non-biodegradable plastic bags. The study was only done on the awareness of the society towards the impact of plastic bag usage and their plastic bag use behaviour change.

In the article, Foolmaun et al. (2021) found that 209 out of 308 of the samples, which translates to 67.86% of the sample population, use plastic bag alternatives after the implementation of the plastic bag ban instead of acquiring bio-degradable plastic bags which are allowed in the border of Mauritius. Besides, four types of alternatives of plastic bags are preferred by the respondents; namely, cloth bags, paper bags, *tente vacous* (carry bag made from Pandanus leaves), and *tente raffia* (carry bag made from nylon). Foolmaun et al. (2021) also stated that, among the plastic carry bag alternatives, cloth bags are favoured by 46% of respondents who use them as as major alternative, followed by paper bags, *tente raffia*, and *tente vacous* which were favoured by 21%, 18%, and 15% of the respondents who uses plastic carry bag alternatives.

Foolmaun et al. (2021) also conducted a study on the respondent's awareness of the environmental impacts and health hazards which are caused by plastic usage, as shown in Figure 5.1. The respondents were tested on their awareness of impacts caused by plastic usage. In the questionnaire, Foolmaun et al. (2021) found that over 80% of the respondents are aware of the environmental impacts and health hazards, followed by 93.18%, 92.53%, 90.26%, 94.48%, 89.29%, and 76.30% of respondents which are aware of the issues of indiscriminate disposal, flooding problems, mosquitoes' proliferation, threats to aquatic life, causing eyesores, and the persistence of plastic in the environment, respectively. This indicates that the implementation of plastic policies is well accepted in Mauritius for the betterment of future generations.

5.2.2 Case Study on Tanzania

Tanzania introduced a plastic ban as early as year 2006. However, the ban was only effective in Zanzibar, Tanzania's semi-autonomous island, and the restriction was only applied on plastic bags which are thinner than the thickness of 30 μm (Bezerra et al., 2021). The ban was reported to prohibit the importation, distribution, and sale of light plastic (The New Humanitarian, 2006). The New Humanitarian (2006) reported that violators of the ban are subjected to a jail-sentence maximum period of

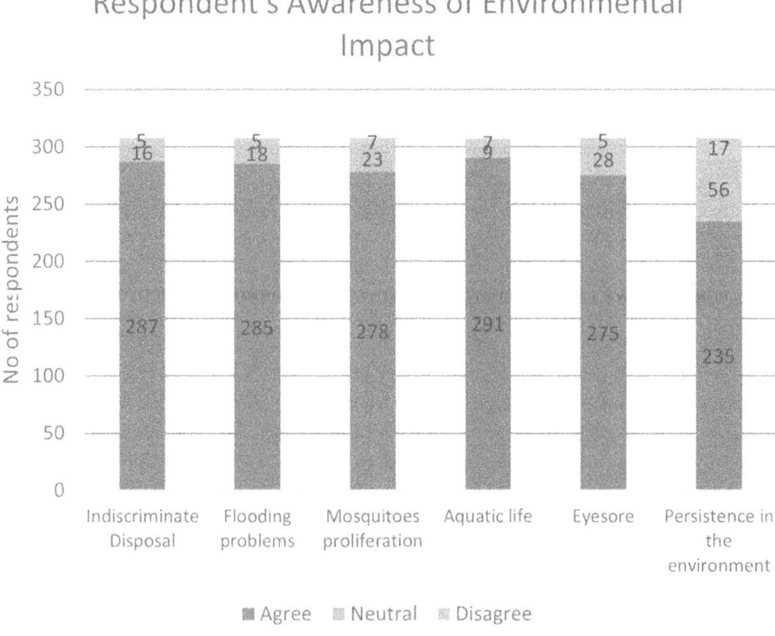

FIGURE 5.1 Respondent's awareness of environmental impact (Foolmaun et al., 2021).

six months or a fine of USD 2,000 or both. A nationwide campaign, "Plastic Wastes Operation," was also organized as a compliment to the ban, aiming to collect wastes, especially light plastics, on the island (The New Humanitarian, 2006). The collected wastes were then transported to mainland Tanzania to be recycled due to the lack of facilities and infrastructure on the island.

The partial ban continued until the enforcement of The Environment Management (Control of Plastic Bags) Regulations, 2015, which was then replaced by The Environmental Management (Prohibition of Plastic Carrier Bags) Regulations, 2019. The act imposes a total prohibition on plastic bags. The prohibiton prohibits the importation, exportation, production, sale, and use of plastic bags, disregarding the thickness of the plastic bags. Under the act, the issuance of licenses which grant permission for individuals and businesses to import, export, manufacture, or sell plastic bags are prohibited by the act. Individuals who import, export, or manufacture exempted plastic packagings are to ensure that the product quality meets the standards prescribed by the Tanzania Bureau of Standards (SUBSIDIARY LEGISLATION to the Gazette of the United Republic of Tanzania No.20. Vol.100 dated May 17, 2019).

The consequences of violating the act are stated clearly in the gazette. Individuals who import, export, distribute, supply, possess, or use plastic bags and plastic wrappings are subjected to different penalties. In the gazette, it is clearly stated that:

a) Individuals who manufacture or import plastic bags and plastic wrappings are subjected to a fine ranging between 20 million shillings and 1 billion shillings or to imprisonment for a term maximum period of two years or both.
b) Individuals who export plastic bags and plastic wrappings are subjected to a fine ranging from 5 million shillings to 20 million shillings or to imprisonment for a maximum period of two years or both.
c) Individuals who store, supply and distribute plastic bags and plastic wrappings are subjected to a fine range between 5 million shillings and 50 million shillings or to imprisonment for a maximum period of two years or to both.
d) Individuals who sell plastic bags and plastic wrappings are subjected to a fine of ranging between 100 thousand shillings and 500 thousand shillings or to imprisonment for a maximum period of three months or to both.
e) Individuals who possess and use plastic bags and plastic wrappings are subjected to a fine range between 30 thousand shillings and 200 thousand shillings or to imprisonment for a term not exceeding seven days or to both.

However, exceptions are given to several situations and cases where several conditions are met (SUBSIDIARY LEGISLATION to the Gazette of the United Republic of Tanzania No.20. Vol.100 dated 17th May, 2019). Firstly, individuals are permitted to sell beverages or other commodities if the commodities require wrappings by plastics by nature. Plastic and plastic packaging which are used for the purpose of medical, industrial goods, agriculture, food processing, and waste-managing are exempted by the gazette.

In the gazette, it is mentioned that imported shipments of prohibited plastic bags which are found will be shipped back to the country of export and the cost shall be paid by the importer, disregarding if the shipments of plastic bags are exempted by the act (SUBSIDIARY LEGISLATION to the Gazette of the United Republic of Tanzania No.20. Vol.100 dated May 17, 2019). Besides, the gazzete also stated that relevant authorities are responisble for ensuring that plastic bags entering the country which are found to violate the act will be confiscated and disposed or recycled. The procees will be done under the supervision of the Council or environmental inspectors, and the cost shall be paid by the offender. The act also assigned responsibilities to product manufacturer or supplier to establish or join a take-back system to ensure that the waste plastic bottles are collected back for the purposes of recycling without incurring additional costs on their products. Lastly, it is also worth taking note that the gazette grants the court the power of issuing orders such as forfeiture of plastic bags, closure of the production unit, cancellation of licenses, or imposing community service on offenders (SUBSIDIARY LEGISLATION to the Gazette of the United Republic of Tanzania No.20. Vol.100 dated May 17, 2019).

The effect of the plastic ban in Tanzania in this study is referred to the findings by Nkya (2020), which were published in the report during 2020 on the effects of the ban of plastic bags in Tanzania. In the report, respondents of the survey had given positive feedback that the plastic ban which was implemented in Tanzania has a positive effect in the country in terms of the environment. It is stated that the number of plastic bags which are visible on the streets of Tanzania is observed to have visibly decreased, as plastic bags are not found in the localities since the ban; in comparison, plastic bags were easily found around the localities of Tanzania and the streets of Tanzania, which sometimes contains wastes and faeces, hence causing environmental pollution and sanitation issues.

In terms of social and environmental impact, it was stated by Nkya (2020) in the report that civilians in Tanzania were observed to have higher awareness towards the importance of environmental protection. The report stated that people in Tanzania were observed to not litter plastic bags in the surroundings. The observation was supported by the phenomena where the plastic bags visible on the streets of Tanzania has rapidly decreased. It was believed that the policy, when the levy is in place on plastic bags, has strengthened the discipline of people in Tanzania to not purchase plastic bags and to not litter plastic bags around.

Besides, the reduction of plastic litter has improved the public cleanliness and health of Tanzania. It was also observed that there was significant reduction of the quantity of visible plastic bags on the street. This has also solved the problem of disposal of faeces in public spaces, hence improving the hygiene of Tanzanian overall. At the meantime, the people in Tanzania were observed to change their carry bag usage habits. In the report, it was stated that people were observed to wash and reuse previously acquired plastic bags. It was also observed that people have begun to opt for non-woven, fabric-made bags, which are known as environmentally friendly bags, as non-woven fabrics are known to be able to decompose. Finally, it was studied by Nkya (2020) that the implementation of plastic usage policy has also impacted the country's society economically. During the interview process, Nkya (2020) noted that several respondents reflected on the situation that there were plastic factories in Tanzania which closed down since the implementation of the ban. As a result of the massive industry closedown, many who worked in plastic-producing factories and packaging-producing factories have lost their jobs and become unemployed. It is also believed that the tax revenues of the Tanzanian government have been reduced as a result of the implementation of the plastic ban, since the implementation of the plastic ban has caused a massive closedown of factories which were related to the plastic manufacturing and packaging industry, which led to an increase in unemployment.

5.2.3 Case Study on Botswana

Another report about the development of Botswana has been conducted by Mogomotsi et al. (2019), which studied the plastic usage behaviour of Botswana locals – specifically, Maun – in 2017. The study was based on approximately 11 years since the implementation of the plastic levy in Botswana during 2006. The research was conducted by analysing the responses which were retrieved from 154 random respondents through questionnaires. In the study, it was concluded that the levy failed to attain its purpose of reducing the usage of plastic bags in the local community. In the study conducted by Mogomotsi et al. (2019), it was revealed that plastic bags are still being frequently used by Botswana locals. The frequency of respondents' plastic bag usage and their frequency of purchase are tabulated in Table 5.3 with the result of analysis showing that 59.7% of the respondents use plastic bags

TABLE 5.3
Frequency of Plastic Bag Usage and Purchase of Respondents (Mogomotsi et al., 2019)

Frequency	Usage	Purchase
Daily	92	27
Every 2 days	18	25
Every shopping trip	32	89
Weekly	12	13
Total	154	154

on a daily basis. According to Mogomotsi et al. (2019), the respondents stated that the plastic bags were used for purposes such as storing items or as litter bags due to the characteristics of plastic bags such as durability, strength, and lightness of the material. However, Mogomotsi et al. (2019) found that only 17.5% of respondents purchase plastic bags on a daily basis.

Besides, it is also observed from Table 5.3 that 21.8% of the respondents use plastic bags on every shopping trip. According to Mogomotsi et al. (2019), plastic bags were mostly used as packaging during respondents' shopping trips. They also found that 57.8% of the respondents purchase plastic bags every shopping trip, where the frequency of shopping trips of the respondents ranges from four times daily to once monthly.

The study also investigated the reasons for the locals' plastic usage. Generally, there are four main reasons which were given by the respondents: 39.6% of the respondents said that plastic bags were being used, as they are readily available; meanwhile, 18.8% of the respondents stated that they are using plastic bags because they used to use plastic bags; while 30.4% of the respondents stated that they use plastic bag because it is cheap; the remaining 11.2% of respondents stated that they use plastic bag because no convenient alternatives are available, which leads them to consume plastic bags on a set path of consumption behaviour, as the benefit of using plastic bags is less than the cost of alternate behaviour (Mogomotsi et al., 2019). Therefore, based on the findings which are stated, it is concluded that the implementation of the plastic levy in Botswana had a minimal effect on the usage of plastic bags of the locals and therefore does not attain the purpose of its implementation. The statement can be further supported by the results which were obtained from respondents, where 86.7% of respondents stated that their consumption has not been reduced after the plastic levy implementation, while only 12.3% of respondents stated that their plastic bag usage has been affected positively (Mogomotsi et al., 2019).

5.3 WESTERN AFRICA

West Africa consists of 16 countries, which are Benin, Burkina Faso, Cape Verde, Gambia, Ghana, Guinea-Bissau, Ivory Coast, Liberia, Mali, Mauritania, Niger, Nigeria, Senegal, Sierra Leone, and Togo. Out of the 16 countries, it was found that 12 countries are implementing plastic reducing policies, with 11 countries implementing bans on plastic bags or plastic products and one implementing a tax on semi-finished and raw plastic materials. The most common plastic reduction method in West Africa is by implementing legislation which bans the production, importation, possession, selling, and usage of plastic bags or non-biodegradable plastic bags, with a few types of plastics for specific purposes listed as exceptions. Some countries implemented or attempted to implement a similar measure, where plastic bags below a specified thickness are allowed, whereas levies are implemented on the usage of plastic bags which have a thickness thicker than the specified thickness. The method of tax implementation is least seen in West Africa as a method of plastic reduction policy, and it was only employed by Ghana in 2014.

Unlike other countries in Africa which usually employ legislative bans and levies on plastic bags and plastic products, Ghana employs strategies which are market-based. The government's effort to combat plastic pollution was first seen in 2011, when Act 512 was introduced and executed in Ghana. Act 512 was described as a type of polluter-pay-principle type of tax and it was reported that the tax rate was set at 20% (Nyavi, 2019). The Ghanan government then introduced and executed Act 863, the Customs and Excise (Duties and Other Taxes) (Amendment) Act (2013), which replaces Act 512. Under Act 512, a 10% tax rate of Environment Excise Tax was implemented on plastic product manufacturers, and the collected tax was intended to be used for the purposes of handling and recycling of plastic waste (Nyavi, 2019). However, it was pointed out that collected funds were stored and not used as intended (Linnenkoper, 2019) until a new fund was established during 2019. The new fund was said to be powered by recycling fees paid by plastic importers and manufacturers.

Aside from collection of tax, it was also discovered that the government of Ghana is also proactive in working on projects and programmes with several constitutions such as Global Plastic Action Partnership (GPAP) and such (Meyerhoff, 2020). According to Meyerhoff (2020), the objectives of the projects are generally to clean up the environment, create jobs in plastic value chain, make resources available to manage plastics, and ensure no one gets left behind. One of the efforts put in was the establishment of a digital system for Ghana's recycling system. The establishment of the digital system was known to benefit the recycling system and the economy of Ghana by establishing a database, which tracks the volume and types of plastic wastes, prices paid across the value chain, etc. The database is foreseen to enable the Ghanan government to collect more accurate data on the problem, which helps the Ghanan government to estimate the severeness of the problem and hence enable the government to make better decisions on the allocation of resources to tackle plastic pollution such as building of recycling plants. The database is also foreseen to help waste pickers to improve their incomes as waste pickers; they were always exploited before the database's establishment by the middlemen of waste collection and the recycling industry, mainly due to the lack of access to information on the market prices of the plastic wastes collected. The scheme is also foreseen to achieve sustainable development, as it can achieve several Sustainable Development Goals (SDG) such as SDG 1, 4, 7, 8, 15, and 17, which are no poverty, quality education, affordable and clean energy, decent work, economic growth, life on land, and partnerships for goals. Though there are no legislative bans implemented in Ghana, it is found that a temporary, non-legislative ban was placed during 2015. The temporary, non-legislative ban was placed after the occurrence of a flash flood, which killed 150 people in Accra, Ghana (Adam et al., 2020). It was discovered that the cause of the flood was clogged storm drains and flood gates (BlueTube, 2015). The ban aimed to ban plastic below 20 μm, but the directive failed to be implemented successfully. For others countries of Western Africa, the policies are list as Table 5.4

5.4 EASTERN AFRICA

Eastern Africa consists of nine countries, which are Burundi, Djibouti, Eritrea, Ethiopia, Kenya, Rwanda, Somalia, South Sudan, and Uganda. Out of the nine countries, eight countries have implemented plastic reducing policies. The majority

TABLE 5.4a
Details of Plastic Usage Policy of Countries in Western Africa

Country	Policy	Year of implementation	Scope	Details
Benin	Partial ban	2018	National	• Acts of importing, manufacturing, possessing, and using of non-biodegradable plastics are prohibited.[*a][*b] • In case of disobedience, fine ranging from 5,000–100,000 CFA francs.[*a]
Burkina Faso	Partial ban	2015	National	• Using of non-biodegradable plastic packages and plastic bags are prohibited.[*a][*c] • In case of disobedience, fine ranging from 100,000 to 10,000,000 CFAF and/or imprisonment of three months to five years.[*a]
Cape Verde	Partial ban	2017	National	• Acts of importing, producing, selling, and using of plastic bags are prohibited.[*a] • In case of disobedience, businesses are subjected to fine range from 50,000 to 800,000 thousand Escudos.[*a]
Gambia	Partial ban	2015	National	• Acts of importing, producing, selling, and using of plastic bags are prohibited.[*a][*b] • In case of disobedience, fine ranging from 1,000–500,000 Gambian dalasi or a jail sentence of minimum six months, maximum of 12 months, or both[*a]

Notes:
[*a] (Adam et al., 2020)
[*b] (Excell et al., 2018)
[*c] (UNEP, 2018c)

of the countries have opted to introduce a ban on plastics by banning the production, importation, selling, and usage of plastic bags. Some implied bans are only on plastics thinner than a specified thickness, and some even applied a levy on plastics thicker than the specified thickness. However, Ethiopia opted to prohibit the issue of permits to produce plastic bags which are thinner than 30 μm while not banning the importation, selling, and usage of plastics thinner than 30 μm. In general, the details of the plastic usage policy in Eastern Africa countries are listed in Table 5.5.

TABLE 5.4b
Details of Plastic Usage Policy of Countries in Western Africa (Cont.)

Country	Policy	Year of implementation	Scope	Details
Ghana	Tentative non-legislative ban	N/A	National	• Non-legislative ban on plastics below 20 μm was tried to be implemented on July 30, 2015.[*a*c] • The ban was the immediate reaction of the flooding incident which caused 150 deaths in Accra. However, the directive failed to be implemented.[*a] • Consultation process on the ban on plastic bags have been ongoing since 2017.[*a]
	Tax	2014		• Implements 10% tax on imported semi-finished and raw plastic materials. The tax is to be paid by the importers and manufacturers.[*a] • The tax revenue is planned to be used for funding the processes such as recycling of plastic waste, and production of plastic waste bins and bags, and production of biodegradable plastic.[*a]
Guinea-Bissau	Partial ban	2013	National	Acts of importing, manufacturing, owning, and selling of non-biodegradable plastics are prohibited.[*a *b]

Notes:
[*a] (Adam et al., 2020)
[*b] (Excell et al., 2018)
[*c] (BlueTube, 2015)

TABLE 5.4c
Details of Plastic Usage Policy of Countries in Western Africa (Cont.)

Country	Policy	Year of implementation	Scope	Details
Ivory Coast	Partial ban	2013	National	• Acts of importing, manufacturing, commercializing, possessing, and using of non-biodegradable plastic bags with thickness thinner than 50 μm.[*a*b] • In case of disobedience, a fine ranging from 100,000 to 1,000,000 CFA francs or a jail period of minimum 15 days and maximum six months.[*a]
Liberia	N/A	N/A	N/A	• Disposal of plastic is regulated.[*c] • No ban is implemented.[*a] • No levy is implemented.[*a]

(Continued)

TABLE 5.4c
Continued

Country	Policy	Year of implementation	Scope	Details
Mali	Partial ban	2012	National	• Acts of importing, manufacturing, owning, selling, and using of non-biodegradable plastic bags are prohibited. • In case of disobedience, subject to fine range from 100,000 to 500,000 CFA or a jail period minimum of three months to maximum of one year.
Mauritania	Partial ban	2013	National	• Acts of importing, manufacturing, and using of plastic bags are prohibited.[*a*b] • In case of disobedience, subject to fine range from 4,630 to 1,653 799 FCAF or a jail period of maximum one year.[*a]

Notes:
[*a] (Adam et al., 2020)
[*b] (UNEP, 2018c)
[*c] (Excell et al., 2018)

TABLE 5.4d
Details of Plastic Usage Policy of Countries in Western Africa (Cont.)

Country	Policy	Year of implementation	Scope	Details
Niger	Partial ban	2013	National	• Acts of importing, manufacturing, selling, owning, and using of low-density plastic and packaging bags are prohibited.[*a*b] • In case of disobedience, fine ranging from 100–1,000,000 francs or a jail period minimum of three months and maximum of six months.[*a]
Nigeria	Ban	N/A	National	• Acts of importing, manufacturing, using and, owning of plastic bags and plastic packaging are prohibited.[*b*c] • In case of disobedience, subject to a fine of maximum ₦500,000 or to jail period maximum of three years or both.[*c] • Any company or organization found guilty are be subjected to a fine of maximum ₦5,000,000. • Bill to ban plastic bags is being considered[*a*c]

Notes:
[*a] (Adam et al., 2020)
[*b] (UNEP, 2018c)
[*c] (Nwafor and Walker, 2020)

TABLE 5.4e
Details of Plastic Usage Policy of Countries in Western Africa (Cont.)

Country	Policy	Year of implementation	Scope	Details
Senegal	Partial ban	2016	National	• Acts of importing, manufacturing, owning, and distributing plastic bags with a thickness < 30 μm are prohibited.[*a] • In case of disobedience, fine range from 10,000 to 20,000 000 CFA or a jail period of maximum two years or both.[*a]
		2020		• Ban on SUPs and all types of plastic bags.[*b] • Bans imports of plastic waste.[*b]
Sierra Leone	N/A	N/A	N/A	• No plastic usage policy implemented in the country.[*a]
Togo	Partial ban	2011	National	• Acts of importing, manufacturing, owning, and distributing non-biodegradable plastic bags are prohibited.[*a*c] • In case of disobedience, fine range from 5 million to 10 million FCAF or a jail period of minimum two months to maximum two years.[*a]

Notes:
[*a] (Adam et al., 2020)
[*b] (Physc.org, 2020)
[*c] (Excell et al., 2018)

TABLE 5.5a
Details of Plastic Usage Policy of Countries in Eastern Africa

Country	Policy	Year of implementation	Scope	Details
Burundi	Ban	2020	National	• Acts of manufacturing, importing, possessing, selling, and using of plastic bags and plastic packaging.[*a] • Exceptions on bio-degradable plastic bags and plastic packaging for the purpose of medical, industrial, and pharmaceutical.[*a]
Djibouti	Ban	2016	National	Acts of producing, importing, and marketing of non-biodegradable plastic bags are prohibited.[*b]
Eritrea	Ban	2005	National	• Acts of producing, importing, selling, and using of plastic bags thickness < 100 μm.[*b] • In case of disobedience, fine up to USD 40,000.[*b]

(Continued)

TABLE 5.5a
Continued

Country	Policy	Year of implementation	Scope	Details
Ethiopia	Ban	2007	National	• Prohibited to grant permit for the manufacture or importation of any nonbiodegradable plastic bags thickness < 30 μm.[*b] • In case of disobedience, fine range from \$5,000–\$20,000 Birr.[*b]

Notes:
[*a] (New Straits Times, 2018)
[*b] (Excell et al., 2018)
[*c] (UNEP, 2018c)

TABLE 5.5b
Details of Plastic Usage Policy of Countries in Eastern Africa (Cont.)

Country	Policy	Year of implementation	Scope	Details
Kenya	Ban	2017 (Aug)	National	• Acts of producing, importing, selling, and using of plastic bag are prohibited.[*a] • In case of disobedience, subject to fine of between 2 million to 4 million shillings or imprisonment of between one to four years or both.[*a]
Rwanda	Ban	2008	National	• Acts of producing, importing, selling, and using of plastic bags and SUP are prohibited.[*b] • Travellers into Rwanda prohibited to carry plastic bags.[*b] • Imported goods with plastic packaging are applied to levy.[*b] • Carry plastic—fine sentence, jail sentence, or making public confessions.[*b] • Smuggle plastic—six months of jail sentences.[*b] • Company executive—jail sentence up to one year.[*b] • Stores—shut down, license suspend, fined.[*b] • Plastic bag manufacturer—fine up to 10 million Rwandan francs.[*b]

Notes:
[*a] (Excell et al., 2018)
[*b] (Government of Rwanda, 2019)

TABLE 5.5c
Details of Plastic Usage Policy of Countries in Eastern Africa (Cont.)

Country	Policy	Year of implementation	Scope	Details
Somalia	Ban	2015	Somaliland, Somalia	Ban on disposable plastic bags.[*c]
South Sudan	Ban	2015	National	• Acts of importing and using plastic bags are prohibited.[*a]
Uganda	Ban	N/A	National	• Attempts to ban plastic bags were announced; however, they are not implemented.[*b] • Latest announcement shows the interest to place a ban on the importation, manufacturing, selling, distribution, and usage of plastic bags < 30 μm. The ban was then delayed and not implemented.[*b] • Ban delayed due to the instrumental power of plastic industry.[*b]

Notes:
[*a] (Juba Monitor, 2018)
[*b] (Behuria, 2021)
[*c] (UNEP, 2018c)

5.4.1 Case Study on Rwanda

Rwanda is found to have employed a plastic usage policy as early as 2004. In the year of 2004, country-wide campaigns were aimed to increase the awareness of the country's people about the negative effects of plastic pollution such as soil pollution, blockage of drainage systems, and endangerment to animals. In the year of 2005, the Rwanda government officials banned the importation and use of plastics thinner than 100 μm (Behuria, 2021). However, the government of Rwanda can still grant permission to allow domestic manufacturers in Rwanda to import plastic items to be used as their packaging. According to Behuria (2021), the official legislation of plastic banning started in 2008, which banned the import of non-biodegradable packaging bags. At the same time, the government of Rwanda also gave an incentive to companies and manufacturers which invest in plastic recycling equipment and equipment which manufactures environmentally friendly bags (Bezerra et al., 2021).

Besides, this legislation was recently reviewed and put into effect immediately in the year of 2019; namely, Law N° 17/2019 of 10/08/2019. Under the reviewed law, individuals are prohibited to use plastic bags and single-use plastic (SUP) items in Rwanda (Government of Rwanda, 2019). Individuals who intend to manufacture, import, and export or use such items are required to obtain permission from exceptional authorities under the law. However, the grantee is responsible for collecting and segregating plastic bags and SUP-made items from wastes and transporting the

TABLE 5.6
Consequences of Offending Law N° 17/2019 of 10/08/2019 (Government of Rwanda, 2019)

Offence	Consequence
Manufacturing plastic carry bags and single-use	Subject to a fine of ten million Rwandan francs
Imports plastic bags and single-use plastic items	• Subject to dispossession of plastic bags and such items • Subject to fine equivalent to ten times the value of imported items
Wholesale of plastic bags and single-use plastic items	• Subject to dispossession of plastic bags and such items • Subject to fine of 700 thousand Rwandan francs
Retailing of plastic bags and single-use plastic items	• Subject to dispossession of plastic bags and such items • Subject to fine of 300 thousand Rwandan francs
Piling or disposing of plastic carry-bag-waste and other single-use plastic items in unauthorized place	• Subject to fine of 50 thousand Rwandan francs • Subject to responsible to remove the waste and repair damages done
Piling or disposing of plastic bag waste and single-use plastic items in unauthorized place (for person who has exceptional authorization from government)	• Subject to fine of 5 million Rwandan francs • Subject to suspension or withdrawal of the authorization

segregated items to recycling facilities. It was also noticed that a levy was placed on goods packaged of plastic material and SUP-made products. The consequences of breaching of the law are stated in the gazette and are tabulated in Table 5.6. The gazette also stated that the amount of the fine will be doubled on the second and consequent offences.

The effects of the plastic ban in Rwanda in this study is mostly referred to the findings by Danielsson (2017), which was published in the report during 2017 which reflects the author's study on the local procedures and the successful outcomes of banning plastic bag in the region. However, the study focuses mainly on the perspective of the government and the perspective of industry within the border of Rwanda. In terms of social and environmental impact, it was stated in the report by Danielsson (2017) that the streets and waterways of Rwanda were observed to be cleaner, and the difference was visibly observed when compared before and after the implementation of the plastic ban. Though non-biodegradable plastic bags can still be seen on the borders of Rwanda, most of them are mainly used by several known sectors which are permitted by the legislation. Besides, it was also mentioned by Danielsson (2017) in the report that recycling technology is also developing in Rwanda, which contributes to the reduction of the number of plastic bags being littered in the streets. However, it was also mentioned that law enforcers in Rwanda have been constantly arresting personnel who possess and use plastic bags which are prohibited by the law on the border and black market, and those who sell the prohibited plastic bags are known to exist within the border.

In terms of economic effect, business owners and industries which involve the use of packaging to package their products have faced difficulty since the implementation of plastic bag ban. As non-biodegradable plastics are banned from being used in the region, only biodegradable packaging is allowed to be used to package their goods. Hence, business owners and industries are facing challenges where the costs of packaging are increased, as the prices of biodegradable packaging are more expensive when compared to non-biodegradable plastic. Moreover, biodegradable packaging is said to be less functional due to the reason that the shelf life of fresh products such as bread, which are packaged using biodegradable materials, are unable to be extended as long as products are packaged in the conventional plastic packaging. Combining the issues of increase in cost and products having shorter shelf life, the issue then extends and causes the low competitiveness of businesses and industries which produce products and rely heavily on their packaging to extend their shelf life in the region. The issue occurs when products are exported to countries which do not implement such bans, as the products are required to be sold at higher prices due to the higher cost of biodegradable packaging and consumed within a shorter period, as the shelf life is relatively short when compared to similar products which are packaged in plastic packaging. This causes the exported products to be less favoured and hence have a lower sale. The statement is said to be supported by Hakuzimana (2021), who stated that a decrease of competitiveness occurred, and it is mainly due to the price hike of packing.

Another economic impact which is observed in Rwanda is the development of recycling industries in the region. In the report, Danielsson (2017) mentioned that one of the Rwandan main manufacturer of plastics, Soimex Plastics, has picked up the opportunity to diversify its businesses by developing the company in the direction of plastic recycling in order to reduce the loss of the company's business which are prohibited by the law. Besides, it was reported by Hakuzimana (2021) that the Rwandan government promoted ecotourism in 2014, which was six years after the implementation of the ban. Hakuzimana (2021) also reported that solutions to turn plastic waste into useful items are being developed by a professional waste management and transport company, Company for Environment Protection and Development (COPED). Amongst the solutions which are being developed, Hakuzimana (2021) listed a few such as recycling old plastic products to remake it into new plastic products bottles, fibres, and cloth threads and to melt polyethylene bags to be used as bonding agent during the brick manufacturing process. The phenomenon should be seen as a good start of the green economy and circular economy in the country, as business owners and industries will soon realize that the economy mode is profitable while bringing less or no harm to the environment. The phenomenon should also be seen as an opportunity to increase the awareness of businesses owners and industries on the issue of environment pollutions.

5.4.2 Case Study on Kenya

Kenya implemented a plastic usage policy as early as 2007. In 2007, Kenya attempted to ban plastic bags which are thinner than 0.03 mm and impose a plastic-use-tax which was rated at 120% (Goitom, 2017). Goitom (2017) also stated that Kenya attempted a plastic ban for a second time in 2011. The ban prohibited plastic bags which were thinner than 0.06 mm, but the bans were not implemented. In 2017, the government of Kenya published a gazette—namely, "Gazette Notice No. 2356"—which places a ban on the usage, manufacturing, and importation of plastic bags for commercial use and household packaging (Republic of Kenya, 2017). Under the ban, plastic carrier bags and flat bags were banned, except for the case where flat bags are used to pack garbage and hazardous wastes. The exception comes with the condition where flat bags require labelling for the use of packing garbage or to be labelled, colour-coded, and incinerated together with the waste if they are used to pack hazardous waste (Abdallah and Macharia, 2017). It was also reported by Abdallah and Macharia (2017) that the ban did not include duty-free shop bags, with the condition that the bags are not brought into the border to Kenya. The individual who breaches the law will be subjected to one of the following penalties (Excell et al., 2018):

a) Fine of minimum 2 million shillings and maximum 4 million shillings.
b) Imprisonment for a period more than one year but less than four years.
c) Both the fine and imprisonment stated previously.

Following up the plastic ban, during September 2017, there was a non-legislative ban of disposable PET bottles in the protected areas of Kenya. The protected areas included Karura Forest and national reserves and game parks (IMPLEMENTATION PLAN FOR THE BAN OF SINGLE USE PLASTICS IN PROTECTED AREAS February 2020). Then, in the 2019/2020 national budget statement of Kenya, the Kenyan government stated a plan to give incentives to support the growth of the plastic recycling industry. The incentive includes exemption of value-added tax (VAT) for all services related to plastic recycling facilities and reduction of corporation tax from 30% to 15% for newly established plastic recycling plant in the first five years (IMPLEMENTATION PLAN FOR THE BAN OF SINGLE USE PLASTICS IN PROTECTED AREAS February 2020). Recently, Kenya has prohibited visitors to bring single-use plastic items such as plastic water bottles, plastic cups, disposable plates, cutlery, or straws into protected areas (UNEP, 2020). According to Republic of Kenya (2017), the list of protected area includes national parks, national reserves and wildlife sanctuaries, national monuments, biosphere reserves, world heritage sites, beaches, protected forests, and ramsar sites, which includes Lake Nakuru, Lake Naivasha, Lake Elmentaita, Lake Baringo, and Tana River Delta.

The first effect of implementation of plastic ban in Kenya is the reduction of plastic bags which are being used in the country, which leads to less litter and less pollution within the borders of Kenya. The effect is discussed in the study conducted by Wahinya and Mironga (2020). In the study, 552 individuals from different demographics such as cereal traders, fruit vendors, vegetable vendors, country environment officers, and park wardens were targeted and identified as the population of the study and 167 individuals were then picked by the application of simple random and census techniques as the sample of study. However, only 125 responses are valid to be studied and analysed. In the study, the response has a mean score of 4.55 and a standard deviation of 0.499 on the statement where the plastic ban eliminated plastic bags, which indicated that less litter and pollution were inflicted. The high mean score and low standard deviation of the responses indicated that respondents highly agree on the statement.

The second effect of the plastic ban which was observed was a decrease of environmental pollution due to plastic litter. The decrease of environmental pollution was observed in a few ways such as improvement of the drainage system, improvement of the aesthetics of city, and reduction of landfills required in the region. The improvement of the drainage system was correlated to the decrease of environmental pollution, as plastic litter was identified as one of the factors which causes clogging of the drainage system, leading to flash floods. The improvement of aesthetics can also be correlated to the decrease of environmental pollution, as it can be understood that the decrease of plastic litter in the country leads to less plastic rubbish visible on the streets or in the ambient environment, hence leading to improved aesthetics. In the meantime, the demand of landfilling in the region has been reduced, indicating that the trash output volume of the region is reduced, which indirectly indicates that plastic bag usage in the region is pursued in a decreasing trend. In the results of the study conducted by Wahinya and Mironga (2020), which is tabulated in Table 5.7, it can be concluded that the respondents highly agree on the statements where the ban has improved drainage systems by reducing plastic litter, improving environmental aesthetics, and reducing landfills which are arose from plastic wastes in the region, as the statements have high mean values of 4.56, 4.52, and 4.46, respectively, and low standard deviations of 0.498, 0.502, and 0.501, respectively.

The third effect of the implementation of the plastic bag ban is the increase of awareness of plastic pollution. The increase of awareness can be expressed in generally two ways, which are recognizing the effects of plastic pollution which are inflicted on the environment as well as actions taken to reduce plastic usage. In the study of Wahinya and Mironga (2020), it was found that the respondents have been rated on a few statements, which are listed in the questionnaire at a high mean and low standard deviation, and hence, it can be understood that the respondents highly

TABLE 5.7
Descriptive Statistics on the Effectiveness of the Implementation of a Plastic Bags Ban (Wahinya and Mironga, 2020)

Effectiveness of the implementation of a plastic bags ban	Mean	Standard deviation
Improved drainage infrastructures as a result of elimination of plastic litter which often clogs drainage systems, causing unnecessary flooding	4.56	0.498
Improved aesthetic beauty of the environment	4.52	0.502
It has led to the reduction of landfills arising from heaps of plastic waste	4.46	0.501

agree on the following statements. The statements which are rated by the respondents stated that the plastic ban implemented in Kenya has:

a) Eliminated plastic bags, which leads to reduced litter and pollution.
b) Reduced contamination of water bodies, and hence, marine life will improve.
c) Improved the draining system in the country by reducing plastic litter, which often clogs drainage systems and caused unnecessary flooding.
d) Decreased the number of breeding grounds for the mosquito population, which spreads malaria.
e) Improved the aesthetic beauty of the environment.
f) Led to the reduction of landfills arising from heaps of plastic waste.

Based on the foregoing statement and the agreement of the respondents towards the statements, it can be deduced that the respondents are aware of the negative effects which are caused to the environment, which are as follows:

a) Usage of plastic bags can cause pollution to the environment.
b) Littering of plastic bags into water bodies causes pollution in the water bodies and negative effects to the marine life.
c) Littering of plastic bags can cause drainage systems to be clogged, which then contributes to the occurrence of flash floods.
d) Littering of plastic bags on the ground can create potential breeding ground for mosquitoes, which might then cause health problems, as malaria can spread through mosquitoes.
e) Littering of plastic bags causes plastic pollution and can negatively affect the aesthetic beauty of environment.
f) Usage of plastic bags contributes to the output of waste, and land pollution can occur, as plastic bags are used to package wastes, which is then buried in landfill.

Hence, it can be concluded that the residents in Kenya have high awareness of plastic pollution. The statement can be further supported by observing the behavioural change of Kenyan locals. A survey, the target of which was respondents of youth in Kenya, was conducted by Oguge et al. (2021), and it found that strong willingness was expressed by the respondents to reduce the consumption of single-use plastic. In the study, the attitudes of respondents towards single-use plastics and the measures taken to reduce single-use plastic pollution were investigated. In terms of respondents' attitude towards single-use plastics, 94.8% of the respondents expressed their desire to switch from single-use plastic to reusable alternatives, and 73% of respondents were found to be willing to pay extra for the use of reusable alternatives. It was also found that 94% of the respondents were willing to promote and use alternative materials if monetary incentives are given. For instance, the acceptable alternative materials—cardboard, glass bottles, food containers, water cans, paper bags, and cloth bags—are the most preferred reusable alternatives by the respondents. The distribution of respondents was observed at 3.7%, 16.2%, 22.5%, 23%, 12%, and 22.6% for the reusable alternatives, respectively.

In the study of Oguge et al. (2021) on the measures taken by the respondents to reduce single-use plastic pollution, there are generally four ways which are employed by the respondents to reduce single-use plastics, including recover wastes, stop throwing away plastic wastes, donate for campaigns to reduce plastics, and participate in campaigns to reduce plastics, respectively. The distribution of respondents on the measures taken were 14.4%, 25.3%, 15.6%, 23.8%, respectively, whereas the remaining 20.8% of the respondents did not take any of the mentioned measures to reduce single-use plastic pollution. It was also observed that 73% of the respondents stated that they are willing to return single-use plastics to designated points of collection for the convenience of recycling the plastic products.

Another observed behavioural change was about the habits of carrier bag usages. The use habits can be observed in two parameters, which are the change of reusable bag ownership, along with their types and the usage of the reusable bags during shopping, respectively. These parameters were investigated in a study conducted by Omondi and Asari (2021). The study found that the average number of reusable bags per household has increased from 4.32 bags per household before the implementation of the ban to 12 bags per household after the implementation of the ban; it was found that, out of 12 bags, 7.7 bags are plastic-based bags. Hence, it is concluded that plastic-based, reusable bags are replacing single-use, plastic-made bags in Kenya. The usage of reusable bags during shopping was also studied by Omondi and Asari (2021), and observation shows that Kenyan people have generally good habits, opting for reusable bags. The study showed that 49% of the sample take their own reusable bags, while 21% of the sample buy reusable bags from stores. The remaining 30% of the respondent sometimes take their own reusable bags on shopping trips or buy new, reusable bags from stores on shopping trips, with a 7% margin for respondents who rarely forget to bring their own shopping bags during shopping trips. The study also reveals that 8% of the sample always forget to bring their own reusable bags, and 9% of the respondents always bring insufficient reusable bags during shopping trips, which shows that Kenyan people have adopted the habits of using reusable bags as carrier bags.

5.5 CENTRAL AFRICA

Central Africa consists of six countries, which are Cameroon, Central Africa Republic, Chad, Equatorial Guinea, Gabon, and Sao Tome and Principe. Out of the six countries, five countries have implemented plastic reduction policies. Four countries have implemented bans on plastic bags. Central Africa Republic, however, has chosen to only implement regulations on the disposal of plastic waste as the country's plastic usage policy, while Chad's plastic usage policy only prohibits the import of non-biodegradable plastic packaging. Therefore, in this aspect, Chad's plastic usage policy is unlike to other countries' plastic usage policy, where the production, selling, distribution, and usage of plastics are also prohibited.

Out of six countries in the region, plastic usage policy does not exist in two countries, while plastic usage policy is announced but not implemented in one country. In the remaining three countries which implemented a plastic usage policy, all three countries have chosen to ban the usage of non-biodegradable plastic bags, but

TABLE 5.8
Details of Plastic Usage Policy of Countries in Central Africa

Country	Policy	Year of implementation	Scope	Details
Cameroon	Ban	2014	National	• Acts of importing, exporting, selling, owning, distributing and, using non-biodegradable plastic bags < 60 μm are prohibited.[*a]
Central African Republic	N/A	N/A	N/A	• Disposal of plastic is regulated.[*b] • No ban is implemented.[*b] • No levy is implemented.[*b]
Chad	Ban	2010	National	• Prohibition of import of non-biodegradable plastic packaging throughout the nation.[*b] • Prohibition of possessing and supply of plastic bag. People found with plastic bags are subjected to fine range between 50,000 to 300,000 CFA francs.[*c] • Shops found with plastic bags will be ordered to close temporarily.[*c]
Equatorial Guinea	N/A	N/A	N/A	No plastic usage policy implemented in the country.[*b]
Gabon	Ban	2010	National	• Acts of importing and selling of non-recyclable plastic bags are prohibited.[*b] • Collection, sorting, storage, transport, recovery, reuse, recycling, and disposal of all types of waste is required.[*b]
Sao Tome and Principe	Ban	N/A	National	Ban production, import, marketing, possession, and usage of plastic bags[*d]

Notes:
[*a] (Ndimuh, 2017)
[*b] (Excell et al., 2018)
[*c] (The New Humanitarian, 2010)
[*d] (FAOLEX, 2020)

with different standards. Cameroon bans non-biodegradable plastic bags which are thinner than 60 μm, Chad bans the importation of non-biodegradable plastic packaging, and Gabon bans on the importation and selling of non-recyclable bags. In general, the details of the plastic usage policy in Central Africa countries are listed in Table 5.8.

5.6 NORTHERN AFRICA

Northern Africa consists of five countries, which are Algeria, Libya, Morocco, Sudan, and Tunisia. Out of five countries, four countries have implemented plastic usage policies. Morocco has implemented bans on the act of producing, importing, distribution, and selling of plastic bags, while Tunisia has implemented a ban on acts of producing, importing, and using plastic bags which are thinner than 50 μm.

TABLE 5.9
Details of Plastic Usage Policy of Countries in Northern Africa

Country	Policy	Year of implementation	Scope	Details
Algeria	Ban	N/A	National	Restrictions are placed on the importation of plastic bags.[*a]
Libya	N/A	N/A	National	• Disposal of plastic is regulated.[*a] • No ban is implemented.[*a] • No levy is implemented.[*a]
Morrocco	Hybrid of ban and levy	2016	National	• Plastics of certain usages are authorized; however, it is required to be marked for their intended use.[*b] • Acts of producing, importing, and distributing of single-use plastic bags are prohibited.[*b] • Consumers are required to pay for shopping bags.[*b] • Plastic bag manufacturers are subjected to fine range between MAD 200,000 to 1 million.[*b] • Persons who possess plastic bags which are intended to be sold are subjected to fine between MAD 10,000 to 500,000.[*b]
Sudan	N/A	N/A		N/A
Tunisia	Hybrid of partial ban and levy	2020 (March)		Acts of producing, importing, and using of plastic bags $< 50\ \mu m$ and place a levy on thicker plastic bags.[*c]

Note:
[*a] (Excell et al., 2018)
[*b] (Ennaji, 2019)
[*c] (UNEP, 2018c)

Tunisia has also implemented a levy on plastic bags which are thicker than 50 μm. Algeria is observed to have only placed ban on the importation of plastic bags, while Libya only regulates the disposal of plastics. In general, the details of the plastic usage policy in Northern African countries are listed in Table 5.9.

5.7 CHALLENGES AND SUGGESTIONS

As mentioned in the previous chapter, the most frequent challenge met by Africa countries' governments in the process of implementing plastic usage policy is the resistance induced by the business power of the plastics industry. The resistance is mainly formed as the industry stakeholders apply pressure on the governments with unemployment and disinvestments, as plastic usage policy is not favourable to plastic-industry businesses. Although the pressure is not applied directly on the government by the business firms, as it might be obvious and can be potentially mistaken as a threat to government, business firms usually apply pressure through the use of mass media and social media, spreading the polarizing information which causes the citizens of the countries to voice objections towards policies which are not favourable to business firms. The potential of using social media as a tool to alter people's political decisions is proven and discussed by Lewandowsky et al. (2020) in the study conducted. It was discussed that pressure is formed in four stages, which are attention economy, choice architectures, algorithmic content curation, and misinformation and disinformation, respectively. The study also discussed the theoretical framework of execution, and hence, it is shown that the usage of social media to alter people's political decisions is viable and could potentially be abused by organizations to pressure authorities to make decisions which are favourable to the organizations.

Therefore, to successfully overcome the first challenge mentioned, African countries' governments should set up and achieve three objectives, which are to educate the citizens about the negative effects of plastic bags usage, to look for possible substitutions of plastic products, and to decrease the economic dependency of countries on the plastics industry. The three objectives are stated, as they are also identified as the challenges which are hindering the success of plastic usage polices' implementation in Africa countries during the stage of literature review. As mentioned previously, citizens should be educated about the harm plastic wastes cause to the environment. This is to ensure that citizens opt not to use plastic-made products in their daily activities by their voluntary actions, hence reducing the demand of plastic-made products. People who are aware of the negative impacts which plastics have on the environment have a sense of social pressure; hence, the pressure causes the individuals to have a lower usage of plastic bags. The statement is supported by the findings of the study conducted by Oguge et al. (2021). In the study conducted by Oguge et al. (2021), it was reported that individuals who have more knowledge of the impact of plastic pollution have higher willingness to perform actions such as switching to single-use plastic alternatives, paying extra to switch to the alternatives, promoting recovery and recycling, and giving up single-use plastics, which aids in the reduction of single-use-plastics consumption. Besides, the African government should ensure that the mass sensitization is conducted for a reasonable period, as it is important for the citizens to develop awareness and sensitization to the issue of plastic pollution to ensure that more citizens will support the effort of the government towards fighting the plastic pollution issue.

Public awareness can be raised in several ways such as giving informational talks about the effects of plastic pollution, beach cleaning activity, and exhibitions of materials which correlate to the effects of plastic of pollution. In these activities, talks are given to inform the public about plastic ban policies and how private sectors can help beat plastic pollution. A coffee break which is plastic-free was also held during the event. The event is believed to be able to raise awareness of the effects on the environment and to educate the public about how plastic-free practises can be practiced and applied in our daily lives. Besides, beach-cleaning activities were also believed to have raised the awareness of the public on the effects of plastic pollution which are implied to the marine ecosystem, as the participating individuals are able to obtain information on how to manage garbage and also to witness the amount and types of plastic garbage which are being dumped into the sea. Besides, by increasing the awareness of the people in the region, it is also expected to be able to tackle hidden issues which are hindering the effort of the government to address the plastic pollution issue. For example, the outdated waste management in Zimbabwe was identified as one of the hidden hindering factors which is hindering the efforts of plastic management, and the government of Zimbabwe has been enacting relevant laws to tackle the weaknesses (Nyathi and Togo, 2020).

Besides, governments should ensure that substitutions are widely available and affordable to citizens. The statement is made as the study conducted indicated that citizens have higher willingness to accept eco-friendly alternatives when the prices of the alternatives are lower. Indeed, due to reasons such as no convenient alternatives and plastics are readily available everywhere at cheaper price compared to the alternatives, such consumers tend to choose plastics and resist changing their plastic usage behaviour. This is also the reason why the reluctance to adopt eco-friendly, reusable alternatives as well as the culture of bringing the eco-friendly reusable alternatives to local society in a shorter period of time, phasing out plastic carry bags in a shorter period of time, or making the culture a sustainable culture are always great challenges to African society.

In order to ensure that the goal of making alternatives widely available and affordable to the public can be achieved successfully, African governments can consider the approach to implement stimulus programmes to support the growth of waste recycling industries and new types of industries which can reduce the net plastic usage in the countries. For example, industries manufacture materials and products from plastic wastes and industries which manufacture biodegradable bags from natural substances. A few examples of the mentioned industries are listed in Table 5.10. With the rise of the new industry together with availability of plastic substitution materials in line with the implementation of the plastic ban, it is believed that the usage of plastic products will decrease in the countries and plastic industries will gradually shift without negatively impacting the economy of the country while being able to reach their goals.

TABLE 5.10
Examples of Companies Manufacturing Goods from Plastic Waste

Company	Location	Example process	Reference
Nelplast	Ghana	Manufactures plastic bricks from recycled plastics	(RFI, 2021)
Gjenge Makers	Kenya	Manufactures plastic bricks from recycled plastics	(Kenya Architecture News, 2021)
Miniwiz	Taiwan	• Manufactures plastic casings for wireless charger from used medical face masks *a • Manufactures modular, adaptable, convertible hospital ward from recycled plastics and materials *b	*a (Euronews and AP, 2022) *b (Pfeifer, 2021)
Company for Environment protection and Development	Rwanda	Manufactures affordable construction materials from plastic wastes	(Hakuzimana, 2021)

Besides that, the enforcement of the plastic usage policies by the African governments should also be more inclusive. Although the use of plastic bags is illegal in several countries on the African continent, more focus should exist on the enforcement of the relevant laws to arrest the illegal importers, distributors, and vendors of the banned plastic products and to seize plastic products before entering the countries. Such a way will be more resource-efficient to combat the issue of illegal plastic usage in the countries, as it can prevent and reduce the amount of illegal plastic products from entering the countries' markets more efficiently. However, there are reports which state that, though illegal usage of plastic products in the countries has decreased significantly, they are not eliminated completely, as there are smugglers who smuggle plastic bags into the countries and sell them on the local black markets. The statement of plastic bags being smuggled into countries which have placed plastic under ban and the establishment of a black market can be supported by a report of UNEP (2018b), where plastic bags were reported to be seen illicitly in the marketplace and on the ground in the town of Koyonzo, Kenya, despite plastic bags having been officially banned in 2017. Another report by Nofuru (2015) reflected similar issues which were observed in Cameroon one year after the implementation of plastic ban in the country. Both the reports reflected the situation where plastic bags or goods packed in plastics, which are illegal in the regions, are being smuggled and sold secretly on the black market. In both reports published by UNEP (2018b) and Nofuru (2015), they reported that the smuggled plastic bags can be obtained easily, as the countries are connected to Uganda, which does not have a ban placed to prohibit the use and manufacture of plastics.

In addition, the high traffic of border-crossing activity is believed to increase the difficulty of inspection and seizure of plastics which are hidden and intended to be smuggled along with the goods imported to the regions from Uganda. Therefore, a possible solution which can be implemented is by regional integration, or preferably, utilizing an existing one such as the African Union to combat the issue stated. By doing so, the member countries of the organization can unify their plastic usage policies, and hence, reduce the risk of smuggling which was caused by the differences in policy within each country in the region. Besides, the member countries can also benefit from it, as countries will supervise each other on the effort of implementing and enforcing the plastic usage policies, hence ensuring high law enforcement.

Besides, law enforcers should continuously monitor the local situation as well as strengthening continual consumer education about the negative impacts of usage of plastics, while also arresting smugglers and dealers who illegally sell plastic products which are prohibited in the region. Law enforcers can conduct more foot patrols, as it creates chances of interacting with the community. The interactions should be taken as chances to develop connections and, hence, improve the social networks between the law enforcer and the local society. By doing so, important information such as newly implemented policies and negative effects of plastic usage can be delivered to consumers easily, hence increasing the effectiveness of public awareness campaigns which are launched by the government bodies. Law enforcers should also continuously enforce the implemented plastic usage policy to prevent plastics which are prohibited in the region from being manufactured and distributed. The stated suggestion is hoped to resolve the problem of having manufacturing companies manufacture banned plastic bags illegally, during odd hours, and distributing the banned plastic through a systematic and well-established network (Foolmaun et al., 2021).

5.8 CONCLUSION

Although the implementation of the plastic usage policies which are observed in the previous chapter are considered to have succeeded, it is also found that there are several studies which have pointed out the weaknesses of plastic usage policies which have been implemented in several regions. The first weakness of plastic usage policy is the effectiveness of utilizing levy as the only or primary instrument of plastic reduction. The method is reported in numerous studies that has successfully reduced the usage of plastic bags by a drastic quantity. However, research found that the levy which was implemented in Maun, Botswana in 2006 had a minimal effect on the consumption of plastic bags. The reduction of plastic usage only persisted for a short period, and hence, the success of attaining the aim of reducing plastic usage in the region will decrease over time. It is believed the decrease of the attainment of success is because consumers are getting used to paying for the usage of plastic bags across time. The failure of attainment is also related to the nature of consumers, who are paralyzed by the status quo or otherwise known as the current state of manner, and hence, do not change their habits of purchasing and continue using plastic bags as their carrier bags. As the levy is implemented with the intention to reduce the plastic demand of consumers in Botswana, the levy is said to have failed to attain its objective as well. Besides, government bodies should also form mechanisms to retrieve the levies which are paid by the customers, as the levies are mostly intended to be collected for utilization to solve issues while reducing the plastic usage issue in the region. This issue is highlighted, as many governments do not have any mechanism to monitor the paid levies, and instead, the levies are collected by the local retailers as their profits. Subsequently, the amount is not properly channelled to resolve issues which are caused by plastic pollution on its borders. Also, in order to ensure that the plastic usage in the region is controlled, it is recommended that governments should cooperate local merchants to promote the use of environmentally friendly, reusable bags, hence minimizing the demand and use of plastic consumption on the border.

Another issue which is identified is that the local stakeholders are mostly not involved in the implementation of the plastic usage policies. It was noticed that most of the plastic reduction legislation is implemented through a top-down approach without prior consultations with local stakeholders. Hence, the plastic usage policies' resistance in the public can be noticeable, as the plastic usage policies are often not in favour of the stakeholders, since the policies might induce economic loss to their businesses. In order improve the public acceptance of the policies, stakeholders should be adequately involved in the drafting and implementation of the plastic usage policies. With the involvement of stakeholders in the process of policies implementation, drafting policies should be done using a better approach which is beneficial to both the efforts of the government to combat plastic pollution and stakeholder's interests. Also, with adequate involvement of the stakeholders in the process of plastic usage policies implementation, a sense of belonging needs to be formed in the mind of the stakeholders, and hence, all parties are committed to the successful implementation of the plastic usage policies. With the commitment of the local stakeholders in the implementation of plastic usage policy, it is believed that the efficiency of the implementation is improved, with the local stakeholders having high willingness to get involved in promoting the usage of environmentally friendly, reusable bags as well as consistently getting involved in activities which raise public awareness on the issue of plastic pollution, hence accelerating the process of phasing out single-use plastics and the process to switch to their substitutes. It is also identified that the sensitization which is carried out by the African governments prior to implementing plastic usage policies is insufficient. The insufficient of sensitization prior to the implementation of the policies is unable to raise the awareness of the public, hence causing the implementation and enforcement of the plastic usage policy to meet resistance and challenges. The insufficient of sensitization also reduces the efficiency of the implemented plastic usage policy, as the public will continue using the prohibited plastic products. Hence, after understanding the resistance which might be caused as the result of insufficient of sensitization prior to the implementation of plastic usage policy, mass sensitization should be planned in such way that the sensitization is sufficiently aggressive, with a reasonable time frame, to ensure that a vast majority of the society will be alert to the issues caused by plastic pollution, hence smoothening the process of implementing plastic usage policies and increasing the efficiency of the implemented plastic usage policy.

ACKNOWLEDGEMENTS

The authors would like to express their sincere appreciation to Loh Weng Keen for their support in the preparation for this chapter.

REFERENCES

Abdallah, A. and Macharia, F. (2017). Legal Alert | Kenya's Ban on the Use, Manufacture and Importation of Plastic Bags. [online] Africa Legal Network. Available at: <www.africalegalnetwork.com/legal-alert-kenyas-ban-use-manufacture-importation-plastic-bags/> [Accessed 2 September 2021].

Adam, I., Walker, T.R., Bezerra, J.C. and Clayton, A. (2020). Policies to reduce single-use plastic marine pollution in West Africa. Marine Policy, [online] 116, p. 103928. https://doi.org/10.1016/j.marpol.2020.103928.

Behuria, P. (2021). Ban the (plastic) bag? Explaining variation in the implementation of plastic bag bans in Rwanda, Kenya and Uganda. Environment and Planning C: Politics and Space, [online] p. 239965442199483. https://doi.org/10.1177/2399654421994836.

Bezerra, C.J., Walker, T.R., Clayton, C.A. and Adam, I. (2021). Single-use plastic bag policies in the Southern African development community. Environmental Challenges, [online] 3, p. 100029. https://doi.org/10.1016/j.envc.2021.100029.

BlueTube (2015). Plastic Bags Kill 150 in Ghana. [online] Available at: <www.bluetubebeach.org/blog/plastic-bags-kill-150-in-ghana/> [Accessed 2 September 2021].

Danielsson, M. (2017). The Plastic Bag Ban in Rwanda: Local Procedures and Successful Outcomes. [online] Available at: <www.diva-portal.org/smash/get/diva2:1067480/FULLTEXT01.pdf>.

Dikgang, J., Leiman, A. and Visser, M. (2010). Analysis of the Plastic Bag Levy in South Africa. [online] Available at: <www.econrsa.org/system/files/publications/policy_papers/pp18.pdf>.

Ennaji, K. (2019). Zero mika: The vision of plastic-free Morocco. Morocco World News. [online] 19 September. Available at: <www.moroccoworldnews.com/2019/09/283009/zero-mika-vision-plastic-free-morocco>.

Environment Protection Act 2016 (Act 18 of 2016). Supplement to Official Gazette. [online] Available at: <https://seylii.org/akn/sc/act/si/2020/81/eng@2020-07-01>.

Euronews and AP. (2022). COVID waste: This taiwanese firm is turning used face masks into phone chargers. [online] Euronews. Available at: <www.euronews.com/next/2022/02/07/covid-waste-this-taiwanese-firm-is-turning-used-face-masks-into-phone-chargers> [Accessed 18 February 2022].

Excell, C., Salcedo-La Viña, C., Worker, J. and Moses, E. (2018). Legal Limits on Single-Use Plastics and Microplastics: A Global Review of National Laws and Regulation. Nairobi: United Nations Environment Programme.

FAOLEX. (2020). Law no. 8/2020 approving measures to reduce the use of plastic bags in São Tomé and Príncipe. [online] FAOLEX Database. Available at: <www.fao.org/faolex/results/details/en/c/LEX-FAOC198723/> [Accessed 2 September 2021].

Foolmaun, R.K., Chamilall, D.S., Munhurrun, G. and Sookun, A. (2021). Was Mauritius really successful in banning plastic carry bags, after promulgation of the regulation prohibiting plastic bags usage? Environment, Development and Sustainability, [online] 23(8), pp. 11660–11676. https://doi.org/10.1007/s10668-020-01134-w.

Goitom, H. (2017). Kenya: Notice outlawing plastic bags issued. [online] Library of Congress. Available at: <www.loc.gov/item/global-legal-monitor/2017-03-31/kenya-notice-outlawing-plastic-bags-issued/> [Accessed 2 September 2021].

Government of Mauritius (2015). Legal Supplement to the Government Gazette of Mauritius No. 81 of 6 August 2015. Government Notices 2015, [online] pp. 1175–1177. Available at: <https://tsapps.nist.gov/notifyus/docs/wto_country/MUS/full_text/pdf/MUS5(english).pdf>.

Government of Mauritius (2020). Legal Supplement to the Government Gazette of Mauritius No. 89 of 18 July 2020. Government Notices 2020, [online] pp. 465–473. Available at: <https://environment.govmu.org/Documents/Legislations/B.%20Regulations/22(i).%20GN%20156%20of%202020%20-%20%20Environment%20Protection%20(Control%20of%20single%20use%20plastic%20products)%20Regulations%202020.pdf>.

Government of Rwanda (2019). LAW N° 17/2019 OF 10/08/2019 Relating to the Prohibition of Manufacturing, Importation, Use and Sale of Plastic Carry Bags and Single-Use Plastic Items. Official Gazette no. 37 bis of 23/09/2019, [online] pp. 3–24. Available at: <https://elaw.org/system/files/attachments/publicresource/Law_relating_to_the_prohibition_of_manufacturing__importation__use_and_sale_of_plastic_carry_bags.pdf>.

Hakuzimana, J. (2021). Break free from plastics: Environmental perspectives and lessons from Rwanda. Journal of Pollution Effects & Control, [online] 9(3). https://doi.org/10.35248/2375-4397.20.9.276.

Jambeck, J., Hardesty, B.D., Brooks, A.L., Friend, T., Teleki, K., Fabres, J., Beaudoin, Y., Bamba, A., Francis, J., Ribbink, A.J., Baleta, T., Bouwman, H., Knox, J. and Wilcox, C. (2018). Challenges and emerging solutions to the land-based plastic waste issue in Africa. Marine Policy, [online] 96, pp. 256–263. https://doi.org/10.1016/j.marpol.2017.10.041.

Juba Monitor (2018). Plastic Pollution Remains Biggest Environmental Problem. [online] Available at: <www.jubamonitor.com/7633-2/>.

Kenya Architecture News (2021). Kenyan startup founder Nzambi Matee recycles plastic to make bricks that are stronger than concrete. [online] World Architecture. Available at: <https://worldarchitecture.org/article-links/egmeg/kenyan-startup-founder-nzambi-matee-

recycles-plastic-to-make-bricks-that-are-stronger-than-concrete.html> [Accessed 18 February 2021].

Lewandowsky, S., Smillie, L., Garcia, D., Hertwig, R., Weatherall, J., Egidy, S., Robertson, R.E., O'Connor, C., Kozyreva, A., Lorenz-Spreen, P., Blashkle, Y. and Leiser, M. (2020). Technology and democracy: Understanding the influence of online technologies on political behaviour and decision-making. Publications Office of the European Union. https://doi.org/10.2760/709177.

Linnenkoper, K. (2019). Ready or not: Plastics recycling 2.0 in Ghana. Recycling International. 19 July. Available at: https://recyclinginternational.com/plastics/ready-or-not-plastics-recycling-2-0-in-ghana/27004/

Magoum, I. (2020). MOZAMBIQUE: Some plastic bags to be banned from 2021. Afrik21. [online] 20 August. Available at: <www.afrik21.africa/en/mozambique-some-plastic-bags-to-be-banned-from-2021/>.

Mensah, H. and Ahadzie, D.K. (2020). Causes, impacts and coping strategies of floods in Ghana: A systematic Review. SN Applied Sciences, 2, p. 792.

Meyerhoff, R. (2020). Ghana's ambitious plan to minimize plastic waste. [online] Forbes. Available at: <www.forbes.com/sites/sap/2020/10/22/ghanas-ambitious-plan-to-minimize-plastic-waste/?sh=7d541a38a327> [Accessed 2 September 2021].

Mogomotsi, P.K., Mogomotsi, G.E. and Phonchi, N.D. (2019). Plastic bag usage in a taxed environment: Investigation on the deterrent nature of plastic levy in Maun, Botswana. Waste Management & Research, [online] 37(1), pp. 20–25. https://doi.org/10.1177/0734242X18801495.

Ndimuh, S. (2017). Plastics dilemma in cameroon four years after ban. [online] Green Vision. Available at: <www.greenvision.news/plastics-dilemma-in-cameroon-four-years-after-ban/> [Accessed 2 September 2021].

New Straits Times (2018). Burundi Plans Plastic Bag Ban. [online] 14 August. Available at: <www.nst.com.my/world/2018/08/401309/burundi-plans-plastic-bag-ban>.

Nkya, E.O. (2020). Assessment on the Effects of Banning Plastic Bag Carrier in Tanzania: A Case of Ilala Municipal Council. [online] Available at: <http://scholar.mzumbe.ac.tz/bitstream/handle/11192/4463/MBA-CM-DCC_Elieshi oberlin_2020.pdf?sequence=1>.

Nofuru, N. (2015). Cameroon struggles to enforce plastic bag ban as black market supports demand from retailers. [online] Global Press Journal. Available at: <https://globalpressjournal.com/africa/cameroon/cameroon-struggles-to-enforce-plastic-bag-ban-as-black-market-supports-demand-from-retailers/> [Accessed 13 December 2021].

Nwafor, N. and Walker, T.R. (2020). Plastic bags prohibition bill: A developing story of crass legalism aiming to reduce plastic marine pollution in Nigeria. Marine Policy, [online] 120, p. 104160. https://doi.org/10.1016/j.marpol.2020.104160.

Nyathi, B. and Togo, C.A. (2020). Overview of legal and policy framework approaches for plastic bag waste management in African Countries. Journal of Environmental and Public Health, [online] 2020, pp. 1–8. https://doi.org/10.1155/2020/8892773.

Nyavi, G.A., 2019. Release funds from environmental excise tax for its intended purpose – Plastic manufacturers to govt. Graphic Online. [online] 7 May. Available at: < https://www.graphic.com.gh/news/general-news/ghana-news-release-funds-from-environmental-excise-tax-for-its-intended-purpose-plastic-manufacturers-to-govt.html> [Accessed 2 January 2024].

Oguge, N., Oremo, F. and Adhiambo, S. (2021). Investigating the knowledge and attitudes towards plastic pollution among the youth in Nairobi, Kenya. Social Sciences, [online] 10(11), p. 408. https://doi.org/10.3390/socsci10110408

Omondi, I. and Asari, M. (2021). A study on consumer consciousness and behavior to the plastic bag ban in Kenya. Journal of Material Cycles and Waste Management, [online] 23(2), pp. 425–435. https://doi.org/10.1007/s10163-020-01142-y.

Pfeifer, H. (2021). A hospital ward made from trash highlights Arthur Huang's mission to revolutionize recycling. [online] Cable News Network. Available at: <http://edition.cnn.

com/style/article/arthur-huang-miniwiz-hospital-ward-c2e-spc-intl-hnk/index.html> [Accessed 11 April 2022].

Physc.org. (2020). Senegal Bans Most Single-use Plastics. [online] Available at: <https://phys.org/news/2020-04-senegal-single-use-plastics.html> [Accessed 2 September 2021].

Princewill, N. (2021). Malawi's Landscape is clogged with plastic waste that could linger for 100 years. One woman has taken on plastic companies and won. [online] Cable News Network. Available at: <https://edition.cnn.com/2021/06/15/africa/malawi-landscape-plastic-pollution-cmd-intl/index.html> [Accessed 18 April 2022].

Republic of Kenya (2017). Gazette Notice No. 2356. [online] Available at: <http://kenyalaw.org/kenya_gazette/gazette/notice/181293> [Accessed 2 September 2021].

RFI (2021). Ghana's Plastic House: A Step Towards Dealing with the Country's Pollution. [online] Available at: <www.rfi.fr/en/africa/20210626-ghana-s-plastic-house-one-idea-to deal with the country s pollution africa environment recycling> [Accessed 18 February 2022].

Richard, N. and Ribet, A. (2021). Plastic ban in mauritius. Business Magazine Mauritius. [online] March. Available at: <www.dlapiperafrica.com/en/mauritius/insights/2021/plastic-ban-in-mauritius-2021.html>.

Sambyal, S.S. (2018). Five African Countries among top 20 Highest Contributors to Plastic Marine Debris in the World. [online] 23 May. Available at: <www.downtoearth.org.in/news/waste/when-oceans-fill-apart-60629>.

South Africa Government Gazette 7348 Vol 443 (2002). [online] Available at: <www.environment.gov.za/sites/default/files/gazetted_notices/eca_plasticbags_regulations_g23393rg-7348gon543_0.pdf>.

SUBSIDIARY LEGISLATION to the Gazette of the United Republic of Tanzania No.20. Vol.100 dated 17th May, 2019. [online] Available at: <https://fbattorneys.co.tz/wp-content/uploads/2019/05/GN-394-of-2019-The-Prohibition-Of-Plastic-Carries-Bags-Regulations-2019.pdf>.

The New Humanitarian (2006). Zanzibar Implements Ban on Plastics. [online] 9 November. Available at: <www.thenewhumanitarian.org/report/61519/tanzania-zanzibar-implements-ban-plastics>.

The New Humanitarian (2010). CHAD: Just Say No to Plastic Bags. [online] 24 November. Available at: <https://reliefweb.int/report/chad/chad-just-say-no-plastic-bags>. [Accessed 16 September 2023]

Thikusho, M. (2019). Plastic Levy Now Mandatory. [online] 13 August. Available at: <www.namibian.com.na/191856/archive-read/Plastic-levy-now-mandatory>.

UNEP (2018a). Africa Is on the Right Path to Eradicate Plastics. [online] Available at: <www.unep.org/news-and-stories/story/africa-right-path-eradicate-plastics> [Accessed 2 September 2021].

UNEP (2018b). How Smuggling Threatens to Undermine Kenya's Plastic Bag Ban. [online] Available at: <www.unep.org/news-and-stories/story/how-smuggling-threatens-undermine-kenyas-plastic-bag-ban> [Accessed 2 September 2021].

UNEP (2018c). Single-Use Plastics: A Roadmap for Sustainability. Single-use Plastic: A Roadmap for Sustainability. New York: UNEP.

UNEP (2020). Kenya Bans Single-Use Plastics in Protected Areas. [online] Available at: < https://www.unep.org/news-and-stories/story/kenya-bans-single-use-plastics-protected-areas> [Accessed 2 September 2021].

US Embassy in Tanzania (2019). Prohibition of Plastic Bags Effective [online]. 1 Jun. Available at: <https://tz.usembassy.gov/prohibition-of-plastic-bags-effective-june-1–2019/> [Accessed 2 September 2021].

Wahinya, M.K.P. and Mironga, J. (2020). Effectiveness of the Implementation of Plastic Bags Ban: Empirical Evidence from Kenya. IOSR Journal of Environmental Science, Toxicology and Food Technology (IOSR-JESTFT), [online] 14(16), pp. 53–61. https://doi.org/10.9790/2402-1406025361.

6 World Organization Plastic Initiatives and Plastic Leakage Prevention

6.1 INTRODUCTION

It is undeniable, the impacts plastic pollution have had on the worldwide ecosystem, especially the fact that microplastics have entered food chains, and prolonged exposure can lead to severe health problems. The efforts to tackle plastic pollution, initiated many years ago, are not promising without the cooperation of all the countries in the world as well as the stakeholders. Hence, a couple of organizations, initiated by developed countries, have been headed to set up global policies and campaigns to minimize and prevent microplastic pollution. United Nations, World Bank United Nations Environmental Program (UNEP), World Economic Forum, Organization for Economic Co-operation and Development (OECD), etc., have initiated negotiations, drafting and setup policies and authorities and organizing forums and campaigns to bring all the stakeholders to act prominently on this pressing pollution problem. The main aim is the reduce the usages of plastic, promote recycling, as well as improve the management of plastic wastes. In other words, the world needs to promote a circular economy of plastic consumption and emphasize the education efforts of communities to participate in the curbing of plastic pollution.

6.2 THE UNITED NATIONS BASEL CONVENTION'S GLOBAL PLASTIC WASTE PARTNERSHIP

Historically, the Basel Convention was made to provide solutions to control exports and imports of hazardous wastes generated from households or industry. The main intention of the Basel Convention was to prevent and minimize speculation that happens by exporting hazardous wastes to developing countries for disposal, subsequently causing serious environmental pollution to the countries. Consider the example of the incident of Toxic Ship of Karin B, as reported in the *New York Times* on September 3, 1988 (Section 1, Page 4), which carried toxic waste originally transported from Italy and wandered the seas of France, Britain, Spain, West Germany, and the Netherlands, who refused the ship landing (*New York Times*, 1988). The waste once landed in Nigeria, yet the locals protested, and finally, the Italian government was ordered to retrieve the cargo. In the context of plastic wastes, the recent year's incident happened in Malaysia in May 2019, where 3,000 tonnes of plastic

DOI: 10.1201/9781003387862-6

FIGURE 6.1 Former Malaysian Minister of Energy, Science, Technology, Environment, and Climate Change (MESTECC), Yeo Bee Yin, shows samples of a plastic wastes shipment before sending it back to the country of origin in Port Klang, Malaysia on May 28, 2019.

wastes were found at Port Klang, Malaysia (Figure 6.1). The 60 containers included cables from the United Kingdom, electronic and household wastes from North America, Japan, Saudi Arabia, and China, compact discs from Bangladesh, and contaminated milk cartons from Australia (Newsweek, 2019). On November 2019, once again the Malaysian government detected containers with plastic wastes at Penang Port, Malaysia. The British government, with a joint statement of the Malaysian government, announced that those containers did not possess any import papers and were illegally sent to Malaysia (Kyodo News, 2019). Britain agreed to take back the 42 containers afterwards.

As such incidents happened regularly, the Basel Convention on the Control of Transboundary Movements of Hazardous Wastes and their Disposal was adopted on March 22, 1989 by the Conference of Plenipotentiaries in Basel, Switzerland. The Convention entered into force on May 5, 1992. It covers 190 parties as of April 2023. Later amendments were made to improve coverage as well as implementation, such as "Band Amendments," Plastic Waste Amendments," and "E-waste Amendments." To date, this environmental treaty is considered to be the most comprehensive to deal with hazardous and other wastes requiring special attention. Since its adoption, the Basel Convention has seen a number of significant developments. Under the Basel Convention, the summarized general obligations of the Parties are as follows:

1) Unless granted "Prior Informed Consent" (PIC), the trans-boundary movements of hazardous and other wastes are not allowed. The related Parties are required to make appropriate national or domestic legislation to prevent and punish illegal traffics of hazardous and other wastes.
2) When managing hazardous and other wastes, the related Parties need to ensure that all the hazardous and other wastes are managed and disposed with Environmental Sound Manner, i.e., all the wastes are needed to be generated, moved, and disposed as minimum as possible, while possible treatment process, reuse, recycling, and recovery should be carried out prior to final disposal, depending on the capability and capacity of the Parties.
3) In cooperation in prevention activities with other Parties and interested organizations, directly and through the Secretariat, including the dissemination

of information on the transboundary movement of hazardous wastes and other wastes, in order to improve the environmentally sound management of such wastes and to achieve the prevention of illegal traffic.

6.2.1 The Plastic Waste Amendments

In reaction to the problem of plastic and microplastic pollution, a proposal was submitted by the government of Norway in June 2019 to amend the annexes of the Basel Convention to address plastic waste within its provisions. As the result, the amendments to Annexes II, VIII, and IX were adopted in the 14th meeting of the Conference of the Parties to the Basel Convention (COP-14, 29 April—10 May 2019) in order to improve the existing transboundary movements of plastics waste and further define the scope of the Conventions in regards to the plastic waste (Wingfield and Lim, 2021). The following is the summary of the amendments, made effective January 1, 2021:

A) Insertion of new entry A3210 in Annex VIII to define the scope of plastic wastes that is hazardous is required prior informed consent (PIC) procedure.
B) New entry of B3011 to replace existing entry B3010 to decide plastic waste to be non- hazardous that do not require PIC produce.
C) Insertion of new entry Y48 in Annex II, which covers mixtures of plastic wastes of A3210 and B3011.

In general, all plastics wastes and their mixtures are required to undergo PIC procedure EXCEPT the following (Basel Convention, 2019):

a) Plastic waste that is hazardous waste pursuant to paragraph 1 (a) of Article 1 of Basel Convention.
b) Plastic waste, listed as follows, provided it is destined for recycling in an environmentally sound manner and almost free from contamination and other types of wastes:

   - Plastic waste almost only consisting of one non-halogenated polymer, including but not limited to the following polymers:

     i. Polyethylene (PE).
     ii. Polypropylene (PP).
     iii. Polystyrene (PS).
     iv. Acrylonitrile butadiene styrene (ABS).
     v. Polyethylene terephthalate (PET).
     vi. Polycarbonates (PC).
     vii. Polyethers.

   - Plastic waste almost only consisting of one cured resin or condensation product, including but not limited to the following resins:

     i. Urea formaldehyde resins.
     ii. Phenol formaldehyde resins.

iii. Melamine formaldehyde resins.
iv. Epoxy resins.
v. Alkyd resins.

- Plastic waste almost only consisting of one of the following fluorinated polymers:

i. Perfluoroethylene/propylene (FEP).
ii. Perfluoroalkoxy alkanes: Tetrafluoroethylene/perfluoroalkyl vinyl ether (PFA), Tetrafluoroethylene/perfluoromethyl vinyl ether (MFA).
iii. Polyvinylfluoride (PVF).
iv. Polyvinylidenefluoride (PVDF).

c) Mixtures of plastic waste, consisting of polyethylene (PE), polypropylene (PP), and/or polyethylene terephthalate (PET), provided they are destined for separate recycling of each material and in an environmentally sound manner and almost free from contamination and other types of wastes.

6.2.2 Plastic Waste Partnership

In order to improve the mutual cooperation of the spectrum of stakeholders from governments, international organizations, and industries in managing the plastic wastes from polluting environments, the Conference of Parties has established Plastic Waste Partnership with the goal, scope, and overall tasks, adapted in detailed from the Terms of Reference for the Basel Convention Partnership on Plastic Waste, as in Table 6.1, with four working groups, formed to address the issue and move forward to solve plastic pollution problem. Overall, the amendments on the Basel Conventions are positive steps for the global arena to battle with plastic pollution for a better future for coming generations.

6.3 WORLD BANK

World Bank plays important roles in supporting programmes related to combatting plastic pollution through PROBLUE, which is an umbrella multi-donor trust fund. PROBLUE supports the sustainable and integrated development of marine and coastal resources for healthier oceans. One of the important projects supported by PROBLUE is ASEAN Regional Action Plan for Combatting Marine Debris (ASEAN APCMD). This initiative was started in November 2017 during the ASEAN Conference on Reducing Marine Debris held in Thailand to develop an integrated land-to-sea policy in the ASEAN region. Subsequently, during the 34th ASEAN Summit in June 2019, it adopted the Bangkok Declaration on Combatting Marine Debris in the ASEAN Region and ASEAN Framework of Action on Marine Debris (see Tables 6.2 and 6.3). Through extensive consultation and collaboration led by Thailand, with a wide spectrum of stakeholders as well as the support from World Bank, the ASEAN APCMD was proposed to be implemented in phases over the five year period from 2021 to 2025. The list of action plans are as follows (ASEAN, 2021):

TABLE 6.1
Goal, Scope, Overall Tasks, and Formation of Working Group of Plastic Waste Partnership

Goal	The goal of the Partnership is to improve and promote the environmentally sound management of plastic waste at the global, regional, and national levels and prevent and minimize their generation so as to, among other things, reduce significantly and, in the long-term, eliminate the discharge of plastic waste and microplastics into the environment, in particular the marine environment.
Scope	The Partnership covers all plastic waste, including wastes containing plastics, generated nationally and disposed of at the national level as well as those which are imported or exported for disposal operations, taking into account the entire lifecycle of plastics.
Tasks	The overall tasks of the Partnership are the following: (a) Collect information and undertake analysis on environmental, health, economic, and social impacts of global, regional, and national policy frameworks and strategies relevant to prevention, minimization, collection, and environmentally sound management of plastic waste. (b) Identify the gaps and barriers to the prevention, minimization, collection and, environmentally sound management of plastic waste and identify best practices, lessons learnt, and possible solutions to the same. (c) Promote the development of policy, regulation, and strategies on the prevention and minimization of plastic waste, in particular, in relation to single-use plastics, inter alia, via better design and innovation to improve durability, reusability, repairability, and recyclability of plastics and to avoid hazardous substances in plastics and on environmentally sound management of plastic waste, taking into account the entire lifecycle of plastics. (d) Advance the prevention, minimization, collection, and environmentally sound management of plastic waste. (e) Undertake pilot projects which support the delivery of the other overall tasks. (f) Collect, analyse, and consider possibilities to improve information on transboundary movements of plastic waste. (g) Facilitate knowledge sharing, capacity building, technical advice, and technology transfer to strengthen and implement policies, strategies, public-private initiatives for the prevention, minimization, collection, and environmentally sound management of plastic waste. (h) Undertake and/or contribute to outreach, education, and awareness-raising activities to widely disseminate the information and knowledge gathered and generated through the activities of the Partnership. (i) Encourage and promote relevant innovation, research, and development.
Working Group	a) Plastic waste prevention and minimization. b) Plastic waste collection, recycling, and other recovery, including financing and related markets. c) Transboundary movements of plastic waste. d) Outreach, education, and awareness-raising.

Source: Adapted from Terms of reference for the Basel Convention Partnership on Plastic Waste and workplan for the working group of the Partnership on Plastic Waste for the biennium 2020–2021, 2019.

TABLE 6.2
Bangkok Declaration on Combatting Marine Debris in ASEAN Region: Key Objectives

Bangkok Declaration: Key Objectives

1. STRENGTHEN actions at the national level as well as through collaborative actions among the ASEAN Member States and partners to prevent and significantly reduce marine debris, particularly from land-based activities, including environmentally sound management.
2. ENCOURAGE an integrated land-to-sea approach to prevent and reduce marine debris and strengthen national laws and regulations as well as enhance regional and international cooperation including on relevant policy dialogue and information sharing.
3. PROMOTE inter-sectoral coordination between ASEAN sectoral bodies to effectively address the multi-dimensional and far-reaching negative effects as well as sources of marine debris pollution.
4. ENHANCE the multi-stakeholder coordination and cooperation to combat marine debris, including implementing joint actions and partnerships for addressing such a challenge.
5. PROMOTE private-sector engagement and investment in preventing and reducing marine debris, including partnerships between public and private sector through various mechanisms and incentives.
6. PROMOTE innovative solutions to enhance plastics value chains and improve resource efficiency by prioritizing approaches such as circular economy and 3R (reduce, reuse, recycle) and welcome capacity building and exchange of best practices among ASEAN Member States as well as support from external partners in this regard.
7. STRENGTHEN research capacity and application of scientific knowledge to combat marine debris; in particular, to support science-based policy and decision making.
8. ACCELERATE advocacy and actions to increase public awareness and participation and enhance education, with the aim to change behaviour towards preventing and reducing marine debris.

Source: Adapted from ASEAN, 2021.

TABLE 6.3
ASEAN Framework of Action on Marine Debris

FRAMEWORK I: POLICY SUPPORT AND PLANNING

A. Promote regional policy dialogue on prevention and reduction of marine debris from land- and sea-based activities by highlighting the issue, sharing information and knowledge, and strengthening regional coordination.
B. Mainstream multi-sectoral policy measures to address marine debris in national and ASEAN's development agenda and priorities.
C. Encourage ASEAN Member States to implement relevant international laws and agreements related to waste management such as MARPOL Annex V ship generated waste, Basel Convention, and UN Environment Assembly resolutions 3/7 on Marine Litter and Microplastics.
D. Develop a regional action plan on combatting marine debris in the ASEAN Region by applying integrated land-to-sea policy approaches.

FRAMEWORK II: RESEARCH, INNOVATION, AND CAPACITY BUILDING

A. Compile regional baseline on status and impacts of marine debris in the ASEAN Region.
B. Strengthen regional, national, and local capacities to develop and implement national action plans/initiatives.
C. Enhance scientific knowledge, transfer marine technology, and promote innovative solution to combat marine debris.
D. Promote integration and application of scientific knowledge to enhance science-based decisions and policies on marine debris prevention and management.

(Continued)

TABLE 6.3
Continued

FRAMEWORK III: PUBLIC AWARENESS, EDUCATION, AND OUTREACH
A. Promote public awareness on status and impacts of marine debris and microplastics.
B. Accelerate advocacy strategy/programme to promote behaviour change to combat marine debris and to incorporate marine debris issue into ASEAN's Culture of Prevention Initiative.
C. Promote platforms for knowledge sharing, innovative solutions, and best practices to combat marine debris.
FRAMEWORK IV: PRIVATE SECTOR ENGAGEMENT
A. Promote collaborative actions with private sector and industry associations to implement measures to address marine debris issues.
B. Encourage private sector investment in and contribution to combat marine debris.

Source: Adapted from ASEAN, 2021.

1. Develop Regional Guidebook on Financial Mechanisms for Investments in Plastic Waste Management.
2. Develop Guiding Principles for Phasing out select Single-use Plastics (SUPs).
3. Develop a Regional Guidebook on Standards for Responsible Plastic Waste Trade, Sorted Plastics Waste, and Recycled Plastics.
4. Elaboration of Best Practice Manual for Development of Minimum Standards and Technical Requirements for Plastic Packaging and Labelling.
5. Undertake Regional Stocktaking of Green Public Procurement.
6. Develop Best Practice Manual for Reducing, Collection, and Treatment of Sea-Based Litter.
7. Develop Guidebook for Common Methodologies for Assessment and Monitoring of Marine Litter.
8. Strengthen ASEAN Regional Knowledge Network on Marine Plastics.
9. Conduct a Regional Study on Microplastics.
10. Coordinate Regional Training Programmes on Plastics and Waste Management.
11. Develop a Behavioural Change Communication Strategy Playbook.
12. Enhance Regional Awareness for Consumers of Labelling of Plastics and Packaging.
13. Establish a Regional Platform for EPR (Extended Producer Responsibility) Knowledge and Implementation Support.
14. Establish a Regional Platform to Support Innovation and Investments in Plastics and Plastic Waste Management.

Meanwhile, the relevant working groups are formed to implement the above ASEAN APCMD action plans are:

- ASEAN Working Group on Environmentally Sustainable Cities (AWGESC).
- ASEAN Working Group on Chemicals and Waste (AWGCW).
- ASEAN Working Group on Environmental Education (AWGEE).

- ASEAN Consultative Committee on Standards and Quality (ACCSQ).
- ASEAN Sectoral Working Group on Fisheries (ASWGFi).
- ASEAN Maritime Transport Working Group (MTWG).
- ASEAN Business Advisory Council (ABAC).
- ASEAN Coordinating Committee on Micro, SMEs (ACCMSME).

Besides that, World Bank also involves financing and advisory on solid-waste management projects via a spectrum of instruments, including traditional loans, results-based financing, development policy financial, as well as technical advisory. The scope of World Bank financing includes the complete lifecycle of waste, starting from the generation to collection, transportation, and finally, treatment and disposal. The following is the target areas of projects and investment of the World Bank (Table 6.4):

TABLE 6.4
Target Areas of World Bank Financing

Target Area	Scope
Infrastructure	Capital investments to build or upgrade waste sorting and treatment facilities, close dumps, construct or refurbish landfills, and provide bins, dumpsters, trucks, and transfer stations are under financeable by World Bank.
Legal structures and institutions	Cost of projects on policy measures and coordinated institutions for the municipal-waste management sector.
Financial sustainability	For projects, help governments improve waste cost containment and recovery such as the design of taxes and fee structures and long-term planning.
Citizen engagement	Programme to promote behaviour change and public participation to achieve functional waste-system.
Social Inclusion	Programme to elevate waste-picker livelihoods with integration of strategies into the established system like provision of safe working conditions, social safety nets, child labour restrictions, and education. In fact, most of the recycling or recovery works in most developing countries rely heavily on informal workers to carry out the work to collect, sort, and recycle 15%–20% of generated waste.
Climate change and the environment	Financing on programme that improves the waste-disposal structure by minimizing greenhouse-gas emission through food loss and waste reduction, organic waste diversion, and the adoption of treatment and disposal technologies that harvest biogas and landfill gas. Any waste projects that make waste-disposal management more resilience by reducing waste disposal in waterways, addressing debris management, and safeguarding infrastructure against flooding.
Health and safety	Projects to improve focus on public health and livelihoods to battle against open burning, mitigating pest and disease vector spreading, and preventing crime and violence.
Knowledge creation	Programmes that enable governments plan and explore locally appropriate solutions through analysis, technical exploration, and data generation to benefit the community at large.

Based on the information retrieved from World Bank website, the amount the World Bank has committed since 2000 has reached more than $4.7 billion, which has involved over 340 solid-waste-management programmes worldwide. Some of the examples shown below in Table 6.5.

6.3.1 International Finance Corporation: Blue Finance

International Finance Corporation (IFC) is a member of World Bank Group. The IFC was found to invest in impactful projects for economic development in developing countries and improve the lives of people in line with the growth of private sector. IFC has founded a financial scheme called Blue Finance. As, globally, more attention has been focused on economic activities to address the climate change

TABLE 6.5
Countries' Waste Management Programmes under the Commitment of World Bank

Region	Countries
East Asia and Pacific	*Indonesia*
	World Bank has supported $100 million loan for the waste management project involving 70 participating cities for around 50 million people. These projects are about improvement on the local policies, technological strengthening of modern, sanitary landfills with landfill gas collection facilities, while reducing, closing, and rehabilitating old and informal dumpsites.
	China
	A loan amount of $80 million was allocated for building of modern anaerobic digestion facility with fermentation and energy recovery capability from household kitchen waste separation for a 3 million population.
	Vietnam
	Investment in flood prevention in the city of Can Tho with better solid-waste management.
	Philippines
	Investment project to reduce flood risk in the Metro Manila to avoid solid waste accumulation in waterways. The improvements include better collection systems by providing incentives to reduce marine litter, especially in Manila Bay.
Europe and Central Asia	*Belarus*
	World Bank provided a loan of $15 million to initiate regional-based solid-waste management with building of waste management facilities in line with the closure of dumpsites.
	Azerbaijan
	Loans are provided to establish state-owned waste management companies to cater to a large, growing population formal solid-waste management, helping to achieve the reuse and recycle rate of 25%.
	Bosnia and Herzegovina
	World Bank financed development of formal waste-management system with rate of access from 25% to 66 % of population.

Region	Countries
Latin America and the Caribbean	*Argentina* Through partnerships with food banks and retainers, a loan of $40 million was used to help to reduce food waste, which led to closure of 70 dumpsites after the construction of 11 waste facilities. *Sint Maarten* A $25 million grant was provided by World Bank to develop integrated sectoral waste-management for better debris management. *Jamaica* World Bank has run results-based financing and infrastructure investments to encourage community participation and waste collection services in 18 communities. Such waste activities have created more jobs, subsequently preventing crimes from happening.
Middle East and North Africa	*Morocco* World Bank has supported a loan for policy development at the total amount of $500 million to strengthen private sector partnerships so that 20,000 informal workers social and lives can be improved with increased fee collection. *West Bank* The loan from World Bank aims to develop sustainable livelihoods of waste pickers, with linked payments for better service delivery. Three landfill sites were built to serve 2 million residences to enable dump closure.
South Asia	*Nepal* World Bank supported $4.3 million on a results-based financing project to improve waste collection and restructure fee collection in five municipalities benefitting 800,000 residents. *Pakistan* The $5.5 million projects resulted in reductions of 150,000 tonnes of carbon dioxide equivalent and expansion of daily compost production, with a volume from 300 to 1000 tonnes per day in a compositing facility in Lahore.
Sub-Saharan Africa	*Liberia* World Bank allocated a $10.5 million loan to improve waste collection and construct a new sanitary landfill and transfer stations. *Burkina Faso* Since 2005, a total of $67 million was loaned to support waste sector planning and construction of two landfills. For instance, Ouagdougou is a capital city able to collect an average of 78% of waste generated, which is significantly higher than the 46% average in Sub-Saharan Africa.

issue, many investors have shown interest in participating in a sustainable economy. Blue Finance offers financing such as Blue Bonds and Blue Loans to raise funds related to water and wastewater management, ocean plastic pollution minimization, marine ecosystem restoration, sustainable transportation, eco-tourism, or renewable energy. IFC financed more than $1 billion dollars in the form of loans and bonds to private sector financial institutions and corporates to enable recycling to take place and subsequently reduce marine plastic pollution in Africa and Asia. For instance, the first Blue Bonds, issued in the East Asia Pacific region by BDO Unibank Inc (largest bank in Philippines) with IFC, has subscribed for $100 million to cater marine plastics issues which threaten the country's public economy in specific areas, while in Thailand, IFC also subscribed $300 million bond issued by TMBThanachart Bank Public Company Limited for Indorama Ventures. The bonds help Indorama Ventures to increase recycling capacity in Thailand, Indonesia, Philippines, India, and Brazil. Indorama Ventures is aiming to produce a minimum of 750,000 metric tonnes of recycled PET (rPET) globally by 2025. Indorama Ventures is also involved on a Waste Heat Recovery (WHR) project at PET and fibre manufacturing facility in Indonesia. The WHR project enables better energy savings by improving energy efficiency, subsequently reducing the production facility of carbon footprints to 25%, whereby energy efficiency (EE) measures are expected to reduce the facility's carbon footprint by as much as 25%. For Latin American, IFC allocated $150 million loan to Companhia de Saneamento Basico de Estado de Sao Paulo (SABESP) to improve water quality and expand sewage collection and treatment in the poorest neighbourhoods, which will reduce pollution of the Pinheros River. There is another $40 million agreement between IFC and Banco Internacional to fund the first private-sector blue bond in Ecuador and Latin America.

For the application guidelines of Blue Finance, IFC has published Guidance for Financing the Blue Economy, Building on the Green Bond Principles, and the Green Loan Principles on January 2022. According to Blue Finance Guidance Frameworks, only projects related to Sustainable Development Goals (SDG) of 6: Clean Water and Sanitation and 14: Life below water will be funded, while they do not cause risk to other SDGs, including 2: No hunger, 3: Affordance and clean energy, 12: Climate action, and 13: Responsible consumption and production. All projects should fulfil internationally accepted sustainability standards such as IFC Performance Standards and the World Bank Environmental, Health, and Safety guidelines, or similar and including also industry-specific sustainability standards as well as certain specific product standards, which may also be applied for a blue investment above national requirements. Finally, these are the following types of activities that can be Blue financed (International Finance Corporation, 2022):

A. Water supply: investments in the research, design, development, and implementation of efficient and clean water supply.
B. Water sanitation: investments in the research, design, development, and implementation of water treatment solutions.
C. Ocean-friendly and water-friendly products: investments in the value chain, including production, packaging, and distribution of environmentally-friendly products that avoid water or ocean pollution.
D. Ocean-friendly chemicals and plastic-related sectors: investments in the research, design,

    1. development, and implementation of measures to manage, reduce, recycle, and treat plastic, pollution, or chemical wastes in coastal and river basin areas.

E. Sustainable shipping and port logistics sectors: investments in the research, design, development, and implementation of water and waste management and reduction measures in shipping vessels, shipping yards, and ports.

F. Fisheries, aquaculture, and seafood value chain: sustainable production and waste management and reduction measures that meet, keep, or exceed the Marine Stewardship Council certification standards or equivalent.
G. Marine ecosystem restoration.
H. Sustainable tourism services.
I. Offshore, renewable energy facilities.

In short, IFC group has committed to providing financial facilities to countries to transform the traditional activities towards sustainable business benefits to global economies by establishing a sustainable economic environment.

6.4 WORLD ECONOMIC FORUM

For addressing the issue of plastic pollution, Global Plastic Action Partnership (GPAP) was officially launched in September 2018 during the World Economic Forum's Sustainable Development Impact Summit. GPAP brings together interested stakeholders of governments, businesses, and civil society to take drastic action on the plastic pollution problem. GPAP has drawn the six key areas, as follows, driving change of the plastic pollution issue with deliveries in the past three years as shown in Table 6.6.

TABLE 6.6
Key Areas and Deliveries of GPAP from 2020–2022 (Global Plastic Partnership Annual Report, 2020, 2021, 2022)

Key Areas	Examples of deliveries
Inform policy— Identify policies and regulatory frameworks which can provide supportive environment to tackle plastic waste and pollution.	*Year 2020* Launched locally governance structures in Ghana and Indonesia. *Year 2021* The Government of Vietnam amended Law of Environmental Protection. More attention is paid to transformation of the sector towards a circular economy model. *Year 2022* The National Analysis and Modelling (NAM) tool is successfully launched to cater to the needs of analysis by the countries' prior implementation of plastic waste management scheme, including aspects of waste generation, plastic waste export and import, collection and sorting rates, recycling rates, disposal rates, mismanaged waste, costs, greenhouse gases emission, jobs, and revenues. Vietnam also formed new recycle polyethylene terephthalate standards and revised EPR legislation on packaging.

(Continued)

TABLE 6.6
Continued

Key Areas	Examples of deliveries
Unlock Financing – Seeking partnership to finance transformation of economy in circular manners of plastic industries, recycling, and waste reduction.	*Year 2020* Involvement of Asian Development Bank to finance Indonesia to reduce marine plastic leakage by 70% by 2025. Develop circular economy framework with USD 77 million in Government of Ghana. *Year 2021* Multiple stakeholders run the Project STOP Nestlé, SYSTEMIQ and Borealis in Jembrana, Bali to develop an integrated waste-management system to serve 140,000 people reduce 3,000 tons of plastic annually. *Year 2022* Launched Toolkit for Investment, which is an instrument of analysis on the readiness of a project to undergo financing based on case studies for recovery and recycling of plastic waste. Ghana's Financing Roadmap was launched.
Transform Behaviour – Engaging programme and activities helps consumers and businesses use and manage plastic responsibly towards sustainability.	*Year 2020* Collaboration with the Ministry of Education and Culture and the non-government organization Aliansi Zero Waste Indonesia to establish a behaviour-change task force to educate the community on the responsible usage of plastics. Ghana also granted USD 7 million on the behaviour change pilot project. *Year 2021* The recycling points in Ghana were mapped and operated by young entrepreneurs in Ghana with more than 116 recycling points in the capital city. This enables the recycling process to be easily accessible. *Year 2022* Launched by in-store refill points in Mexico City, with 30 dispensers, to refill household products that can be found across the city.
Boost Innovation – Engagement of technological innovation in the plastic value chain, enabling the transition to a circular economy.	*Year 2020* Engagement of technology developer SAP in the Waste Recovery Innovation Challenge in Ghana. Set up a platform via World Economic Forum's Uplink initiative to assist entrepreneurs with mentorship and investment bridging. *Year 2021* Suntory and Indorama Ventures have awarded a $5,000 grant to the 12 innovators under the programme of Incubation Network and the Ocean Plastic Prevention Accelerator (OPPA). These recipients were selected from the Uplink innovation challenge, which called for idea to improve the traditional, informal waste-management economy in Indonesia. *Year 2022* Launched Global Plastic Innovation Network as an open collaboration platform. In Indonesia, a company called TrashCon, with a unique plastic recovery and recycling solution, raised funds from investors to scale up business.

Key Areas	Examples of deliveries
Harmonise Metric – Develop and use assessment tools to generate data of analysis in order to obtain reliable data to understand the status of plastic pollution.	*Year 2020* Pew Charitable Trusts and SYSTEMIQ developed the first plastic pollution model with better data-analysis capability. *Year 2021* Vietnam has released the results of plastic waste statistics for analysis through Vietnam Zero Waste Alliance's Waste Audit, the World Bank's Plastics Circularity Market Study, and the International Union for the Conversation of Nature's Guidance for Plastic Pollution Hotspotting. *Year 2022* Based on the promising analysis performed on Indonesia, Ghana, and Vietnam, the National Analysis and Modelling (NAM) tools are available for other countries to perform to cater the different model solution, trade-offs, and target settings.
Promote Inclusivity – Diverse and inclusive of opinions, alternatives, methods of the members to tackle plastic pollution problems.	*Year 2020* Launched a global gender-guidance document for all parties of interest to embed gender equality in the plastic value-chain. *Year 2021* Ghana has conducted an investigation on gender roles in the plastics and waste-management value chains, including women as regulators, market actors, workers, consumers, and community members, and these findings are used as references for other countries as well. *Year 2022* Local organization in Nigeria and Maharashtra, India has provided critical training and capacity building, personal hygiene products, and sustainable tools for local waste pickers for over 2,200 people, with 1,890 being women.

6.5 ORGANIZATION FOR ECONOMIC CO-OPERATION AND DEVELOPMENT (OECD)

The Organization for Economic Co-operation and Development (OECD) is an inter-government organization with a current 38 member countries. OECD was founded in 1961. Most of the OECD members are high-income countries. To address the plastic global issue, OECD has published Global Plastics Outlook Policy Scenarios to 2060 to provide inputs and recommendations on challenges to curb the plastic pollution via transition to sustainable and circular usage of plastics. It is undeniable that certain policies are needed to control plastic usage in these countries. When environmental benefits are protected, the economy and living behaviours of communities can be impacted. The publication of OECD has analysed the pros and cons accordingly.

According to the publication by OECD, mismanaged of waste is the main source of the leak of macroplastics to the terrestrial and aquatic environment. However, through the year 2060, the leakage of microplastics will be dominant and the growth may be quick in coming decades. The projection of leakage will increase from 6.1 million tonnes in 2019 to 11.6 million tonnes in 2060 (OECD, 2022). When marine activities and microplastics leakage increase significantly, there is uncertainty whether the leakages may be several times higher than the projection. By 2060, there will be 145 million tonnes of plastics accumulated in the ocean. Moreover, when the land transportation volume is higher, there is also the tendency of more airborne microplastics that are contributed from the wear of tyres and brake pads as aerial microplastic pollution in the form of particulate matter. This is another serious source of unsolvable microplastic pollution, which requires further efforts to look into any possible solutions in the near future.

While an OECD publication has reported that the current policies on reducing greenhouse gases emission make plastics seems to be more environmentally friendly, yet, plastic lifecycle emissions are projected to increase from 1.8 giga tonnes CO_2 in 2019 to 4.3 Gt CO_2 in 2060. One of the main reasons that cause the projection is that the policies that promote bio-based plastic have reduced greenhouse gases, yet there are concerns about higher demand for agricultural land to grow biomass as the input for bio-based plastics. Agricultural land will cause the loss of natural areas, and more agricultural activities can make eutrophication happen substantially. Hence, production of bio-based plastics to substitute for fossil-type plastics is not a smart solution to curb plastic impacts on the environment.

OECD has underlined two most important types of policies to curb plastic consumption: (I) increase production lifespans with better quality, repairable, and reusable features; (II) improve waste management and recycling facilities accessibility while managed by highly responsible personnel to reduce plastic leakage to the environment. The implementation of two types by policies is known as *Regional Action* policy, which should be done in a package; otherwise, effectiveness may be low. This is because the package is to restrain plastic demand and production, enhance recycling, and close leakage pathways. It is anticipated that implementation of a plastic tax can effectively reduce the plastic demands for almost 20% in the OECD countries, where high taxing rates can significantly reduce single-use plastic. However, the rate of reduction will not be so promising or slower after the partial elimination of single-use plastic, with the substitution of the prolonged lifespan of plastic products. This is expected because the prolonged lifespan of plastic products will reduce the turnover of products in the market, while recycling and the supply of recycled plastics increased significantly after the implementation of the Regional Action policy. Since plastics are linked to many economic activities, changes in plastic demand can cause macroeconomic impacts. However, the extent of impacts can be less for some countries like China with established, highly accessible, and affordable recycling technology in the countries, while they also tend to receive good demands of machinery to other countries who are in the stage of setting up recycling facilities. In the opposite respect, capital expenditures can be costly for recycling technology importers like Sub-Saharan Africa and non-OECD European Union countries. In addition, according to OECD, waste treatment is another factor that needs to be considered, with the additional investment anticipated to be USD 320 billion, where OECD countries mainly spend on recycling related facilities, where non-OECD countries need to invest USD 100 billion in recycling, together with another USD 60 billion in waste-collection infrastructure. All the previously mentioned efforts are mainly to reduce the worsening rate of plastic pollution problem; nonetheless, the stocks of plastics in the environment continue to build up. OECD has also been urged to implement more stringent policies to eliminate the leakage of plastics to the environmental as much as possible by upgrading to *Global Ambitious* policies package/instruments. The comparison of both commitment of policies are given by OECD, as shown in Table 6.7. The main difference of Regional Action and Global Ambitious is the level of stringency and coverage of policies/ instruments are based on economic development of countries. Nevertheless, the plastic pollution problem remain a global issue, and these problems will remain unsolved without the full cooperation of all the parties. For the implementation of Global Ambitious, OECD also expected that they have synergistic interaction with climate mitigation policies involving carbon tax and decarbonation of the global power sector because plastics are linked to fossil-based derived products. The synergy between plastics and climate mitigation policies maximizes the benefits of both areas of environmental policy. According to projections, the Global Ambition and Climate Mitigation scenario is expected to achieve a significant reduction in greenhouse gas (GHG) emissions from plastics throughout their lifecycle. By 2060, it is estimated that these emissions will decrease by 67% from 4.3 Gt CO_2e to 1.4

TABLE 6.7
Comparison of Regional Action Policies and Global Ambitious Policies

Pillar	Policy instrument	*Regional Action* scenario	*Global Ambition* scenario
	Packaging plastics tax	*EU*: USD 1,000/tonne by 2030, constant thereafter *Rest of OECD*: USD 1,000/tonne by 2040, constant thereafter *Non-OECD*: USD 1,000/ tonne by 2060	*Global*: USD 1,000/ tonne by 2030, doubling by 2060
Restrain plastics production and demand and design for circularity (hereafter, restrain demand)	Non-packaging plastics tax	*OECD*: USD 750/tonne by 2040, constant thereafter *Non-OECD*: USD 750/ tonne by 2060	*Global*: USD 750/ tonne by 2030, doubling by 2060
	Ecodesign for durability and repair	*Global*: 10% lifespan increase, 5–10% decrease in demand for durables, increase in demand for repair services such that ex ante total expenditures are unchanged	*Global*: 15% lifespan increase, 10–20% decrease in demand for durables, increase in demand for repair services such that ex ante total expenditures are unchanged

(Continued)

TABLE 6.7
Continued

Pillar	Policy instrument	*Regional* *Action* scenario	*Global* *Ambition* scenario
Enhance recycling	Recycled content target	*OECD*: 40% recycled content target *Non-OECD*: 20% recycled content target	*Global*: 40% recycled content target
	EPR for packaging, electronics, automotive, and wearable apparel	*OECD + EU*: 20% points increase in recycling, tax on plastics inputs—USD 300/tonne by 2030, constant thereafter, subsidy on waste sector such that the instrument is budget neutral	*Global*: 20% points increase in recycling, tax on plastics inputs—USD 300/tonne by 2030, constant thereafter, subsidy on waste sector such that the instrument is budget neutral
	Enhance recycling through waste management	*EU, Japan & Korea*: 60% recycling rate target by 2030, 70% by 2060 *Rest of OECD, the People's Republic of China (hereafter 'China')*: 60% recycling rate target by 2060 *Rest of non-OECD*: 40% recycling rate target by 2060	*EU, Japan and Korea*: 60% recycling rate target by 2030, 80% by 2060 *Rest of OECD, China*: 80% recycling rate target by 2060 *Rest of non-OECD*: 60% recycling rate target by 2060
Close leakage pathways	Improved plastic waste collection	*OECD*: full reduction of mismanaged waste shares* *Non-OECD*: halving of mismanaged waste shares*	*Global*: full reduction of mismanaged waste shares*
	Improved litter collection	*High-income countries'* collection rates increase 5%-points; *middle income countries* income-scaled increase	*Low-income countries* collection rates increase 10%-points; *high income countries* collection rates increase 5%-points; *middle income countries* income-scaled increase

Source: * Waste streams from uncollected litter and from markings and microbeads are not included in this policy, as they are not managed as waste. (Adapted from OECD Global Plastics Outlook: Policy Scenarios to 2060 @ Table 1 B. 1. Details on the implementation of the circular plastics scenario.)

Gt CO_2e, surpassing the emissions levels of 2019. Plastics policies primarily contribute to this reduction by minimizing the use of plastics and the generation of waste. Moreover, climate change mitigation policies go a step further by enhancing the GHG-intensity of both plastics production and waste management, leading to additional decreases in overall GHG emissions. Overall, the suggestions of implementation of Regional Action and Global Ambitious policies/instruments can take seriously references by the policymakers when drafting the policies in the respective countries for localized and globalized benefits.

6.6 UNITED NATIONS ENVIRONMENTAL PROGRAMME

On March 2, 2022, United Nations Environmental Programme (UNEP) in the United Nations Environmental Assembly at Nairobi adopted Resolution 5/14, titled "End plastic pollution: Towards an international legally binding instrument." This is a historical move for the members of the United Nations to develop an international, legally binding instrument on plastic pollution as well as the marine environment, with negotiations to end in 2024. In fact, the United Nations has been working thoroughly for many decades to manage plastic debris and marine litter problems. The summary of timelines has been recorded in the publication of UNEP in 2021, titled "From Pollution to Solution, A Global Assessment of Marine Litter and Plastic Pollution" in Figure 6.2. To address the UNEP Resolution 5/14, an Open-Ended Working Group was met in Dakar from May 30 to June 1, 2022 with the decision to form Intergovernmental Negotiation Committee (INC) to drive the development of an international legally binding instrument on plastic pollution, including in the marine environment. The first and second INC meeting were held in Uruguay (November 28–December 2, 2022) and France (May 29 to June 2, 2023) (shown in Figure 6.3). After the second round of negotiation meeting, around 170 countries agreed to develop a first draft by November 2023 that will be the first global treaty to curb plastic pollution by the end of 2024. The coming third INC, fourth INC, and fifth INC will be held in Kenya (13–November 17, 2023), Canada in Spring 2024, and the Republic of Korea in autumn 2024, respectively. The INC aims to complete the negotiations by the end of 2024 with the comprehensive approach to manage the full life cycle of plastic.

In general, the leakage of plastic wastes to the environment and oceans can be prevented via the following actions, which are urged at different hierarchies, as shown in Table 6.8. UNEP has identified that both technological and education approaches are equally important to effectively curb plastic pollution globally. Effective, tailored social- and behaviour-change campaigns and initiatives need to be conducted regularly to create awareness on responsibility consumption and disposal. The behaviour change of consumers is important through formal and informal education events to strengthen support and compliance with plastic reduction policies. In addition, educational programming, communication, and advocacy campaigns serve as effective tools for generating public support and engagement around plastic pollution. These initiatives create a sense of environmental responsibility, which is essential as a prerequisite for taking action smartly to meet the expected outcomes. Indeed, tailored campaigns and communication initiatives that highlight the availability of viable alternatives and provide accessible pathways to access them are

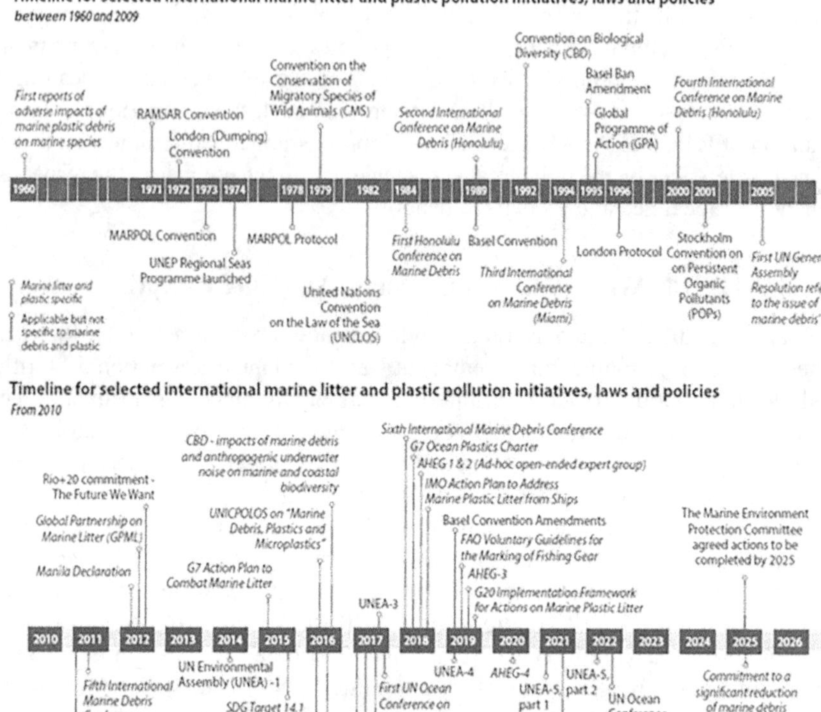

FIGURE 6.2 Timelines of related international marine litter and plastic pollution initiatives, laws, and policies from 1960 onwards.

Source: Adapted from UNEP (2021), "From Pollution to Solution, A Global Assessment of Marine Litter and Plastic Pollution."

more likely to translate awareness and concern for plastic pollution into tangible action, subsequently unlocking transformative patterns of social and behavioural change. Hence, UNEP has suggested the effective public campaign should include the following elements:

a) **Customizable elements:** The aims and goals should fit the localized and political context and relate to the target audiences' gender, age, and education levels.
b) **Emphasize positive norms among communities:** Audiences tend to adopt plastic reduction and safe disposal of plastics by demonstrating that such behaviour is a positive social norm which is widely practised by the majority communities for the sustainable future.

FIGURE 6.3 Photo Credit by Stephanie Lecocq from Reuters for second INC Meeting at UNESCO Headquarters in Paris, France, May 29, 2023 (Reuters, 2023).

c) **Emphasize benefits:** Promote social and behavioural change by emphasizing the benefits for individuals and highlighting how individual actions can contribute to collective efforts, ultimately leading to improved livelihoods on a global scale. Thoughtfully crafted campaigns have the potential to create lasting impacts on both individual and societal consumption patterns by showcasing real-life instances of how redirecting purchasing habits and promoting reuse behaviours can serve as an inspiration for society, accelerating the adoption of sustainable consumption and production practices.
d) **Target and action:** Set achievable tasks to progress with forward action such as refusing avoidable, harmful, and unnecessary plastic products, especially single-use items that cannot be recycled or have excessive or unnecessary plastic packaging.
e) **Communication progression:** In regions where regulations regarding plastic pollution are lacking, citizen behaviour campaigns prove to be powerful advocacy tools, motivating individuals to voice their concerns and exert pressure on governments and businesses for legislative action or the provision of plastic-free alternatives.

6.7 BIOPLASTICS AND BIODEGRADABLE PLASTICS

Nowadays, synthetic polymers such as polyethyelene, polypropylene, polystyrene, poly(vinyl chloride), polyethylene terephathalate, etc., are vastly used in daily activities. However, the drawbacks of petroleum-based plastics are because they are being

TABLE 6.8
Actions at Different Hierarchies to Curb Leakage of Plastics to the Environment

Hierarchy	Actions
Community level	Education and knowledge on recycling, including training on cleaning, identifying, segregating, as well as economic incentives using low-cost infrastructure. Dynamic and continuous awareness raising efforts are required.
Manufacturing level	Develop completely biodegradable plastic products, innovation on design to make the products easy to separate and recycle by the industry. Upon the end of life, the users are provided incentives and facilities to return the plastic items, including extended producer responsibility
Local and national government level	Develop multi-stakeholder processes, evidence-based policy frameworks, enforcement to address illegal, unreported and unregulated plastic disposal, and setup incentive schemes for a closed-loop economy to encourage the collection, sorting, recycling, reuse, and repair. Impose strict regulation on single-use plastics.
Regional and international level	Coordination is needed to harmonize regulatory standards, define common methodologies to assess the scope, sources, and impacts of plastic products/wastes; share knowledge, good practices, and guidelines on responsible recovery, management, and prevention of plastic wastes; harmonize products, categorize and promote enforcement of existing laws and regulations; and invest in cleaning and recycling technologies while scaling good practices. Governments can leverage existing partnerships and mechanisms such as the Regional Seas and Coastal to share knowledge, data, and good practices as well as build capacity on plastic products.

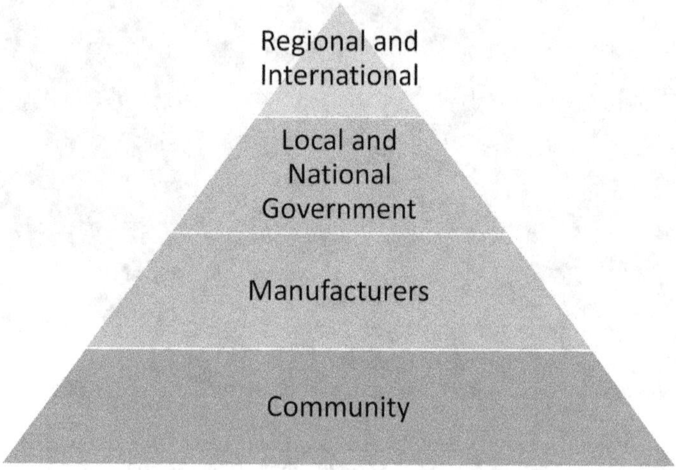

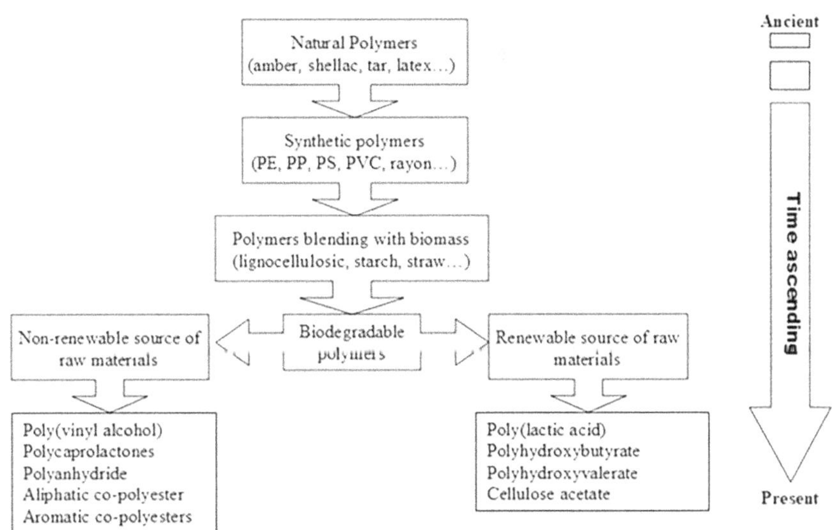

FIGURE 6.4 Trends of polymer development.

Source: Adapted from Sin and Bee (2019), with permission of Elsevier.

used to produce a variety of types of single-use products such as food packaging, decorative packaging, food containers, disposable utensils, electrical and electronic appliances, etc. Almost all the petroleum-based plastics are non-degradable, and they are barely recycled. Therefore, scientists recommend bioplastics as the potential candidates to replace petroleum-based and non-degradable plastics. As the prefix of "bio," this type of polymer is biodegradable, while they are preferably derived from renewable sources, especially using plant-based inputs. As summarized by Sin and Bee (2019), the domestic plastics development shall follow the trends as in Figure 6.4.

Sin and Bee (2019) provided an overview of biodegradable polyester derived from both renewable and non-renewable sources in Figure 6.5. This illustration demonstrates that biodegradable polymers, primarily belonging to the polyester family, contain oxygen elements in their chemical chains. The presence of oxygen is crucial, as it facilitates the initiation of chain scissioning reactions when exposed to moisture, resulting in the breakdown of the chains into smaller fragments that can be easily consumed by microorganisms. Additionally, there is another category of plastics products, known as oxo-degradable plastics, commonly utilized in the production of plastic bags or single-use utensils and containers which undergo disintegration within a short period of time. These oxo-degradable polymers typically incorporate pro-degradant additives into fossil-based polymers like polyethylene, polypropylene, and polystyrene to initiate chain scissioning. It is important to note

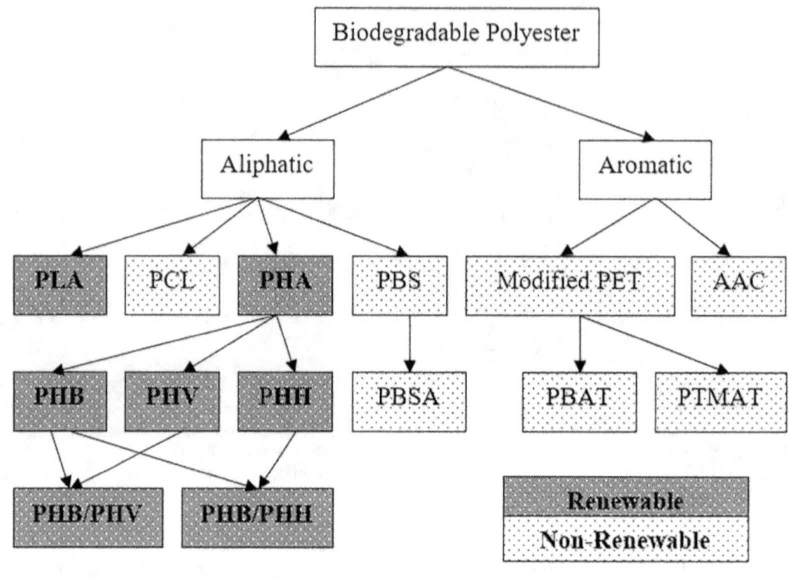

AAC – Aliphatic-Aromatic Copolyesters
PBAT – Poly(Butylene Adipate/Terephthalate)
PET – Poly(Ethylene Terephthalate)
PBS – Poly(Butylene Succinate)
PBSA – Poly(Butylene Succinate/Adipate)
PCL – εPolyCaproLactone
PLA – Poly(Lactic Acid)
PHB – Poly(Hydroxy Butyrate)
PHH – Poly(Hydroxy Hexanoate)
PHV – Poly(Hydroxy Valerate)
PTMAT – Poly(TetraMethylene Adipate/Terephthalate)

FIGURE 6.5 Biodegradable polyester family.

Source: Adapted from Sin and Bee, 2019, with permission from Elsevier.

the advantages and disadvantages of oxo-degradable polymers. On the positive side, only small amounts of pro-degradant additives are required, which do not significantly affect the production parameters in the factory, and the material costs are relatively insignificant. However, there are drawbacks to consider. The pro-degradant additives, composed of transition metal compounds, carry the risk of causing unforeseen pollution to the environment. Additionally, the disintegration of plastic chains can generate microplastics, leading to environmental pollution once again.

Table 6.9 presents a comprehensive list of major producers of biopolymers in the global market. Over the past decade, the demand for biopolymers has steadily increased as consumers become more aware of the importance of using biodegradable polymers to mitigate plastic waste pollution. However, the price of biodegradable polymers remains a significant barrier to their widespread adoption by a broader range of consumers. Compared to petroleum-based non-degradable polymers, biodegradable polymers can be at least twice as expensive. Consequently,

TABLE 6.9
Players in Biodegradable Polymer/Packaging Industry

Company	Country
BASF	Germany
Biomatera	Canada
Biome Bioplastics Ltd	United Kingdom
Biomer	Germany
BIOP Biopolymer Technologies Ag	Netherlands
Biotec- Biologische Naturverpackungen GmbH & Co.	Germany
Cereplast	Italy
Corbion	France
Danimer Scientific	United States
Fkur Plastics Corp.	United States
Futerro	Belgium
Galactic SA	Belgium
Huhtamaki Group	Finland
Mitsui Chemicals	Japan
Natureworks LLC	United States
Novamont S.p.A.	Italy
Plantic Technologies Ltd	Australia
Rodenburg Biopolymers B.V.	Netherlands
Synbra Technology B.V.	Netherlands
Teijin Limited	Japan
Teknor Apex	Singapore
Tianan Biologic Material Co. Ltd.	China
Tianjin Guoyun Biological Materials Co. Ltd.	China
Toray Industries, Inc.	Japan
Toyobo Co., Ltd.	Japan
Zhejiang Hisun Biomaterials Co., Ltd.	China

Source: Adapted from Sin and Bee, 2023. With permission of Elsevier.

many developing and underdeveloped countries are not inclined to utilize biodegradable polymers. Figure 6.6 clearly illustrates that only developed regions with high gross domestic product (GDP) have a presence in the biodegradable industry. This indicates a lack of penetration of biodegradable polymers in regions with middle and low GDP. Ironically, these developing and underdeveloped countries are grappling with severe plastic pollution, with overflowing landfills and rivers filled with tonnes of plastic waste floating on the surface. In cases where the use of biodegradable plastics is economically impractical due to their high cost, the most suitable approach is to educate communities about proper plastic waste management to reduce long-term environmental impacts.

The biopolymers market has experienced significant growth in recent years, particularly in Europe, driven by the support and awareness of the European community for sustainable products. Additionally, many countries have implemented regulations and bans on single-use plastics, such as food packaging and straws,

FIGURE 6.6 Countries with major biodegradable polymers players until 2020.

Source: Adapted from Sin and Bee, 2023. With permission of Elsevier.

which have further increased the demand for alternatives to non-degradable polymeric materials. According to the IHS Markit Chemical Economics Handbook: Biodegradable Polymer Report, the market value of biodegradable plastics reached \$1.1 billion in 2019 and is projected to reach \$1.7 billion by 2023. In terms of consumption, biodegradable polymer usage amounted to 484.7 kilo tonnes in 2019 and is expected to reach 984.8 kilo tonnes in 2022, with a compound annual growth rate of 15.2% between 2017 and 2022.

Despite the COVID-19 pandemic causing a decline in global economic activity and consumer product demand, the demand for plastic products, especially those used for hygiene packaging in food service, has remained resilient. For example, many airlines have implemented individual packaging for in-flight food to ensure safety and reduce contact. Similarly, event organizers for conferences, meetings, celebrations, and weddings have shifted to pre-packaged food serving instead of buffet-style arrangements to minimize contact. These new practices have increased the demand for biodegradable polymers as alternatives to single-use plastics.

Figure 6.7 illustrates that Western Europe currently represents the largest consumer of biodegradable polymers, followed by Asia, Oceania, and North America (specifically, the United States and Canada). This is primarily due to the efforts of governments in these regions to educate their communities about responsible handling and disposal of single-use plastics. Consumer acceptance in these regions is higher compared to other parts of the world (see Figure 6.8). Moreover, these countries have developed robust environmental protection frameworks that highlight the benefits of choosing biodegradable plastics for packaging. Consumers are well-informed that biodegradable and compostable plastics can be transformed into energy through incineration or composting processes, rather than ending up in landfills. For example, Japan has encouraged the use of biodegradable plastics by

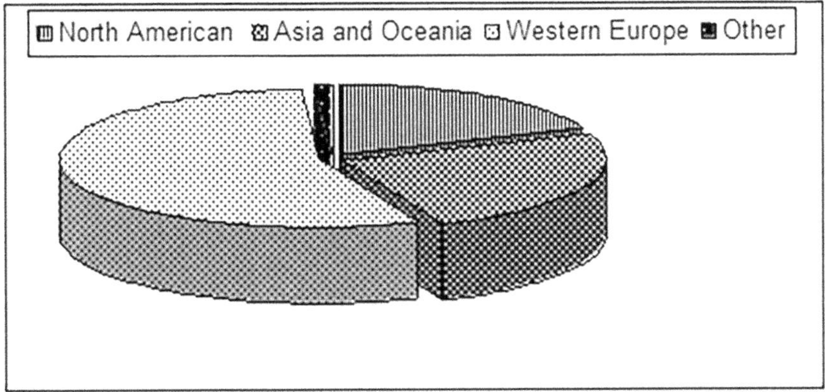

FIGURE 6.7 Distribution of world consumption of biodegradable polymers in 2018.

Source: Adapted from Sin and Bee, 2023. With permission of Elsevier.

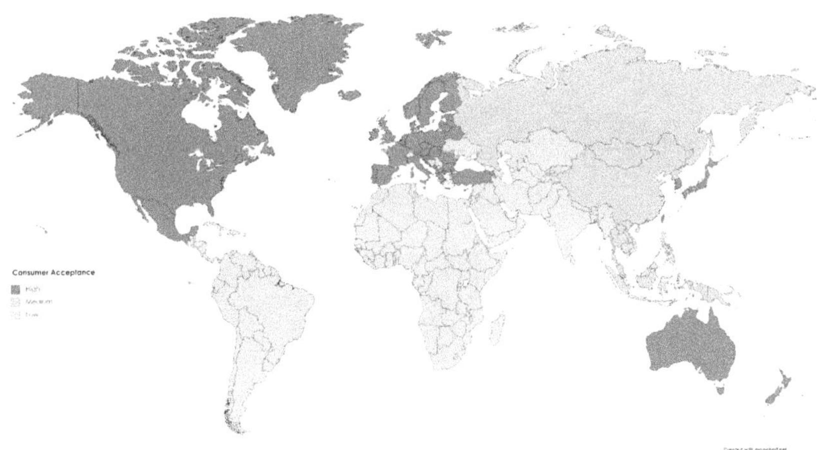

FIGURE 6.8 Consumer acceptances of biopolymers in accordance to regions.

Source: Adapted from Sin and Bee, 2023. With permission of Elsevier.

food producers and provided collection points to separate biodegradable polymers from other commodity plastics like polyethylene, polypropylene, and polystyrene. This specialized management allows biodegradable plastics to be transformed into value-added products. The strong initiatives taken by governments in these countries have driven the demand for biodegradable polymers, leading to the growth and global leadership of the biopolymer industry in clean technology.

Biodegradable polymers can be produced from both petroleum-based and renewable sources, and both types have gained significant attention in the industry. Renewable biodegradable polymers are not only biodegradable but also sourced from

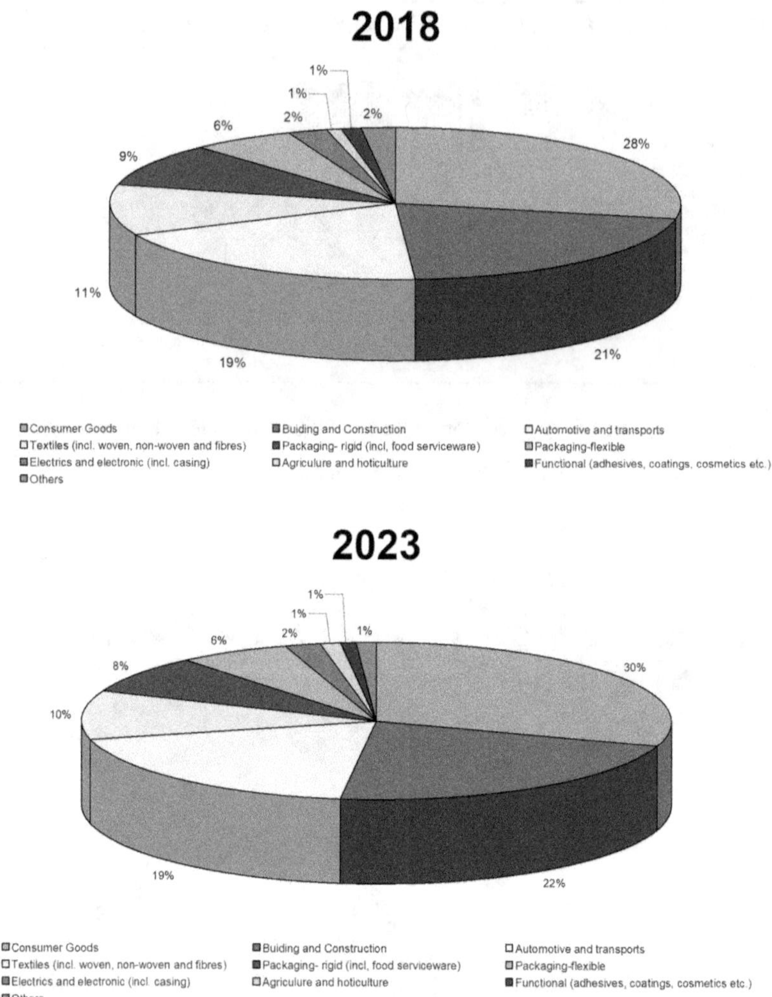

FIGURE 6.9 Percentage of market share of biopolymers in 2018 and 2023.

environmentally friendly plants, while petroleum-based biodegradable polymers help address the issue of non-degradable plastic waste accumulation. As shown in Figure 6.9, consumer goods represent the largest share of biopolymers, followed by the building and construction industry. Over the next five years, from 2018 to 2023, the market shares of consumer goods and the building and construction industry are expected to increase by 1% and 2%, respectively. This growth can be attributed to government policies and successful education campaigns that encourage consumers to replace fossil-based plastics with biopolymers. Many countries have implemented regulations and bans on non-degradable plastics for environmental protection, leading to increased demand for biopolymers. For example, China, with a population of

1.44 billion, has banned the use of plastic bags in the restaurant industry and prohibited the production and sale of plastic bags less than 0.025 mm thick as well as plastic straws. By 2025, no single-use plastic items will be offered. Major supermarkets in China no longer provide free plastic bags to customers, resulting in savings of at least 60 million barrels of oil per year. Similarly, in Europe, Directive 94/62/EC on Packaging and Packaging Waste has set requirements for plastic and packaging waste. Plastic and packaging waste must meet the European standard EN 13432 and be declared as compostable before being marketed to the public. Countries like Ireland and the United Kingdom have implemented plastic bag levies, resulting in a significant reduction in plastic bag usage per capita. Although the use of plastic bags is not entirely avoidable in modern life, the production of reusable plastic bags made from compostable materials is recommended to reduce the burden on the environment. Several countries impose high charges on single-use plastic products, which are several times higher than the cost of materials for biodegradable polymers. This has led consumers to choose between bringing their own bags or paying a higher cost for biodegradable bags from supermarkets. As a result, the production of biodegradable polymers remains profitable and well accepted by consumers.

In Southeast Asian countries like Malaysia and Singapore, supermarket operators charge 20 cents per carrier bag. Some Malaysian states have implemented a soft-landing strategy, initially imposing plastic bag charges only on Saturdays for the first two years, followed by daily charges after two years, based on positive consumer response. As awareness of biopolymer packaging grows, many companies have expanded their product range to include biopolymer materials. However, the biodegradability and compostability of these so-called "eco-plastic" products remain questionable. Therefore, these eco-plastic products need to undergo testing according to standards to verify their biodegradability and compostability. In the European Union, compostable packaging must meet the requirements of EN 13432, while other countries have their own standards for allowing the use of a compostable logo (see Table 6.10). There are also labels for biomaterials that fulfil the criteria of environmentally friendly products, as listed in Table 6.11. These labels can be used for bio-based products, such as bio-based polyethylene and polypropylene, which are derived from renewable resources but are not biodegradable or compostable materials. Overall, biodegradable plastics are a sort of technological innovation which can help to partially substitute petroleum based non-degradable plastics, while the availability and price of the biodegradable plastics remains the main concern for them to be widely used by consumers.

6.8 CONCLUSION

Management of plastic pollution has gained global attention since decades ago. Importantly, curbing plastic pollution required long-term efforts and strategies, resources, understandings, agreements, negotiations, implementation, and education to address the problem comprehensively. All nations need to play their roles to minimize single-use plastic and manage plastic waste in an appropriate manner to avoid leakage to the environment. Global citizens are no longer working in the dark because plenty of policies, guidelines, suggestions, or even treaties have been

TABLE 6.10
Certification of Compostable Plastic for Respective Countries

Certification body	Standard of reference	Logo
Australia Bioplastics Association (Australia) www.bioplastics.org.au	EN 13432: 2000	
Association for Organics Recycling (UK) www.organics-recycling.org.uk	EN 13432: 2000	
Polish Packaging Research and Development Centre (Poland) www.cobro.org.pl/en	EN 13432: 2000	
DIN Certco (Germany) www.dincertco.de/en/	EN 13432: 2000	
Keurmerkinstituut (Netherlands) www.keurmerk.nl	EN 13432: 2000	
Vincotte (Belgium) www.okcompost.be	EN 13432: 2000	
Jätelaito-syhdistys (Finland) www.jly.fi	EN 13432: 2000	
Certiquality/CIC (Italy) www.compostabile.com	EN 13432: 2000	
Biodegradable Products Institute (USA) www.bpiworld.org	D 6400–04	
Bureau de normalisation du Québec (Canada) www.bnq.qc.ca	BNQ 9011–911/2007	

Certification body	Standard of reference	Logo
Japan BioPlastics Association (Japan) www.jbpaweb.net	Green Plastic Certification system	
Biodegradable Products Institute (North America) www.bpiworld.org	D6400 or ASTMD6868	
Nordic Ecolabeling www.nordic-ecolabel.org/	EN 16640:21 and EN 16785–1:2015	
DIN-Geprüft www.dincertco.de/	AS 5810 and NF T51–800	  

Source: Adapted from Sin and Bee, 2023. With permission of Elsevier.

TABLE 6.11
Certification for Sustainable Eco-Plastic or Related Products

Certification body	Label
EU Ecolabel	

(Continued)

TABLE 6.11
Continued

Certification body	Label
Roundtable on Sustainable Biomaterials	
International Sustainability & Carbon Certification	
REDcert	
Blue Angel	

Source: Adapted from Sin and Bee, 2023. With permission of Elsevier.

developed, while some are in the pipeline. Currently the most important issue is that all the nations and their communities need to extend their full cooperation to curb the issues. Consumers should adopt behaviour change to minimize usage of plastic products, particularly single-use plastic, while practising recycling for circular economy in daily activities. Finally, all of us should be confident that minimizing plastic usage will not lead to inconvenience, yet this behaviour change can guarantee a better and sustainable future for coming generations.

REFERENCES

ASEAN (2021). ASEAN Regional Action Plan for Combating Marine Debris in the ASEAN Member States. ASEAN.

Basel Convention (2019). On the Control of Transboundary Movements of Hazardous Wastes and Their Disposal. Protocol on Liability and Compensation For Damage Resulting from Transboundary Movements of Hazardous Wastes and Their Disposal. Available at https://www.basel.int/Portals/4/Basel%20Convention/docs/text/BaselConventionText-e.pdf. Accessed 4 January 2024.

Global Plastic Action Partnership (2020). Annual Report 2020. Global Plastic Action Partnership.

Global Plastic Action Partnership (2021). Annual Report 2021. Global Plastic Action Partnership.

Global Plastic Action Partnership (2022). Annual Report 2022. Global Plastic Action Partnership.

International Finance Corporation (2022). Guidelines Blude Finance- Guidance for Financing the Blue Economy, Building on the Green Bond Principles and the Green Loan Principles. International Finance Corporation

Kyodo News (2019). Malaysia to Send 42 Containers of Illegal Plastic Waste Back to Britain. https://english.kyodonews.net/news/2019/11/eb81cb372c28-malaysia-to-send-42-containers-of-illegal-waste-back-to-britain.html. Accessed 19 April 2023.

New York Times (1988). Toxic Waste Boomerang: Ciao Italy! www.nytimes.com/1988/09/03/world/toxic-waste-boomerang-ciao-italy.html. Accessed 19 April 2023

Newsweek (2019). Malaysia Has Started Returning Tons of Trash to the West: 'We Will Not Be the Dumping Ground of the World'. www.newsweek.com/plastic-waste-malaysia-minister-yeo-bee-bin-south-east-asia-trash-1436969. Accessed 19 April 2023.

OECD (2022). OECD Global Plastics Outlook: Policy Scenarios to 2060. www.oecd.org/publications/global-plastics-outlook-aa1edf33-en.htm. Accessed 9 June 2023.

Reuters (2023). Plastic Recycling in Focus as Treaty Talks Get Underway in Paris. www.reuters.com/business/environment/plastic-recycling-focus-treaty-talks-get-underway-paris-2023-05-29/. Accessed 30 May 2023.

Sin, L. T. and Bee, S. T. (2019). Polylactic Acid- A Practical Guide for the Processing, Manufacturing, and Applications of PLA. Second Edition. Elsevier.

Sin, L. T. and Bee, S. T. (2023). Plastics and Sustainability- Practical Approaches. Elsevier.

Terms of Reference for the Basel Convention Partnership on Plastic Waste and Workplan for the Working Group of the Partnership on Plastic Waste for the biennium 2020–2021. (2019). www.basel.int/Portals/4/download.aspx?d=UNEP-CHW.14-INF-16-Rev.1.English.pdf. Accessed 29 April 2023.

UNEP (2021). From Pollution to Solution, A Global Assessment of Marine Litter and Plastic Pollution. UNEP.

Wingfield, S. and Lim, M. (2021). The United Nations Basel Convention's Global Plastic Waste Partnership: History, Evolution and Progress. Bank, M. S. (Eds.) In Microplastic in the Environment: Pattern and Process. Book Series: Environmental Contamination Remediation and Management (ENCRMA). Springer. https://doi.org/10.1007/978-3-030-78627-4_10

Index

Note: Page numbers in **bold** indicate tables on the corresponding page.

For Product Safety Concerns and Information please contact our EU representative GPSR@taylorandfrancis.com
Taylor & Francis Verlag GmbH, Kaufingerstraße 24, 80331 München, Germany

www.ingramcontent.com/pod-product-compliance
Lightning Source LLC
LaVergne TN
LVHW020617110826
845149LV00002B/501

* 9 7 8 1 0 3 2 4 8 2 0 3 3 *